CHEMISTRY

*A Project of the
American Chemical Society*

Volume I, Chapters 1–5

W. H. Freeman and Company
New York

ISBN 0-7167-9670-8

©2003 American Chemical Society
All rights reserved

GenChem Editorial/Writing Team

Dr. Jerry Bell, **Chief Editor**
American Chemical Society

Dr. Stephen Branz
San Jose State University

Dr. Diane Bunce
The Catholic University of America

Dr. Melanie Cooper
Clemson University

Dr. I. Dwaine Eubanks
Clemson University

Ms. Lucy Pryde Eubanks
Clemson University

Dr. Herbert Kaesz
University of California, Los Angeles

Dr. Wayne Morgan
Hutchinson Community College

Dr. Dorit Noether

Dr. Maureen Scharberg
San Jose State University

Dr. Robert Silberman
SUNY–Cortland

Dr. Emmett Wright
Kansas State University

GenChem Advisory Board

Dr. Ronald Archer
University of Massachusetts

Dr. Thomas R. Cech
University of Colorado

Dr. Robert DeHaan
Emory University Medical School

Dr. Slayton Evans
University of North Carolina–Chapel Hill

Dr. Bruce Ganem
Cornell University

Dr. Gordon Hammes
Duke University Medical Center

Dr. Morton Hoffman
Boston University

Dr. Donald E. Jones

Dr. Richard Lerner
Scripps Research Institute

Dr. Joe Miller
DuPont Experimental Station

Dr. Stanley H. Pine
California State University

Dr. Truman Schwarz
Macalester College

Dr. Judith Voet
Swarthmore College

Dr. Vera Zdravkovich
Prince George's Community College

GenChem Editorial Review Panel

Dr. Ronald Archer
University of Massachusetts

Dr. Morton Hoffman
Boston University

Dr. Truman Schwarz
Macalester College

Dr. Vera Zdravkovich
Prince George's Community College

Here's what field testers are saying about...

CHEMISTRY
A Project of the American Chemical Society

"For the first time in more than 20 years, I felt I was teaching my students something real."
—*Jonathan Mitschele, Saint Joseph's College*

"By the end of the semester, I saw a dramatic improvement in my students' critical thinking skills."
—*Kent Chambers, Hardin-Simmons University*

"What drew me to this book was the treatment of entropy in Chapter 8. It is absolutely without parallel and the chapter is supported by an equally marvelous interactive Web tutorial as well."
—*Glenn Keldsen, Purdue University North Central*

"I have enjoyed working with a textbook that incorporates student activities directly into the text. This makes it much easier to enable students to be actively involved in the course . . . this is a unique feature. . . . I look forward to seeing the final product."
—*Laura Eisen, George Washington University*

"The text emphasizes the concepts of chemistry . . . the how and why of chemical phenomena, not the usual teaching of algorithms and applied mathematics."
—*Michelle Dose, Hardin-Simmons University*

"It is hard for an old professor to learn new teaching tricks, but it was worth it to engage in the trial of the ACS GenChem project. The promise of helping students gain a deeper and more intuitive understanding of the concepts of chemistry is very rewarding."
—*Neil Rudolph, Adams State College*

"The text facilitates (even encourages) the use of lecture groups and in-class activities, both of which the students enjoyed. Working effectively in groups is one of the goals/objectives of Adams State's general education program, so this course helps to fulfill that specific goal."
—*Marty Jones, Adams State College*

"I think this text has the potential to change both how we teach and how the students learn chemistry. The students are much more engaged with the material in class and appreciate the balance between activities and lecture. I'm growing in enjoyment of this also."
—*Priscilla Bell, Whittier College*

". . . the experience provided me a challenge that made teaching more fun and exciting."
—*Amina El-Ashmawy, Collin County Community College*

An Innovative Support Package for the Instructor . . .

The ***Chemistry*** *Instructors' Manual* provides instructors with a set of options for integrating active learning into their curriculum. The teaching strategies presented in the *Instructors' Manual* cover the spectrum of general chemistry courses, from small studio sections to large, multilaboratory sections. The *Instructors' Manual* includes information for each type of activity students will encounter. Access the *Instructors' Manual* via our Web site:

www.whfreeman.com/acsgenchem

For *Investigate This* activities, the *Instructors' Manual* includes:

- Goals of activity
- Anticipated results
- Optional modification for the activity to accommodate larger classes
- Set-up and clean-up times
- Detailed lists of reagents, materials, procedures, and classroom options
- Digitized photographs of the activity set-ups and results
- Follow-up discussion and activities

For *Consider This* activities, the *Instructors' Manual* includes:

- Goals of activity
- Detailed solutions
- Follow-up discussion and activities

For *Check This* activities, the *Instructors' Manual* includes:

- Detailed solutions

For the Student . . .

The *Personal Tutor*

The *Personal Tutor* is a print supplement for the ACS Chemistry textbook. Its primary purpose is to give you more guidance and practice with problems and computations. Using the *Personal Tutor* is much like visiting the professor during office hours or going to a human tutor. The *Personal Tutor* is completely integrated in the text.

First, go to the Web site www.whfreeman.com/acsgenchemhome click on "Personal Tutor," and take the diagnostic exam. Once you've completed the exam you will receive feedback on what sections of the *Personal Tutor* you must review.

You may need to take advantage of the *Personal Tutor* frequently or you may need its assistance only a few times during the course. The questions can be a good review when preparing for a test even if you are confident that you know how to solve the problems or do the computations.

The Student Web site

For additional support, visit our Web site at **www.whfreeman.com/acsgenchemhome** where you will find

"Interactive Web Companion" featuring:
- Molecular level animations
- Drag and drop activities
- Integrated review questions

Molecular Database: More than one hundred 3-D rotatable molecules

Images from the text: Every image in full color

The Molecular Structure Model Set

Models of molecules help you to understand their physical and chemical properties by providing a way to visualize the three-dimensional arrangement of the atoms. A superb study aid, the Molecular Structure Model Set uses polyhedra to represent atoms and plastic connectors (scaled to correct bond lengths) to represent bonds. Plastic plates representing orbital lobes are included for indicating lone pairs of electrons, radicals, and multiple bonds—a feature unique to this set.

Student Survey

The feedback that you provide to the writers and editors of *Chemistry* will be analyzed and integrated into the final version of this text. We appreciate your thoughtful comments and insights. ***Please cut out this survey and return it to your instructor.***

Investigate This activities
- a. were useful and/or informative.
- b. were somewhat useful and/or informative.
- c. I tried using, but did not find useful.
- d. I did not try using.

Consider This activities
- a. were useful and/or informative.
- b. were somewhat useful and/or informative.
- c. I tried using, but did not find useful.
- d. I did not try using.

Check This activities
- a. were useful and/or informative.
- b. were somewhat useful and/or informative.
- c. I tried using, but did not find useful.
- d. I did not try using.

Web site tutorials at www.whfreeman.com/acsgenchem
- a. were useful and/or informative.
- b. were somewhat useful and/or informative.
- c. I tried using, but did not find useful.
- d. I did not try using.

What was your overall impression of this text?

How does this text compare with those you've used in other courses?

May we have your permission to use your comments in our marketing materials? If so, please print your name, and the name of your school below.

Name_____

School_____

Career aspiration_____

Year: Freshman ____
        Sophomore ____
        Junior ____
        Senior ____
        Other ____

Brief Contents

Volume One

Chapter 1 Water: A Natural Wonder

Chapter 2 Aqueous Solutions and Solubility

Chapter 3 Origin of Atoms

Chapter 4 Structure of Atoms

Chapter 5 Structure of Molecules

Volume Two

Chapter 6 Chemical Reactions

Chapter 7 Chemical Energetics: Enthalpy

Chapter 8 Entropy and Molecular Organization

Chapter 9 Chemical Equilibrium

Chapter 10 Reduction-Oxidation: Electrochemistry

Chapter 11 Reaction Pathways

Contents

Volume One

Introduction

CHAPTER 1 Water: A Natural Wonder

Section 1.1. Phases of Matter   1-6
  Density
  Solids
  Liquids
  Gases
  Phase changes
Reflection and projection

Section 1.2. Atomic Models   1-11
  Electrical nature of matter
  Atoms and elements
  Electron-shell atomic model

Section 1.3. Molecular Models   1-17
  Molecular bonding and structure

Section 1.4. Valence Electrons in Molecular Models: Lewis Structures   1-20
  Lewis structures
  Connectivity
Reflection and projection

Section 1.5. Arranging Electron Pairs in Three Dimensions   1-25
  Tetrahedral arrangement
  Modified ball-and-stick model

Section 1.6. Polarity of the Water Molecule   1-29
  Effect of electric charge on polar molecules
  Water molecules are polar
  Bond polarity
  Electronegativity
Reflection and projection

Section 1.7. Why Is Water Liquid at Room Temperature?   1-37
  Induced-dipole attractions
  Attractions among polar molecules
  The hydrogen bond

Section 1.8. Further Structural Effects of Hydrogen Bonding in Water   1-45
  Ice: a hydrogen-bonded network
  Variation of liquid water density with temperature

Section 1.9. Hydrogen Bonds in Biomolecules   1-48
  Proteins
  Protein folding
  Nucleic acids
  Strength of hydrogen bonds
Reflection and projection

Section 1.10. Phase Changes: Liquid to Gas   1-54
  Energy diagrams
  Energy units and conversions
  Correlations with molecular models

Section 1.11. Counting Molecules: The Mole   1-60
  Mole and molar mass
  Molar vaporization energies
  Vaporization energies and molecular attractions
Reflection and projection

Section 1.12. Specific Heat of Water: Keeping the Earth's Temperature Stable   1-68
  Molar heat capacity
  Specific heat
  Temperature
  Thermal energy (heat)
  Explaining the high specific heat of water

Section 1.13. Outcomes Review   1-74

Section 1.14. EXTENSION — Liquid Viscosity   1-76

CHAPTER 2 Aqueous Solutions and Solubility

Section 2.1. Substances in Solution 2-6
Solution nomenclature
The solution process
Favorable and unfavorable factors

Section 2.2. Solutions of Polar Molecules in Water 2-9
Hydrogen bonds among unlike molecules
Nonpolar solutes
Molecular reorganization
Intermediate cases
Like dissolves like
Solutes with multiple polar groups
Reflection and projection

Section 2.3. Characteristics of Solutions of Ionic Compounds in Water 2-16
Solution conductivity
Ionic solids and solutions
Ion-dipole attractions in ionic solutions

Section 2.4. Formation of Ionic Compounds 2-20
Names and Formulas
Why ionic compounds form
Chemical reaction equations
Formation of ionic crystals

Section 2.5. Energy Changes When Ionic Compounds Dissolve 2-27
Solubility: hydration energy and lattice energy
Reflection and projection

Section 2.6. Precipitation Reactions of Ions in Solution 2-31
Formation of precipitates
Identity of the precipitate
Ionic reaction equations
Equilibrium

Section 2.7. Solubility Rules for Ionic Compounds 2-36
Solubility rules for ionic compounds
Reorganization factors in ionic solubility
Exceptions to the rules
Reflection and projection

Section 2.8. Concentrations and Moles 2-41
Concentration
Moles and molarity

Section 2.9. Mass–Mole–Volume Calculations 2-46
Preparing solutions of known molarity
Reflection and projection

Section 2.10. Reaction Stoichiometry in Solutions 2-50
Stoichiometric reaction ratios
Limiting reactant
Amount of product formed
Amount of reactant remaining
Reflection and projection

Section 2.11. Solutions of Gases in Water 2-57
Water as a reactant

Section 2.12. The Acid–Base Reaction of Water with Itself 2-61
The pH scale
Brønsted-Lowry acids and bases

Section 2.13. Acids and Bases in Aqueous Solutions 2-64
Oxyacids
Ionic compounds with oxyanions
Nucleic acids are oxyanions
Carboxylic acids
Extent of proton-transfer reactions
A Brønsted-Lowry base: ammonia
Review of properties of solutions of gases
Brønsted-Lowry acid–base pairs

Stoichiometry of H_3O^+(aq)–OH^-(aq) Reactions
Reflection and projection

Section 2.14. Outcomes Review 2-75

Section 2.15. EXTENSION — CO_2 and Le Chatelier's Principle 2-77
The carbon cycle
Reactions in carbon dioxide–limewater solutions
Le Chatelier's Principle

CHAPTER 3 Origin of Atoms

Section 3.1. Spectroscopy and the Composition of Stars and the Cosmos 3-6
Spectroscopy
Continuous and line spectra
Line spectra of elements and stars
Elemental abundance in the universe
Trends in elemental abundance
Reflection and projection

Section 3.2. The Nuclear Atom 3-14
The nuclear atom
Protons and atomic number
Neutrons and mass number
Isotopes
Ions

Section 3.3. Evolution of the Universe: Stars 3-20
The Big Bang theory
The first nuclear fusions
The life and death of a star
The birth of a new star
Reflection and projection

Section 3.4. Nuclear Reactions 3-25
Emissions from nuclear reactions
Balancing nuclear reactions
Positron–electron annihilation
Radioactive decay
Half life
Radio isotopes and human life

Section 3.5. Nuclear Reaction Energies 3-37
Mass is not conserved in nuclear reactions
Comparison of nuclear and chemical reactions
Nuclear binding energies
Fusion, fission, and stable nuclei
Nuclear chain reactions
Reflection and projection

Section 3.6. Cosmic Elemental Abundance and Nuclear Stability 3-48
Trends in elemental abundance
Correlations between abundance and nuclear binding energy

Section 3.7. Formation of Planets: The Earth 3-51
The age of the earth
The elements of life
Reflection and projection

Section 3.8 Outcomes Review 3-56

Section 3.9. EXTENSION—Isotopes: Age of the Universe and a Taste of Honey 3-58
Age of stars and the universe
Stable isotopes and pure honey

CHAPTER 4 Structure of Atoms

Section 4.1. Periodicity and the Periodic Table 4-6
Origin of the periodic table
Filling gaps in the periodic table
Ionization energy

Section 4.2. Atomic Emission and Absorption Spectra 4-11
Spectrum of the sun: elemental analysis by absorption of light
Reflection and projection

Section 4.3. Light as a Wave 4-14
Wave nomenclature
Superimposed waves and diffraction
Diffraction of light
Electromagnetic waves
Electromagnetic spectrum
Source of electromagnetic radiation

Reflection and projection

Section 4.4. Light as a Particle: The Photoelectric Effect 4-24
Emission from glowing objects
Planck's quantum hypothesis
The photoelectric effect
The dual nature of light

Reflection and projection

Section 4.5. The Quantum Model of Atoms 4-33
Quantized electron energies in atoms
Electron energy levels in atoms

Section 4.6. If a Wave Can Be a Particle, Can a Particle Be a Wave? 4-38
Wavelength of a moving particle
Experimental evidence for electron waves

Reflection and projection

Section 4.7. The Wave Model of Electrons in Atoms 4-41
Waves and atomic emissions
Standing electron waves
Probability picture of electrons in atoms
Orbitals

Section 4.8. Energies of Electrons in Atoms: Why Atoms Don't Collapse 4-45
Kinetic energy of an electron
Potential energy of a nucleus and an electron
Total energy of an atom
Why atoms don't collapse
Ionization energy for the hydrogen atom

Reflection and projection

Section 4.9. Multielectron Atoms: Electron Spin 4-50
The electron wave model for helium
The electron wave model for lithium: a puzzle
Magnetic properties of gaseous atoms
Electron spin
Pauli exclusion principle
Exclusion principle applied to lithium: puzzle solved

Section 4.10. Periodicity and Electron Shells 4-54
Patterns in the first ionization energies of the elements
Electron shell model for atoms
Atomic radii
Electronegativity
A closer look at the electron shell model
Electron spin and the shell model

Reflection and projection

Section 4.11. Wave Equations and Atomic Orbitals 4-64
Schrödinger wave equation
Wave equation solutions: orbitals for one-electron atoms
Electron configurations
Energy levels in multielectron atoms
Electron configurations for multielectron atoms

Section 4.12. Outcomes Review 4-72

Section 4.13. EXTENSION — Energies of a Spherical Electron Wave 4-74
Kinetic energy of an electron wave
Potential energy of an electron wave
Total energy of an atom

CHAPTER 5 Structure of Molecules

Section 5.1. Isomers 5-6

Section 5.2. Lewis Structures and Molecular Models of Isomers 5-9

Atomic connections in isomers
Reflection and projection

Section 5.3. Sigma molecular orbitals 5-15
Molecular orbitals
Localized, one-electron σ (sigma) molecular orbitals
Sigma bonding orbitals
Sigma nonbonding molecular orbitals

Section 5.4. Sigma Molecular Orbitals and Molecular Geometry 5-21
The sigma molecular framework
Molecular shapes
Bond angles
Molecules containing third period elements
Why are there only four sigma orbitals on second period elements?

Reflection and projection

Section 5.5. Multiple Bonds 5-28
Double bonds and molecular properties
Triple bonds
Multiple bonds in higher period atoms

Section 5.6. Pi Molecular Orbitals 5-33
Pi orbital standing wave
Sigma-pi molecular geometry with one pi orbital
Triple bonds: sigma-pi geometry with two pi orbitals

Reflection and projection

Section 5.7. Delocalized pi orbitals 5-39
Bond order
Orbital energies

Section 5.8. Representations of Molecular Geometry 5-47
Tetrahedral representation
Trigonal planar representation
Lewis structures and molecular shape
Condensed structure for 1-butanol
Skeletal structure for 1-butanol

Reflection and projection

Section 5.9. Stereoisomerism 5-56
Cis- and trans-isomers
Stereoisomers
Polarized light and isomerism
Tetrahedral arrangement around carbon

Reflection and projection

Section 5.10. Functional Groups — Making Life Interesting 5-65
Alkanes
Functional groups
Alkenes
Functional groups containing oxygen and/or nitrogen
Alcohols
Carbonyl compounds
Carboxyl compounds
Multiple functional groups

Reflection and projection

Section 5.11. Molecular Recognition 5-72
Noncovalent interactions
Molecular recognition

Section 5.12. Outcomes Review 5-74

Section 5.13. EXTENSION — Antibonding Orbitals: The Oxygen Story 5-76
Paramagnetism of oxygen
Antibonding pi molecular orbitals
Molecular orbital model for oxygen
Ground-state oxygen reacts slowly

Volume Two

CHAPTER 6 Chemical Reactions

Section 6.1. Classifying Chemical Reactions
Identifying reaction products
Classifying chemical reactions

Section 6.2. Ionic Precipitation Reactionss
Calcium–oxalate reaction stoichiometry
Continuous variations

Reflection and projection

Section 6.3. Lewis Acids and Bases: Definitions
Types of Lewis acid–base reactions

Section 6.4. Lewis Acids and Bases: Brønsted-Lowry Acid–Base Reactions
Strong and weak Brønsted-Lowry acids and bases
Protons and electron pairs
Relative strengths of Lewis bases

Reflection and projection

Section 6.5. Predicting Strengths of Lewis/Brønsted-Lowry Bases and Acids
Electronegativity and relative Lewis base strength
Atomic size and relative Lewis base strength
Oxyacids and oxyanions: carboxylic acids
Delocalized p bond in carboxylate
Energetics of proton transfer
Oxyacids and oxyanions of other elements
Comparisons within and between periods
Predominant acid and base in a reaction

Reflection and projection

Section 6.6. Lewis Acids and Bases: Metal Ion Complexes
Colors of metal ion complexes
Calcium–EDTA complex ion formation
Complex ion formation and solubility
Metal ion complexes with four ligands
Porphine-metal ion complexes in biological systems

Reflection and projection

Section 6.7. Lewis Acids and Bases: Electrophiles and Nucleophiles
Nucleophiles and electrophiles
Alcohol-carboxylic acid reaction
Condensation polymers
Biological condensation polymers

Section 6.8. Formal Charge
Rules for formal charge
Interpretation of formal charges
Reactions reduce formal charges

Reflection and projection

Section 6.9. Oxidation-Reduction Reactions: Electron Transfer
Direction of an oxidation-reduction reaction
Charge balance
Oxidation numbers
Interpreting oxidation numbers
An alternative way to assign oxidation numbers

Reflection and projection

Section 6.10. Balancing Oxidation-Reduction Reaction Equations
The oxidation-number method
The half-reactions method
Oxidation-reduction reactions in basic solutions

Section 6.11. Oxidation-Reduction Reactions of Carbon-Containing Molecules   6-00
Oxidation of methanal by silver ion
Yeast fermentation of glucose

Section 6.12. Outcomes Review

Section 6.13. EXTENSION — Titration
Ethanoic (acetic) acid reaction with hydroxide

Titration
Ascorbic acid–iodine reaction

CHAPTER 7 Chemical Energetics: Enthalpy

Section 7.1. Energy and Change
Measuring energy
Energy conservation and chemical changes

Section 7.2. Thermal Energy (Heat) and Mechanical Energy (Work)
Directed and undirected kinetic energy
Chemical reaction energies

Section 7.3. Thermal Energy (Heat) Transfer
Thermal energy transfer by radiation
Thermal energy transfer by contact
Reflection and projection

Section 7.4. State Functions and Path Functions
State functions
Path functions

Section 7.5. System and Surroundings
Open, closed, and isolated systems
Reflection and projection

Section 7.6. Calorimetry and Introduction to Enthalpy
Systems and surroundings
Enthalpy
Constant-pressure calorimeter
Thermal energy change and temperature
Calorimeter heat capacity
Calculating enthalpy change
Improving calorimetric measurements
Reflection and projection

Section 7.7. Bond Enthalpies
Chemical reactions: bond breaking and bond making
Bond enthalpy
Homolytic bond cleavage
Average bond enthalpies
Patterns among average bond enthalpies
Bond enthalpy calculations
Accuracy of bond enthalpy calculations
Reflection and projection

Section 7.8. Standard Enthalpies of Formation

Section 7.9. Harnessing Energy in Living Systems
Coupled reactions
Energy captured as ATP^{4-}
Reflection and projection

Section 7.10. First Law of Thermodynamics: Pressure–Volume Work
Definition of work
Constant volume and constant pressure reactions
Calculating pressure–volume work

Section 7.11. Enthalpy Revisited
Definition of enthalpy
Analysis of constant volume and constant pressure reactions
Pressure–volume work calculation
Reflection and projection

Section 7.12. What Enthalpy Doesn't Tell Us

Section 7.13. Outcomes Review

CHAPTER 8 Entropy and Molecular Organization

Section 8.1. Mixing and Osmosis
Mixing
Osmosis

Section 8.2. Probability and Change

Probability
Number of molecular arrangements

Section 8.3. Counting Molecular Arrangements in Mixtures
Mixing model
Arrangements for the unmixed system
Arrangements for the mixed system

Section 8.4. Implications for Mixing and Osmosis in Macroscopic Systems
Diffusion
Molecular model for osmosis

Reflection and projection

Section 8.5. Energy Arrangements Among Molecules
Model for energy arrangements
Counting energy arrangements

Section 8.6. Entropy
Net entropy: the second law of thermodynamics
Absolute entropy

Reflection and projection

Section 8.7. Phase Changes and Net Entropy
A dilemma
Melting and freezing water
Net entropy change for ice melting
Magnitude of thermal entropy changes
Quantitative expression for thermal entropy change
Dilemma resolved
Factors that influence the direction of a phase change
Phase equilibrium

Section 8.8. Gibbs Free Energy
Focusing on the system
ΔS_{net} as a function of system variables
Shift to an energy perspective
Gibbs free energy and the direction of change
Criterion for equilibrium
Why a new thermodynamic variable?
Why "free" energy?

Reflection and projection

Section 8.9. Thermodynamics of Rubber
Molecular structure of rubber

Section 8.10. Colligative Properties of Solutions
Freezing point of a solution
Entropy change for freezing from solution
Solution-solid equilibrium
Freezing point lowering
Colligative properties
Origin of colligative properties
Quantifying freezing point lowering
Effect of solute ionization on colligative properties
Boiling point elevation

Section 8.11. Osmotic Pressure Calculations
Osmotic pressure
Explaining osmotic pressure
Quantifying osmotic pressure

Reflection and projection

Section 8.12. Thermodynamic Calculations for Chemical Reactions
General equation for calculating entropy changes
Entropy example: glucose oxidation
Free energy change for glucose oxidation
Entropy example: dissolving ionic solids
Entropies of ions in solution
Significance of ΔG and $\Delta G°$

Reflection and projection

Section 8.13. Why Oil and Water Don't Mix
Oriented water molecules
Clathrate formation

Section 8.14. Ambiphilic Molecules: Micelles and Bilayer Membranes
Ambiphilic molecules
Micelle formation
Detergent action
Phospholipid bilayers
Getting through a phospholipid bilayer membrane

Section 8.15. The Cost of Molecular Organization

Section 8.16. Outcomes Review

CHAPTER 9   Chemical Equilibrium

Section 9.1. The Nature of Equilibrium Identifying systems at equilibrium
$Fe^{3+}(aq)$ reaction with $SCN^-(aq)$
Equilibrium in the $Fe^{3+}(aq)$–$SCN^-(aq)$ system
Co^{2+} complexes with water and chloride
Temperature and the Co^{2+}–water–chloride system

Reflection and projection

Section 9.2. Mathematical Expression for the Equilibrium Condition
The equilibrium constant expression
Standard states in solution
Other standard states

Section 9.3. Acid–Base Reactions and Equilibria
The acetic acid–water reaction
Solution pH for an acid of known K_a
More elaborate equilibrium calculations
The autoionization of water
The acetate ion–water reaction
Relationship between K_a and K_b for a conjugate acid–base pair
pK

Reflection and projection

Section 9.4. Solutions of Conjugate Acid-Base Pairs: Buffer Solutions
Conjugate base-to-acid ratio
Acid–base buffer solutions
Buffer pH
Preparing a buffer solution

Section 9.5. Acid–Base Properties of Proteins
Acid–base side groups on proteins
Isoelectric pH for proteins
Electrophoresis
Acids, bases, and sickle-cell hemoglobin

Reflection and projection

Section 9.6. Solubility Equilibria for Ionic Salts
Solubility product
Solubility and solubility product
Common ion effect

Reflection and projection

Section 9.7. Thermodynamics and the Equilibrium Constant
Changes under nonstandard conditions
Entropy change for a gas phase reaction
Free energy change for a gas reaction
The reaction quotient and the equilibrium constant

Section 9.8. Temperature Dependence of the Equilibrium Constant
Reflection and projection

Section 9.9. Thermodynamics in Living Systems
Free energy for cellular reactions
Hydrolysis of ATP
Equilibrium constant for hydrolysis of ATP
Free energy change for cellular ATP reactions
Equilibria in oxygen transport

Section 9.10. Outcomes Review

Section 9.11. EXTENSION—Competing Equilibria
Equilibrium constant for a reaction going in reverse
Multiple equilibria
Combined equilibrium constants

CHAPTER 10   Reduction-Oxidation: Electrochemistry

Section 10.1. Electrolysis
Electrolysis of water
Anode and cathode
Stoichiometry of electrolysis
Applications of electrolysis

Reflection and projection

Section 10.2. Electric Current from Chemical Reactions
Separated half reactions
Salt bridge
Half cells

Section 10.3. Work From Electrochemical Cells
Cell potential, E
The sign of E

Reflection and projection

Section 10.4. Concentration Dependence of Cell Potentials
pH dependence of the silver-quinhydrone cell potential
Le Chatelier's principle and cell potentials
Electrochemical cell notation

Section 10.5. Free Energy and Electrochemical Cells: The Nernst Equation
$[Ag^+(aq)]$ and cell potential
Free energy change and cell potential
The Nernst equation
E^0 and the equilibrium constant, K

Section 10.6. Combining Cell Potentials for Reactions
Combining reactions and free energies
Combining cell potentials

Reflection and projection

Section 10.7. Half-Cell Potentials: Reduction Potentials
The standard hydrogen electrode
Measuring a reduction potential
Signs of reduction potentials
Table of standard reduction potentials
The direction of redox reactions

Section 10.8. Reduction Potentials and the Nernst Equation
Nernst equation for a half reaction
Reduction potentials under biological conditions

Reflection and projection

Section 10.9. Carbon-Containing Reducing Agents: Glucose
The Cu^{2+}–Cu^+ redox system
Glucose is oxidized to gluconic acid
Redox reaction in the Benedict's test
Reducing sugars
A glucose biofuel cell

Section 10.10.   Coupled Redox Reactions
The Blue Bottle redox reaction(s)
Methylene blue couples glucose oxidation to oxygen reduction
Coupled reactions in a biofuel cell
Glucose oxidation in aerobic organisms

Reflection and projection

Section 10.11.   Outcomes Review

Section 10.12. EXTENSION—Cell Potentials and Non-Redox Equilibria

Concentration cells
Cell potentials and complexed ions

CHAPTER 11 Reaction Pathways

Section 11.1. Pathways of Change
Reactant concentration effects
Temperature effect
Catalysis

Section 11.2. Measuring and Expressing Rates of Chemical Change
Rate of reaction
Initial reaction rates
Alternative rate expressions
Rate expressed in molarity

Reflection and projection

Section 11.3. Reaction Rate Laws
Dependence of $H_2O_2(aq)$ decomposition rate on $[H_2O_2(aq)]$
Rate laws
Reaction order

Section 11.4. Reaction Pathways or Mechanisms
Reactions occur in steps
Bottlenecks: rate-limiting steps
Applying the concept of rate-limiting steps
Reaction mechanisms are tentative

Reflection and projection

Section 11.5. More Ways to Analyze Rate Data
Radioactive decay
Progress of a first order reaction with time
Measurement of radioactive decay
Half life and first-order rate constant
Other first order reactions
Kinetics of the acetone iodination reaction
Reaction pathway for acetone iodination
Protonation of carbonyl oxygen
Formation of an enol intermediate
Enol reaction with triiodide

Reflection and projection

Section 11.6. Temperature and Reaction Rates
The frequency factor, A
Activation energy for reaction
Molecular energy distributions

Section 11.7. Light: Another Way to Activate a Reaction
Rate of photochemical reactions
Competing reactions and the steady state
Light and life

Reflection and projection

Section 11.8. Thermodynamics and Kinetics

Section 11.9. Outcomes Review

Section 11.10. EXTENSION — Enzymatic Catalysis
Enzymes and rate laws
Substrate concentration effects
Michaelis-Menten mechanism for enzyme catalysis
Limiting cases for the Michaelis-Menten mechanism
Determining and interpreting K
Enzyme specificity
Enzyme active sites

Introduction

Everything you hear, see, smell, taste, and touch involves chemistry and chemicals (matter). And hearing, seeing, smelling, tasting, and touching all involve intricate series of chemical reactions and interactions in your body. With such an enormous range of topics, chemistry offers you fascinating opportunities to explore and study. At the same time, all these possibilities make chemistry seem a daunting subject to study. Aware of both the fascination and the challenge of studying chemistry, the American Chemical Society chose a team of chemists to consider what concepts would help you open the doors to opportunities that require a knowledge of chemistry without being overwhelming. The team also took up the challenge to develop effective approaches to learning and teaching chemistry. The result of the team's efforts are this textbook, **Chemistry**, and its complementary materials, including project-based laboratory experiments, your molecular model kit, the **Web Companion**, and the **Personal Tutor**.

Learning chemistry, even with a limited range of concepts and content, requires a good deal of effort from both you and your instructors. To facilitate your efforts, we have written **Chemistry** in a conversational tone designed to be accessible and engaging. But you cannot learn chemistry only by reading about it, just as you cannot learn how to write a short story or how to find fossils simply by reading about how others do it. Learning how others do something you want to do is important, but you must also practice doing it yourself. Chemists and other scientists learn about the world through experimenting. They then try, often in collaborative efforts, to develop models of the world at the molecular level that explain their results and allow them to predict the outcomes of other possible experiments. We have tried to incorporate this same approach in this textbook.

Throughout **Chemistry**, we present activities and thought-provoking questions that are intended to promote active small-group and whole-class participation. To encourage your participation and collaborative learning efforts, four features appear often in each chapter:

Intro.1. **Investigate This**

An *Investigate This* usually involves short experiments that introduce the chemical concepts explored in the following paragraphs. The investigations are designed to be carried out in small groups or in the whole class setting.

Intro.2. Consider This

A *Consider This* follows each *Investigate This* and usually asks you to discuss and develop hypotheses or explanations for what you have observed. At other places a *Consider This* will ask you to think about and discuss the consequences of what has just been presented or to anticipate what is to come. The intent in all cases is to involve the class in a discussion.

Intro.3. Worked Example

Each *Worked Example* guides you through the reasoning involved in solving a problem. Thinking about *how* to solve a problem is often more important and more challenging than actually carrying out the solution procedure, so we place an emphasis on this thinking. Almost all Worked Examples include the following components, after the statement of the problem:

Necessary information: What do you need to know, including the information from the problem statement, in order to solve the problem?

Strategy: How do you put the information together in order to solve the problem? What concepts are involved and how are they to be used? With an appropriate strategy (there is often more than one) in hand, the problem is essentially solved.

Implementation: Carry out the strategy using the needed information to obtain the solution to the problem. Calculations, if necessary, are done at this stage.

Does the answer make sense? Once you get an answer to a problem, you should always check to be sure it makes sense. You should also check to be sure you have carried out any numerical calculations correctly, but making *sense* of the answer is a distinct task (and can sometimes flag possible numerical problems). Is the answer about the size you would expect (based on other experiences, for example)? Does it have the expected direction (sign or change from some baseline)? And so on…

Intro.4. Check This

At least one *Check This* follows each *Worked Example* and presents a similar problem or problems so that you can practice the strategy presented in the *Worked Example*. *Check This* problems also appear in other places, where you are asked to practice some technique or answer questions based on what has just been presented in the text. ***Chemistry*** is designed to be used with paper, pencil, calculator, and model kit at hand, so you can try each *Check This* as you come to it.

In addition to these features within each chapter, there is an *Outcomes Review* section near the end of each chapter and *End-of-Chapter Problems* you can use to test your problem-solving skills. Use the *Outcomes Review* to remind yourself of the important ideas from the chapter and the *End-of-Chapter Problems* to check your understanding of these ideas. Some of these problems will give you more practice with the kinds of problems you meet in the *Worked Example* and *Check This* activities throughout the chapter. Other *End-of-Chapter Problems* are included to stretch your thinking and engage you in problem-solving strategies that are combinations of strategies introduced in the chapter or that extend a bit beyond them. Most scientists work cooperatively, and we encourage you to try working on these problems collaboratively as well. Often a group can come up with more and better solutions than an individual working in isolation.

Throughout **Chemistry**, you will find an emphasis on understanding and reasoning, and on models of all kinds: physical, computer, and mathematical models, and analogies. We use models, because it is difficult to observe individual atoms or molecules as they undergo the changes and interactions that lead to the events we can easily observe in nature or in the laboratory. As we try to understand the physical and chemical properties of atoms and molecules and how they cause observable effects we will use three levels of description, which are exemplified by this page from the **Web Companion**:

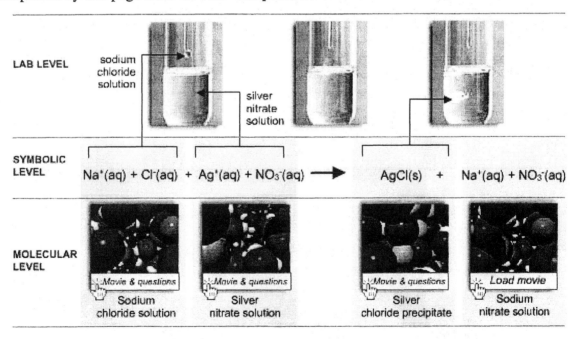

- *Lab Level:* These are the observations you make on macroscopic systems, such as that shown in these frames from a movie of a reaction between two solutions in a test tube or in your Investigate This activities.

- *Molecular Level:* These are our models of what is going on among the particles (atoms, ions, and molecules) that gives rise to the effects observed in the laboratory. These models are animated in the *Web Companion* and are usually shown as less complicated still figures in the text. You will also often use models you build yourself with your molecular model kit.

- *Symbolic Level:* Intermediate between and connecting the Lab and Molecular Levels is the Symbolic Level of description. This is the descriptive level which you probably associate with chemistry. This level is essential, because it combines a great deal of information in a succinct format. It is also the most abstract level of representation, since all the symbols need to be interpreted to make sense of the description. The usual approach in *Chemistry* will be first to try to understand systems at the Lab and Molecular Levels and only then proceed to the Symbolic Level.

The *Web Companion*, from which the above figure is taken, is designed to provide you opportunities to use interactive animations, movies, and other resources that provide a visual (usually moving) means to examine many of the concepts included in the written text. When a

> WEB Chap X, Sect X.5.3-4
> A brief description of what you will find on these pages is given here.

Web Companion page (or pages) is available for some concept, this marginal box appears in the text. To access the *Web Companion*, visit www.whfreeman.com/acsgenchemhome and select "Web Companion." Find the appropriate location in the *Companion* by selecting the chapter and subsection referenced. The Companion may also be available on your institutional computers and, if so, your instructors will tell you how to access it. There are also *Consider This* and *Check This* problems, as well as *End-of-Chapter Problems*, based on the *Web Companion*. These are denoted by this symbol, WEB , and a reference to the chapter and section you will need to access.

A very large percentage of students who take the general chemistry course in a college or university have already had at least one year of a high school chemistry course. We assume that you are in this category and have probably been exposed to a good deal of the nomenclature and methods that are part of the study of chemistry. We take advantage of this background to move quickly into an examination of the properties of water that depends on some familiarity with the properties of atoms and molecules. You probably have also done some of the algebraic and arithmetic calculations that are a part of essentially all beginning chemistry courses. We take advantage of this experience as well, by providing a review of only necessary concepts and then using them to try to answer questions based on our initial studies of water. The brief reviews we provide in the text may, however, not be enough to make you comfortable with the problems we pose, so we have provided a *Personal Tutor*.

The primary purpose of the *Personal Tutor* is to give you more guidance and practice with problems and computations in the areas that seem to give students trouble. Using the Personal Tutor is much like visiting your instructor during office hours or going to a human tutor. Marginal boxes like this one will alert you to a section in the *Personal Tutor* that might be helpful for the topic under discussion. Before using the *Tutor*, for the first time, visit

> **Personal Tutor**
> A brief description directs you to the section you might find helpful for this part of the text.

www.whfreeman.com/acsgenchemhome, select "Personal Tutor," and take the diagnostic exam. When you have completed the exam, you will get feedback on what sections of the *Personal Tutor* would be helpful for you to study. You may need to take advantage of it frequently or you may need its assistance few times or not at all during the course. The questions in the Tutor can be a good review when preparing for tests, even if you are confident that you know how to solve the problems or do the computations. We urge you to take advantage of this resource in whatever ways it can be helpful.

We have outlined above how we designed this textbook and its complementary materials to provide you, your classmates, and your instructors the resources to learn and teach chemistry actively and interactively. Now let us return to the first task the American Chemical Society team considered, what concepts to include in **Chemistry**. Several concepts or "big ideas" recur in one form or another through the book. Brief statements of these concepts are:

- Attractions between positive and negative centers hold matter together and are responsible for chemical reactions.
- The lower its energy, the more stable the system.
- During change, energy is conserved: $\Delta E_{net} = 0$.
- The properties of elements repeat periodically as the atomic number of their atoms increases.
- Electrons in atoms and molecules act like matter waves with quantized energies; the more spread out a matter wave, the lower (more favorable) its energy.
- Change occurs in the direction that increases the number of distinguishable arrangements of particles and/or energy quanta. Entropy, S, is a measure of this number, and in all spontaneous processes, net entropy increases, $\Delta S_{net} > 0$.
- Reactions are at equilibrium when $\Delta S_{net} = 0$ for the change from reactants to products.

- When a reaction at equilibrium is disturbed, the system reacts to minimize the disturbance. This is LeChatelier's principle. Reactions at equilibrium are quantitatively described by a temperature dependent equilibrium constant ratio.

- Electric current can produce reduction-oxidation chemical reactions. Reduction-oxidation chemical reactions can produce an electric current.

- The rate of a chemical reaction depends on the concentrations of species and the temperature of the system. These are a result of the reaction pathway.

Some of the concepts in this list may look familiar and others probably do not. Our goal in structuring *Chemistry* to emphasize active and collaborative learning has been to provide the means for you to understand these concepts. The understanding you gain will allow you to apply the concepts not only to the problems we and your instructors provide, but also to the problems and systems you meet in other courses, and most importantly to interesting and intriguing systems you meet in the world outside the classroom. It has been an enjoyable challenge to write *Chemistry* and develop the complementary materials. We hope it is an enjoyable challenge to use them and learn chemistry.

Chapter 1. Water: A Natural Wonder

Section 1.1. Phases of Matter 1-6
    Density 1-7
    Solids 1-7
    Liquids 1-7
    Gases 1-8
    Phase changes 1-9

Reflection and projection 1-10

Section 1.2. Atomic Models 1-10
    Electrical nature of matter 1-11
    Atoms and elements 1-13
    Electron-shell atomic model 1-14

Section 1.3. Molecular Models 1-17
    Molecular bonding and structure 1-19

Section 1.4. Valence Electrons in Molecular Models: Lewis Structures 1-20
    Lewis structures 1-21
    Connectivity 1-22

Reflection and projection 1-24

Section 1.5. Arranging Electron Pairs in Three Dimensions 1-24
    Tetrahedral arrangement 1-26
    Modified ball-and-stick model 1-27

Section 1.6. Polarity of the Water Molecule 1-28
    Effect of electric charge on polar molecules 1-29
    Water molecules are polar 1-30
    Bond polarity 1-31
    Electronegativity 1-32

Reflection and projection 1-35

Section 1.7. Why Is Water Liquid at Room Temperature? 1-36
    Induced-dipole attractions 1-37
    Attractions among polar molecules 1-38
    The hydrogen bond 1-41

Section 1.8. Further Structural Effects of Hydrogen Bonding in Water 1-44
    Ice: a hydrogen-bonded network 1-44

Variation of liquid water density with temperature 1-46
Section 1.9. Hydrogen Bonds in Biomolecules 1-47
Proteins 1-47
Protein folding 1-48
Nucleic acids 1-50
Strength of hydrogen bonds 1-51
Reflection and projection 1-52
Section 1.10. Phase Changes: Liquid to Gas 1-52
Energy diagrams 1-53
Energy units and conversions 1-56
Correlations with molecular models 1-57
Section 1.11. Counting Molecules: The Mole 1-58
Mole and molar mass 1-59
Molar vaporization energies 1-63
Vaporization energies and molecular attractions 1-64
Reflection and projection 1-65
Section 1.12. Specific Heat of Water: Keeping the Earth's Temperature Stable 1-66
Molar heat capacity 1-68
Specific heat 1-69
Temperature 1-70
Thermal energy (heat) 1-70
Explaining the high specific heat of water 1-72
Section 1.13. Outcomes Review 1-72
Section 1.14. EXTENSION — Liquid Viscosity 1-74

Water in all its forms is essential in many ways for life on Earth and is also inviting for recreation and relaxation. Molecular level illustrations of the three phases of water are shown in the circles and you can see both solid and liquid water in the photograph. Gaseous water is invisible, but the clouds (fog) of tiny water droplets you see are evidence that water molecules leave the surface of the warm water as a gas and then condense as they move into the cold air above the pool.

Chapter 1. Water: A Natural Wonder

> O plunge your hands in water
> Plunge them up to the wrist;
> Stare, stare in the basin
> And wonder what you've missed.
>
> *W. H. Auden (1907–1973),*
> As I Walked Out One Evening

The quotation from Auden describes something we do many times every day, but few of us stop to wonder what we have missed. Water is by far our most familiar chemical compound, and we spend much of our lives in contact with water as a liquid, solid, or gas. The example in the picture on the facing page shows people enjoying a plunge in the pool formed by a warm thermal spring that is surrounded by new-fallen snow. Because it is so familiar, we seldom pause to consider what a unique and remarkable substance water is.

About 70% of the human body mass is water, and everything we drink and much of what we eat is mostly water. Living systems are fundamentally **aqueous** systems and one emphasis of this textbook is the chemistry that is applicable to living systems. We will often use this chemistry

> Aqueous is from Latin, *aqua* = water, which also shows up in "aquarium," "aquatic," and so on.

as a jumping-off point to develop a systematic understanding of chemistry that applies to the chemistry in organisms as well as to chemistry in the laboratory and in industrial processes. While we will concentrate on the important chemistry that occurs in aqueous media, we will apply this understanding to other conditions as well.

We will start with the bulk properties of water and then go on to see how the structure of the water molecule helps to explain these properties. Along the way, as we develop concepts to explain how water acts, we will apply these concepts to other molecules and substances. You are probably familiar with several of the concepts in this chapter, so this will be a review of what you have already learned. Many of these ideas will be developed in more detail in later chapters, so this will also be a look ahead to where we are going. In most cases we will begin with experiments and observations and use these to develop molecular-level explanations that will

> **Personal Tutor**
> Try the on-line diagnostic exam to check your background knowledge and find out how you might best use the **Tutor**. Also, skim Sections 1.1—1.5 of the textbook and try several of the Check This and Consider This problems as well as End-of-Chapter Problems to find out what concepts you may need to spend more time on.

account for the results. In all cases, we will emphasize reasoning from evidence and always look for further evidence to support our explanations.

Section 1.1. Phases of Matter

> ***1.1. Consider This***   *What do you already know about water?*
> Use words and/or drawings to make a list of all the properties of water that you know from experience or have learned about in previous courses. Which properties do you think are essential for life on Earth? Why are they essential? Compare your list and analysis with those of your classmates. What properties of water did you learn about from others in your class?

Your list and analysis from Consider This 1.1 probably confirmed that you already know a lot about the properties of water at both the macroscopic and molecular levels. Did your list include the fact that water is the only common substance found in all three **phases of matter**, solid, liquid, and gas, on the surface of the Earth under normal conditions? The photograph that opens this chapter exemplifies this observation. There you can see the liquid water of the thermal spring pool and the solid water of the snow covering the shrubs, trees, and mountains in the background. You cannot see any gaseous water, because gaseous water, which we usually call water vapor, is invisible. But you have evidence that water vapor is leaving the warm water surface, because you can see the mist or fog of tiny liquid water droplets (water dust) formed as the vapor condenses (liquefies) in the cold air above the pool. You can also observe this effect in the water vapor leaving the spout of a teakettle of boiling water, as shown in Figure 1.1.

> "Steam" is another name for gaseous water near its boiling point. The mist of visible condensed water droplets is often incorrectly called steam, so we usually will not use the word steam.

Figure 1.1. Water vapor and a cloud of liquid water droplets from a teakettle.
Invisible water vapor leaving the spout expands into the cooler surrounding air and condenses to form a visible cloud of tiny water droplets (water dust) a few centimeters from the spout.

Density

Density is a measure of how much matter occupies a given volume. Mass is used to measure how much matter is present, so the density of a substance is expressed as the mass of the substance per unit volume in units of g·mL^{-1} or kg·L^{-1}. Density is one of the physical properties we can use to identify and characterize a pure substance.

Solids

Solids have a definite volume and shape that do not depend upon the size and shape of their container. The density of solid water is 0.917 g·mL^{-1} (at 0 °C). The densities of other pure solids range from about 0.5 to 20 g·mL^{-1}. Figure 1.2(a) shows a molecular-level model of solid water (ice). The molecules in ice form an orderly array with the molecules attracted to one another, close to each other, and fixed in place with respect to one another. If Figure 1.2(a) represents a "snapshot" of the molecules taken at a particular instant, another snapshot taken later would show the same molecules in the same locations.

> **WEB Chap. 1, Sect. 1.1**
> Interactive molecular-level animations of the phases and phase changes of water

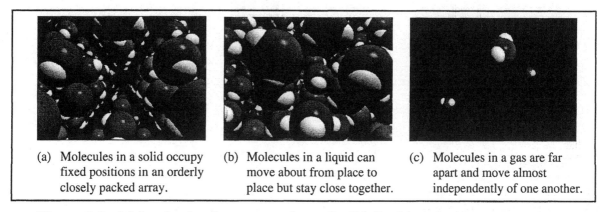

(a) Molecules in a solid occupy fixed positions in an orderly closely packed array.

(b) Molecules in a liquid can move about from place to place but stay close together.

(c) Molecules in a gas are far apart and move almost independently of one another.

Figure 1.2. Molecular-level representations of solid, liquid, and gaseous water.
These are space-filling representations of water molecules, which are discussed in Section 1.6.

Liquids

Liquids have a definite volume, but take the shape of their container; they are fluid (pourable). The density of a pure liquid substance is nearly the same as the density of the same substance in the solid phase. For example, at the melting point of water (the temperature at which solid ice becomes liquid water) the density of liquid water is 1.000 g·mL^{-1}. To explain the density, Figure 1.2(b) shows the molecules in liquid water attracted to one another and close together, as they are in the solid. To explain the fluidity of liquids, our model assumes the molecules in a liquid can move about, so they are shown in a disorderly jumble in the figure. A snapshot taken at a later time would show a different jumble than the one we see in Figure 1.2(b) and would probably show some different water molecules.

Note that the density of liquid water is greater than the density of solid water. Did your list in Consider This 1.1 include the fact that solid water (ice) floats on liquid water? Water is the *only* common substance that exhibits this property. For all other substances, the density of the solid phase is larger than the density of the liquid phase, so the solids sink in the liquids, as shown in Figure 1.3 for a solid and liquid alcohol compared to ice and water.

Figure 1.3. Comparison of the densities of solid and liquid (a) *t*-butanol and (b) water.

1.2. Consider This *How do the volumes of liquid and solid water compare?*

(a) If you freeze 1 L of liquid water, will the volume of the solid ice be larger, smaller, or the same as the liquid? Explain the reasoning for your answer.

(b) [WEB] Chap 1, Sect 1.1.3. Is there evidence on this page of the *Web Companion* to support your answer in part (a)? What is the evidence and how does it support your answer.

Gases

Gases take the volume and shape of their container. The density of a pure gaseous sample varies with the pressure and temperature of the gas because the volume of the sample depends strongly on these variables. To compare gas densities, we choose a standard set of conditions and measure or calculate the density of all gases under these conditions. A common choice for this comparison is the gas at one atmosphere pressure and 0 °C, which is called **standard temperature and pressure (STP)** conditions. The density of gaseous water at STP is 8.03×10^{-4} g·mL^{-1}. Our representation of gaseous water (water vapor), Figure 1.2(c) shows the molecules far apart and moving essentially independently of one another. Since there are many fewer

> The densities of solids and liquids also vary with temperature, but the variation is much smaller than for gases. The variation of solid and liquid densities with pressure is usually negligible.

molecules in a given volume of gas than the same volume of liquid, the density of a pure gaseous substance is much less than the density of the same substance in the liquid or solid phase. The independent motion of the molecules means that they can move about to occupy as much volume as is available to them.

Phase changes

To change a solid to liquid or a liquid to gas requires an input of energy to break the attractions that hold the molecules in place in the solid or together in the liquid. The **melting point** (mp) and **boiling point** (bp) are the temperatures at which, respectively, the solid to liquid and liquid to gas phase changes occur for a pure substance at a pressure of one atmosphere. Melting and boiling point temperatures are characteristic of a substance. The melting and boiling points of water, 0 °C and 100 °C, are familiar examples.

1.3. Check This *Molecular level representation of boiling water*

(a) When water boils at one atmosphere pressure and 100 °C, you observe lots of bubbles forming and rising to the surface of the liquid. Make a sketch representing the gas in a bubble and the liquid surrounding it. Use circles or other geometric shapes to represent molecules and label them to indicate what molecule each represents.

(b) WEB   Chap 1, Sect 1.1.5-6. Compare your representation in part (a) with the one that is animated on these pages.

The change of a liquid to a gas, **vaporization**, can occur at temperatures below the liquid boiling point, but the pressure of the gas that is formed is less than one atmosphere. Did you note in Consider This 1.1 that vaporization (which we also call evaporation) of water from your skin helps cool you down when you exercise? Water absorbs more heat per gram than any other substance when it changes from liquid to gas. Water also requires a great deal of heat just to warm it up. Since we are about 70% water, this capacity to absorb a lot of heat without warming too much helps us maintain a constant body temperature, even on hot days.

> We will discuss these properties in more detail in Sections 1.10, 1.11, and 1.12.

Vaporization of water from the warm thermal spring shown in the chapter-opening photograph explains where the water molecules come from that condense to form the mist above the pool. **Condensation**, the formation of a liquid from its gas (vapor) is the reverse of vaporization, and **freezing**, the formation of a solid from its liquid is the reverse of melting. We often refer to both liquids and solids as **condensed phases**, since their molecules are more tightly packed together than in the gas phase. Energy is released to the surroundings when a gas condenses or a liquid freezes. This is why you can be badly burned by hot water vapor (steam) if it condenses to water on your skin.

Because the range of temperatures at the surface of the Earth is from somewhat below to somewhat above 0 °C, water exists on Earth in all three phases. Water is unique; all other common substances exist naturally in only a single phase. For example, the major gases in the atmosphere, nitrogen (bp = -196 °C) and oxygen (bp = -183 °C) are never found naturally as liquids (or solids). Iron (mp = 1535 °C) and silicon dioxide (quartz, mp = 1610 °C) are never found naturally as liquids (or gases) at the surface of the Earth.

> In localized hot spots, like volcanoes or deep-sea vents, molten rock from deep beneath the surface actually reaches the surface of the Earth.

Reflection and projection

Because water is so familiar, perhaps you had never considered how strange its properties are compared to the large number of other substances you also see around you. Water occurs naturally as a solid, liquid, and gas at the surface of the Earth. Solid water floats on liquid water. A large quantity of thermal energy (heat) is required to raise the temperature of water and to vaporize it. We have shown how the physical phases of water can be visualized at the molecular level to help explain some of the observed properties of solids, liquids, and gases. However, these models are applicable to other substances as well as water and do not explain what is unique about water. We need to have a more detailed picture or model of water molecules and their interactions, in order to explain the properties of water that we can see, feel, and measure. Since molecules are made of atoms, we will begin with a brief reminder about atomic properties and then go on to consider water and other molecules. Briefly review those concepts in the next four sections that are already familiar and spend more time working through those that are new or less familiar.

Section 1.2. Atomic Models

It is common knowledge, even among people who have never studied chemistry, that an atom of oxygen and two atoms of hydrogen are somehow connected to form a water molecule, H_2O, "aitch-two-oh." On the other hand, it often comes as a surprise to learn that the connections result from the electrical nature of matter.

1.4. Investigate This *What electrical effects can you observe?*

Drape a 1 × 15-cm strip of thin plastic sheet from a plastic bag over one finger, so that equal lengths hang down on each side. Place one finger of your other hand *between* the two halves of the plastic strip. Quickly slide your hand down the full length of the free ends of the strip, with the plastic loosely pinched between your fingers and thumb. Immediately release the free ends. Observe what happens. Try rubbing the strips again. Observe what happens.

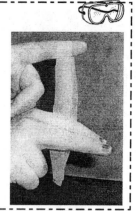

1.5. Consider This *Do electrical effects explain your Investigate This 1.4 results?*

What, if anything, does electricity have to do with your observations in Investigate This 1.4? Have you ever observed other effects similar to what you observed when you rubbed the plastic strip? What were they? What might they have to do with electricity?

Electrical nature of matter

The electrical nature of matter was recognized for centuries before scientists developed a model to explain it. You have probably experienced an electric shock after scuffing across a carpet or sliding across an automobile seat, and then touching some metal object. The phenomenon of objects acquiring electrical charge is illustrated by rubbing a balloon on cloth or your hair. Initially, as Figure 1.4(a) illustrates, the balloon and the cloth are each electrically neutral. The number of positive and negative particles in each is balanced and they just cancel one another out.

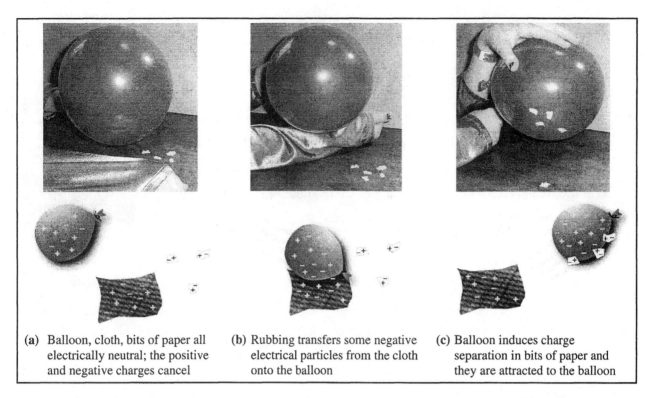

(a) Balloon, cloth, bits of paper all electrically neutral; the positive and negative charges cancel

(b) Rubbing transfers some negative electrical particles from the cloth onto the balloon

(c) Balloon induces charge separation in bits of paper and they are attracted to the balloon

Figure 1.4. Electrostatic attraction between an electrically charged balloon and bits of paper.

Figure 1.4(b) illustrates how negative charges are rubbed off the cloth and onto the balloon. This produces a negative electric charge on the surface of the balloon and leaves behind a net positive charge on the cloth. The charged balloon attracts electrically neutral objects, such as the small bits of paper in Figure 1.4(c), by *inducing* a shift in the charge distribution in the paper, which has no net electrical charge. The surface of the paper nearest the balloon develops a charge that is opposite in sign to the charge on the balloon. **Electrostatic attraction** occurs between electrical charges of opposite sign, and the paper sticks to the balloon. Conversely, electrical charges of the same sign repel one another.

1.6. **Consider This** *Can an electrostatic model explain the Investigate This 1.4 results?*
Make drawings, modeled after those in Figure 1.4, that show how the electrostatic model can explain your observations in Investigate This 1.4 when two strips of thin plastic hanging next to one another are rubbed simultaneously. Compare your explanations and sketches with those of other students. Discuss any differences, and try to reach agreement on an explanation.

Atoms and elements

Our electrostatic model for matter assumes that matter is composed of negative and positive particles. Atoms are made of these positive and negative particles, so we need to consider how the particles are combined in the structure of atoms. Our model for atoms has evolved during the two hundred years since John Dalton (English minister and scientist, 1766-1844) proposed an atomic model based on the *assumption* that all matter is composed of indivisible atoms. Scientists have since learned that every **atom** is made up of a positively charged **nucleus** (plural:

> Atom is derived from the Greek *atomos*, which means indivisible or not able to be cut.

nuclei) surrounded by negatively charged **electrons**, e^-. The nucleus consists of particles called **protons**, p, and **neutrons**, n. Protons and neutrons have nearly identical mass, which is almost 2000 times the mass of an electron. Protons are positively charged and neutrons are electrically neutral, as their name implies. The value of the charge on a proton or electron is called a **unit charge**, and is written as 1+ (said as *"one plus"*) for protons and 1– (said as *"one minus"*) for electrons. Atoms are electrically neutral; they have equal numbers of protons and electrons, so the positive and negative charges balance one another. Numerical values for the mass and charge of these subatomic particles are given in Chapter 3, Section 3.1, Table 3.1.

Elements are pure substances that are composed of all the same kind of atom. The identity of an element is defined by the number of protons in the nuclei of its atoms. Every atom of the same element has the same number of protons, and therefore the same nuclear charge and the same number of electrons to balance the nuclear charge. For example, the heaviest element that occurs on Earth in quantities large enough to mine is uranium, with 92 protons. The lightest element is hydrogen, with 1 proton. At the time of this writing, 116 elements are known; 84 of them occur naturally in measurable quantities. The other 32 exist on Earth only because they have been made

> Processes that result in nuclear transformations are discussed in Chapter 3.

in nuclear reactors and particle accelerators or result from nuclear decay of other elements.

The inside front cover and end paper of the book list names, symbols, and nuclear charges for all the known elements. One listing is in the form of a **periodic table** of the elements. The periodic table arranges all the known elements into rows (called **periods**) and columns (called **groups**) to place elements with similar chemical properties in the same group. It is the chemist's principal tool for organizing information about the elements. Dmitri Mendeleev, (Russian chemist, 1834-1907), devised one of the first periodic tables in 1869. We will frequently use the periodic table (or parts of it) to illustrate trends in atomic properties. The development and structure of the periodic table are central themes of Chapter 4.

Electron-shell atomic model

The current model for the atom is based on the observation that atomic nuclei have radii on the order of 10^{-14} m, while atoms themselves have radii on the order of 10^{-10} m, some ten thousand times larger than the radius of the nucleus. To put these values in perspective, if a nucleus were the size of a pea, the entire atom would be about the size of the baseball stadium in Figure 1.5. For any atom, the region outside the nucleus is the domain of its electrons. The nucleus contains essentially the total *mass* of the atom, but the *size* of an atom is determined entirely by the distribution of its electrons.

> **Web Chap. 1, Sect. 1.2**
> Explore the shell model and an interactive periodic table to investigate atomic properties.

Figure 1.5. Relative size of a nucleus and an atom.
Imagine a pea, the nucleus, at second base. The atom would be about the size of the entire stadium.

A simple atomic model, called the **electron-shell model**, illustrated in Figure 1.6, is sufficient for understanding much of chemistry. Negatively charged electrons are held in atoms by their attraction to the positively charged nucleus, but the locations of electrons are only approximately known. Spherical shells arranged concentrically around the nucleus represent the volumes of space where electrons are most likely to be found. Two categories of electrons are distinguished in Figure 1.6. **Core electrons** are nearest the nucleus and are so strongly attracted to the nucleus that they never interact with other atoms. **Valence electrons** are furthest from the nucleus and are more weakly attracted. Valence electrons are responsible for interactions of atoms with one another.

> Valence, Latin, *valentia* = strength or power, was originally used to characterize the combining properties of atoms.

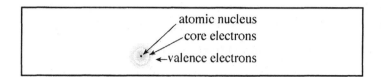

Figure 1.6. Electron-shell model of an atom.

Figure 1.7 illustrates the shell model, as applied to the first twenty elements of the periodic table. The field for each element includes the element name, its one-or-two letter symbol, the units of positive charge in its atomic nucleus (which is called the **atomic number**), and the units of negative charge (symbolized as e^-) associated with its core and valence electrons. The sum of the unit charges for core electrons and valence electrons must equal the units of positive charge for the nucleus. Within any period (row), the number of core electrons is the same for all the elements. The first period has no core electrons, the second period has two, the third has ten, and the fourth has eighteen.

	I 1 valence e^-	II 2 valence e^-	III 3 valence e^-	IV 4 valence e^-	V 5 valence e^-	VI 6 valence e^-	VII 7 valence e^-	VIII 8 valence e^-
1 no core electrons	hydrogen **H** · nucleus: 1+ core e^-: 0– valence e^-: 1–		Scale $\mid\leftarrow\rightarrow\mid$ 100 pm					helium **He** · nucleus: 2+ core e^-: 0– valence e^-: 2–
2 2 core electrons	lithium **Li** · nucleus: 3+ core e^-: 2– valence e^-: 1–	beryllium **Be** · nucleus: 4+ core e^-: 2– valence e^-: 2–	boron **B** · nucleus: 5+ core e^-: 2– valence e^-: 3–	carbon **C** · nucleus: 6+ core e^-: 2– valence e^-: 4–	nitrogen **N** · nucleus: 7+ core e^-: 2– valence e^-: 5–	oxygen **O** · nucleus: 8+ core e^-: 2– valence e^-: 6–	fluorine **F** · nucleus: 9+ core e^-: 2– valence e^-: 7–	neon **Ne** · nucleus: 10+ core e^-: 2– valence e^-: 8–
3 10 core electrons	sodium **Na** · nucleus: 11+ core e^-: 10– valence e^-: 1–	magnesium **Mg** · nucleus: 12+ core e^-: 10– valence e^-: 2–	aluminum **Al** · nucleus: 13+ core e^-: 10– valence e^-: 3–	silicon **Si** · nucleus: 14+ core e^-: 10– valence e^-: 4–	phosphorus **P** · nucleus: 15+ core e^-: 10– valence e^-: 5–	sulfur **S** · nucleus: 16+ core e^-: 10– valence e^-: 6–	chlorine **Cl** · nucleus: 17+ core e^-: 10– valence e^-: 7–	argon **Ar** · nucleus: 18+ core e^-: 10– valence e^-: 8–
4 18 core electrons	potassium **K** · nucleus: 19+ core e^-: 18– valence e^-: 1–	calcium **Ca** · nucleus: 20+ core e^-: 18– valence e^-: 2–						

Figure 1.7. Periodic variation of valence electrons and atomic size for the first 20 elements.

The number of valence electrons increases across a period. The beginning of each period, corresponds to a new electron shell, the **valence shell** for that period. The valence shell for the first element in the period contains one electron, and all the electrons added across the previous period are now core electrons in the new element. Then electrons are added one at a time to the new valence shell as each new element is added across the period. The result is that *the number of valence electrons is the same for every element in a group* (a column) of the periodic table. (Helium in the eighth column is an exception with only two valence electrons; all the other eighth column elements —noble gases— have eight valence electrons.)

> The periodic table of Figure 1.7 has the atoms arranged in eight groups rather than the eighteen shown on the inside front cover. The arrangement of eight groups is often used when discussion is limited to the first twenty elements.

1.7. Check This *Number of valence and core electrons*

(a) Use the periodic table on the inside front cover and the data in Figure 1.7 to determine the number of valence electrons in (i) bromine, (ii) strontium, and (iii) selenium. Explain how you get your answers.

(b) How many electron shells surround the nucleus of a phosphorus atom? How many electrons are in each shell? Are your numbers consistent with those in the electron shell model in the WEB Chap 1, Sect 1.2? Explain your answers.

The spherical models in Figure 1.7 are scaled to be proportional to the relative sizes of the atoms. Compare atomic sizes *within a vertical group* and you will see that the sizes of atoms increase with increasing atomic number. Not surprisingly, a core of ten electrons is larger than a core containing two electrons, and a core of eighteen electrons is larger than a core of ten electrons. Therefore, when a new valence shell is created, its diameter is greater than the valence shell diameter of the element of the same group in the previous period. However, across a period from left to right, the sizes of atoms generally *decrease* despite the fact that the numbers of electrons *increase*. Remember that the nuclear charge is also increasing across a period; the increased charge attracts all the electrons more strongly and draws them closer to the nucleus.

1.8. Worked Example *Trends in atomic size*

Predict the relative sizes of calcium, strontium, and barium atoms.

Necessary information: We need the periodic table on the inside front cover to find the relationship of these three elements to one another.

> *Strategy:* Determine whether these elements are in the same period (row) or the same group (column) and use the trends noted in the previous paragraph to make our prediction.
>
> *Implementation:* Calcium, strontium, and barium are in the second column of the periodic table and increase in atomic number in this same order. Since atoms of elements in the same group increase in size with increasing atomic number, we predict that the sizes of these atoms are in the order: calcium < strontium < barium.
>
> *Does the Answer Make Sense?* The second column of the periodic table in Figure 1.7 shows the second column elements increasing in size from top to bottom; it seems reasonable to expect the trend to continue, as we have predicted.

1.9. Check This *Trends in atomic sizes*

Predict the relative sizes of arsenic, selenium, and bromine atoms. Explain the basis of your prediction.

Section 1.3. Molecular Models

A **molecule** is a collection of atoms that are all strongly attached to one another and move and act together as a single entity. Essentially all of the 116 elements are able to associate with each other to form molecules. Molecules may contain as few as two atoms, or as many as your imagination permits. Biologically important molecules often contain thousands of atoms and some contain millions. Molecules may contain only one kind of atom (O_2, oxygen gas, for example) or more than one kind (H_2O, for example). In living systems, the great majority of molecules contain two or more of the elements carbon, hydrogen, oxygen, nitrogen, sulfur, and phosphorus.

Molecules are difficult to study individually, so much of what we know about their properties we *infer* from observations of the **macroscopic** or **bulk properties** of collections of molecules. We then invent physical and mathematical models of molecules to help interpret these experimental observations. Once suitable **molecular models** have been developed, they can be used to explain new observations and to suggest avenues for further study. Our goal is to help you become comfortable working in both directions—from observations to molecular-level explanations, and from molecular models to predicted properties.

> "Macroscopic" describes properties that we can observe directly with our senses. "Bulk properties" are also associated with enough of a substance for the properties to be directly observable.

Figure 1.8 shows six models or representations of the water molecule. Some may already be familiar to you. You may even have included one or more of these representations in your list from Consider This 1.1 and you should rely on what you have learned previously as you use them. Others of the representations are probably unfamiliar. We will introduce them in this chapter and continue to use them throughout the book.

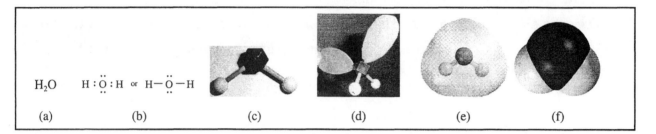

Figure 1.8. Six different models of the water molecule.

You may not think of the **molecular formula** for water, H_2O, Figure 1.8(a), as a molecular model at all. However, the formula tells you how many atoms of hydrogen, H, and oxygen, O, combine to produce a water molecule. A **compound** is a substance whose molecules are all the same; they all have the same molecular formula. The composition of a compound, such as water or sucrose (table sugar) is the same, no matter where you find it, because all its molecules have the same number of atoms of each element.

> **WEB Chap. 1, Sect. 1.3**
> Investigate the relationships of these molecular models to one another.

The molecular formula H_2O is called a **line formula** because it is written on a single line of text. Line formulas often provide no explicit information about the way the atoms are connected to each other, although you can often figure out the linkages by examining the sequence of atomic symbols in the formula. Figures 1.8(b) through (f) each show the connectivity within the water molecule. Figure 1.8(b) presents two representations of what is called an **electron-dot model**, which shows the distribution of valence electrons in the molecule. The remaining representations all show that the water molecule has a bent structure. Representation (c) is called a **ball-and-stick model**. It shows connectivity within the molecule as well as the bent structure. The representations, (d) through (f), provide basic shape information as well as other details about the properties of water molecules. Each of these models is useful within the context of specific discussions of the properties of water. We will explain them as they are used later in the chapter.

Molecular bonding and structure

Each pair of atoms in a molecule is linked by a **covalent bond**, which results from pairs of valence electrons being shared between the two atoms. Figure 1.9(a) is a representation of the formation of a covalent bond between two hydrogen atoms to produce a hydrogen molecule, H_2. As the electron shells of the two atoms overlap, a new distribution of the electrons forms in which the electrons are shared by and attracted to both nuclei. Figure 1.9(b) represents this H–H bond formation with electron-dot structures for the atoms and molecule and Figure 1.9(c) extends this representation to show the shared electron-pair bonds in water. A more detailed discussion of molecular bonding, which explains how a pair of electrons makes up a covalent bond is the subject of Chapter 5.

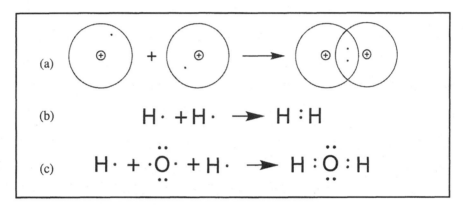

Figure 1.9. Representations of shared pairs of electrons forming covalent bonds.
(a) and (b) show the formation of H_2 and (c) the formation of H_2O. Valence electrons are represented by dots and shared electron pairs are shown in red.

Figure 1.8(c) and Figure 1.10 show a photograph of a ball-and-stick model for the water molecule made with a molecular-model kit like yours. The red ball and two aqua balls in the figures represent, respectively, an oxygen **atomic core**, the nucleus plus core electrons, and two hydrogen atomic cores, protons. Each of the sticks that connect the balls represents a pair of valence electrons, forming a covalent bond. Ball-and-stick models show the **molecular shape**, or geometry, which is *the arrangement of atomic nuclei relative to each other in a molecule*. **Bond lengths** (the distances between pairs of covalently-bonded atomic nuclei) and **bond angles** (the angle formed by the two bonds of three linked atoms) define the shape of a molecule. The numerical values in Figure 1.10 are the results of many experiments, which show that a water

> Ball-and-stick model kits use balls drilled with several patterns of holes to represent atomic centers. Thus, molecules of various geometries can be made. In your model kit, the "balls" (except for hydrogen) are polyhedra, but the kits are still referred to as "ball-and-stick.". Model kits usually use red for oxygen white or aqua for hydrogen, blue for nitrogen, and black for carbon.

molecule has two identical H–O bonds, each of which is 94 picometers, pm (1 pm = 10^{-12} m) long. The H–O–H bond angle is 104.5° so that water is described as being bent, or as a "V-shaped" molecule.

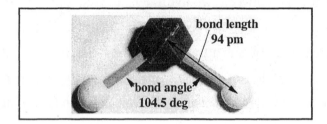

Figure 1.10. Geometry of the water molecule.

Ball-and-stick models provide useful information about molecular shape, but provide no information about valence electrons—beyond the fact that each stick represents a pair of bonding electrons. Before you can build a ball-and-stick model, you need to know how many atoms of each type combine to form the molecules. Then you must consider how atoms use valence electrons to form molecules. We need a model that helps visualize all the valence electrons.

Section 1.4. Valence Electrons in Molecular Models: Lewis Structures

Use the partial periodic table, Figure 1.7, to verify that an oxygen atom has six valence electrons and two core electrons, and that a hydrogen atom has one valence electron. The two oxygen core electrons do not participate in forming chemical bonds. In a water molecule, only the eight valence electrons (six from the oxygen atom and one each from the two hydrogen atoms) have a role in binding the three atoms together and in determining the shape of the molecule. The eight valence electrons in water are represented as *four electron pairs*, as shown by the electron-dot models in Figures 1.8(b), 1.9(c) and 1.11.

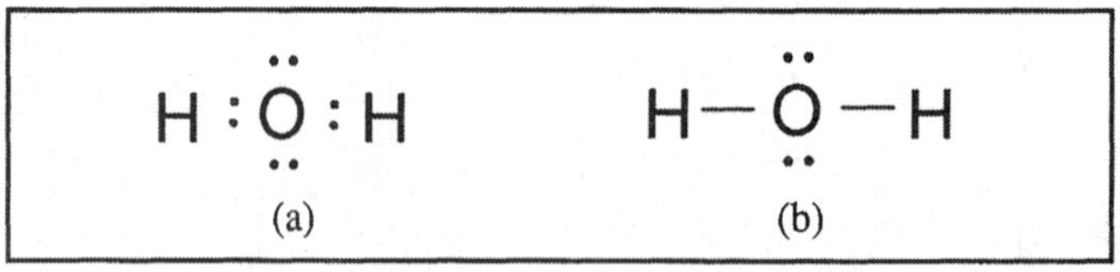

$$H:\overset{..}{\underset{..}{O}}:H \qquad H-\overset{..}{\underset{..}{O}}-H$$

(a) (b)

Figure 1.11. Two representations of the electron-dot model (Lewis structure) of water.
Electron pairs in covalent bonds may be represented as (a) dots or (b) strokes connecting the atomic symbols.

Note carefully that the meanings for the letter symbols for elements vary with context. In electron-dot diagrams like Figure 1.11, "O" represents the *atomic core*, the nucleus plus *core* electrons. The valence electrons are explicitly indicated using dots (or strokes). When *atoms* of oxygen are specified, as in the periodic table of Figure 1.7, the symbol "O" represents the nucleus plus *all* the electrons. The "O" symbol is sometimes used to represent *only* the oxygen nucleus, as you will see in Chapter 3. The inconsistency in the way chemists use atomic symbols should not create confusion if you remain attentive to the context of the discussion.

Lewis structures

Electron-dot models are also called **Lewis structures** to honor G. N. Lewis (American chemist, 1875–1946), who first proposed the electron-pair, covalent bond. Lewis structures are often written with the covalent bonds shown as a pair of dots [the red dots in Figure 1.11(a)] to reinforce the idea that the bond involves a pair of electrons. A Lewis structure may also be written with the covalent bonds shown as a stroke (line) between the bonded atomic cores [the red lines in Figure 1.11(b)]. The strokes are quicker to write, and make it easier to see the relationship between a Lewis structure and a ball-and-stick model. Valence electrons that are written adjacent to only one element symbol (the black dots in Figure 1.11) are associated exclusively with that element and are called **nonbonding electrons.**

When we write Lewis structures, we keep in mind that second-period elements (carbon through fluorine) in molecules typically have eight associated valence electrons. The "rule of eight," or **octet rule**, is useful for figuring out appropriate Lewis structures for molecules that contain these second-period elements. Also, hydrogen always has two valence electrons associated with it in molecules, and the pair of electrons is always written between H and the atomic core to which it is bound.

The first step in writing a Lewis structure is to determine the total number of valence electrons in the atoms that make up the molecule. Then these electrons are distributed among the atomic cores so that each second-period element is surrounded by an octet (four pairs) of electrons. For the purposes of counting these octets, valence electrons written *between* a pair of element symbols are considered to be associated with *both* of the bound atoms. *Electrons shared between a pair of atoms count toward satisfying the octet rule for both atoms.*

1.10. **Worked Example**   *The Lewis structure for methane*

Write the Lewis structure for the methane molecule, CH_4.

Necessary information: We need the number of valence electrons for carbon and hydrogen atoms from the periodic table.

Strategy: Determine the total number of valence electrons in the molecule, arrange them to surround the second period element (carbon) with four electron pairs, and then add the hydrogen atomic cores.

> *Implementation:* From Figure 1.7, we find that the C atom has four valence electrons and each of the four H atoms has one; the total number of valence electrons in CH$_4$ is eight. C is the only second-period element in the molecule, so we surround it with four pairs of electrons:
>
> $:\!\overset{..}{\underset{..}{C}}\!:$
>
> Each H atomic core can share one pair of electrons with another atomic core, so we use each of the four electron pairs on the C to form an electron-pair bond with one of the H atomic cores:
>
> $$H:\overset{\overset{H}{..}}{\underset{\underset{H}{..}}{C}}:H \quad \text{or} \quad H-\overset{\overset{H}{|}}{\underset{\underset{H}{|}}{C}}-H$$
>
> *Does the answer make sense?* All the valence electrons in the methane molecule are accounted for. There are two-electron bonds between all four H atomic cores and the C atomic core and eight electrons surround the carbon, as our Lewis structure rules require. The structure with bonding electron pairs shown as strokes reminds us that there are no nonbonding electrons in this molecule; all the valence electrons are bonding electrons.
>
> > In this chapter we use names of compounds without telling you the origin of the names. For now, please focus on the connections, shapes, and polarities of the molecules, not their names.

1.11. Check This   *Writing Lewis structures*

(a) Write the Lewis structure for the ammonia molecule, NH$_3$.

(b) Write the Lewis structure for the hydrogen fluoride molecule, HF

Connectivity

You can write Lewis structures if you know or can figure out the connectivity (which atoms are linked to which) in the molecule. Figuring out the connectivity is simplified somewhat by the fact that hydrogen can have only two valence electrons. Thus hydrogen is covalently bonded to *only one other atom.* Second-period elements have characteristic numbers of covalent bonds and nonbonding electrons in their molecules, as shown in Table 1.1. We are particularly interested in the second-period elements because many of the molecules you will meet in this book are **biomolecules,** the molecules essential for life. They are composed mainly of hydrogen, carbon, nitrogen, and oxygen. Lewis structures can help you understand and interpret the properties of these molecules. Writing Lewis structures for molecules containing two or more second-

> Out of every 100,000 atoms in your body there are:
> | | |
> |---|---|
> | 60,560 | H |
> | 25,670 | O |
> | 10,680 | C |
> | 2,440 | N |
> | 650 | all others |
>
> The prevalence of H and O is not surprising. Approximately 70% of your body mass is water.

period elements is only slightly more complicated than writing the structures for molecules containing one second-period element.

Table 1.1. Bonding and nonbonding electron pairs for second-period elements in molecules.

Element	C	N	O	F
Covalent bonds	4	3	2	1
Nonbonding pairs	0	1	2	3

1.12. **Worked Example**   *Lewis structure of methanol*

Write the Lewis structure for methanol, CH_4O.

Necessary information: We need the number of valence electrons for carbon, oxygen, and hydrogen atoms from the periodic table and the number of covalent bonds to carbon and oxygen from Table 1.1.

Strategy: Determine the possible connectivity of atoms and the total number of valence electrons in the molecule. Place two valence electrons between each pair of atomic cores and distribute any that are left over to give every second-period element eight electrons surrounding its atom symbol.

Implementation: Hydrogen atoms can bond to only one other atom, so we must connect the C and O atomic cores together and distribute the H atomic cores around them in accordance with Table 1.1:

$$\begin{array}{c} H \\ H\ C\ O \\ H\ H \end{array}$$

From Figure 1.7, we find that a C atom has four valence electrons, an O atom has six, and each of the four H atoms has one, so the total number of valence electrons in CH_4O is 14. We distribute these to make two-electron bonds between each pair of atomic cores:

$$\begin{array}{c} H \\ H:\overset{..}{C}:O \\ H\ H \end{array}$$

Only 10 valence electrons are accounted for in this structure and there are only four electrons around the O atomic core. Table 1.1 shows that an O atomic core usually has two nonbonding electron pairs in addition to the two bonding pairs:

$$\begin{array}{c} H \\ H:\overset{..}{C}:\overset{..}{\underset{..}{O}}: \\ H\ H \end{array} \quad \text{or} \quad \begin{array}{c} H \\ H-\overset{|}{\underset{|}{C}}-\overset{..}{\underset{..}{O}}: \\ H\ H \end{array}$$

All 14 valence electrons are accounted for, so this is the Lewis structure.

> ***Does the answer make sense?*** There are two-electron bonds between all four H atomic cores and the C and O atomic cores, a two-electron bond between the two C and O atomic cores, and the C and O are each surrounded by eight electrons, as our Lewis structure rules require. The structure with bonding electron pairs shown as strokes reminds us that there are both nonbonding and bonding electrons in this molecule. The line formula for methanol is often written as CH_3OH to make the connectivity and structure more apparent.

1.13. Check This *Lewis structure of ethane*

Write the Lewis structure for ethane, C_2H_6. Check your result to see that it is consistent with the information in Table 1.1. Try writing a line formula for ethane that more clearly shows its connectivity.

Reflection and projection

The preceding sections focused primarily on building physical and conceptual models to represent atoms and molecules. The discussion centered on a few second-period elements and hydrogen, all of which form covalent bonds with each other. The physical models are based on the electron-shell description for atoms, which takes into account the size and charge of the nucleus, core electrons that do not take part in chemical interactions, and valence electrons that interact with electrons from other atoms.

Lewis structures for molecules represent all the valence electrons, but the connectivity must be known before the structure can be written, unless the molecule is so simple that the atoms can be assembled in only one way. Constraints on how molecules can be assembled are imposed by (a) the octet rule for second-period elements, (b) hydrogen's limit of two electrons, and (c) the usual number of covalent bonds that each second-period element forms (4 for carbon, 3 for nitrogen, and 2 for oxygen). Ball-and-stick molecular models show the three-dimensional geometry of the nuclei in a molecule, but do not always represent all the valence electrons. The role of the valence electrons in producing the molecular geometry is the topic of the next section.

Section 1.5. Arranging Electron Pairs in Three Dimensions

The Lewis structure of H_2O in Figure 1.11, of CH_4 in Worked Example 1.10, and of those you wrote for NH_3 and HF in Check This 1.11 show four pairs of valence electrons arranged around the second period elements, but the representations are written on a sheet of paper, which has only two dimensions. Two-dimensional Lewis structures cannot unambiguously reveal molecular geometry in three-dimensions. Though the H–O–H diagram shows that water has a central oxygen atom to which two hydrogen atoms are connected, it does not show the 104.5° H–O–H angle. Accurate representations of three-dimensional geometries of molecules reflect the spatial arrangements of electron pairs around the atomic cores. Knowing these geometries is important because the arrangement of electrons and the three-dimensional structure of molecules are responsible for most of their physical and chemical properties. The next step in building a useful three-dimensional model is to figure out how electron pair arrangements produce observed molecular geometries.

To begin, consider that each of the four valence electron pairs around oxygen is strongly attracted to the 6+ positive charge of the atomic core (the 8+ positive nucleus and the two core electrons, 2– negative charge). As we will discuss in Chapter 5, the four pairs of electrons around oxygen occupy four separate regions of space. The strength of the attraction between positive

> We often say that core electrons "shield" valence electrons from the full nuclear charge. This shielding is not 100% effective, so the valence electrons are attracted by a charge that is somewhat larger than just the core charge.

and negative charges increases as the distance between the charges decreases. Therefore, the most favorable attraction occurs when each valence-electron pair is located as close as possible to the positive charge. Finally, we have to consider the shape of the region occupied by the electron pairs. For simplicity, think of each of the four regions of space occupied by valence-electron pairs as roughly spherical, like four tennis balls or round balloons.

1.14. **Investigate This** *What arrangement of balloons around a point is stable?*

For this investigation, use four 8- or 10-inch round balloons. Inflate all four balloons to the same size and tie off their stems so they remain inflated. Attach two of the balloons together by wrapping their stems together. Attach a third and then a fourth balloon in the same way. Try to make different geometries of the four attached balloons. For the most stable arrangement you can find, describe and record the geometry of the centers of the balloons with respect to their mutual point of attachment.

1.15. **Consider This** *What is the geometry of four valence electron pairs?*

Imagine lines connecting the center of each of the balloons in Investigate This 1.14 to the mutual point of attachment. For the most stable arrangement, try drawing this set of four lines on a piece of paper. Find an atom center in your molecular model kit that has holes in the correct geometric arrangement and place connectors in the holes to help visualize the lines you are trying to draw.

Tetrahedral arrangement

In Investigate This 1.14 you probably found an arrangement of four balloons that places their centers all the same distance from a central point, as shown for a set of tennis balls in Figure 1.12. Here, three balls are arranged in a triangle, and the fourth ball is placed on top of the triangle. This arrangement is called **tetrahedral**, because lines joining the centers of the balls form a regular **tetrahedron**, the four-sided geometric figure shown in red in the figure. The tetrahedral arrangement puts all the balls as close as possible to the central point (the mutual point of attachment for the four balloons in Investigate This 1.14). Four valence electron pairs always form a tetrahedral arrangement around second-period elements because that arrangement places the valence electron pairs nearest the nucleus.

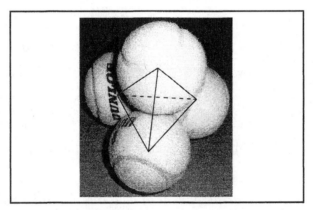

Figure 1.12. Tetrahedral arrangement of four balls.
The red lines represent connections between the centers of the balls that form a regular tetrahedron.

To visualize the angles between lines from the center of a tetrahedron to its corners, use your ball-and-stick model kit that has some polyhedra with four holes drilled tetrahedrally. The black polyhedra (carbon) are tetrahedral, and nitrogen (blue) and oxygen (red) are drilled the same way. Put a stick of the same length in each of the four holes, as in Figure 1.13. The angle formed by any two of these sticks is called the **tetrahedral angle**, which is 109.5°. If the arrangement of valence-electron pairs around the oxygen atom in water were exactly tetrahedral, the H–O–H bond angle would be 109.5°. The value of 104.5°, see Figure 1.10, obtained experimentally, is close to this predicted angle. The tetrahedral arrangement is very symmetric. Any two sticks can be chosen for the H–O covalent bonds, and the same "V-shaped" molecular model results.

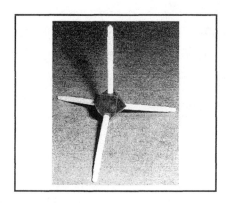

Figure 1.13. Ball-and-stick tetrahedral array of sticks.

Modified ball-and-stick model

Your model kit has parts that you can use to visualize all valence electrons in a molecule—both bonding and nonbonding electrons. Figure 1.14 shows a ball-and-stick model for water made with sticks and "paddles." In this model, each stick represents a pair of electrons in a covalent bond. Each paddle represents a pair of nonbonding electrons. Think of the paddles as cross-sections of volumes of space (like your balloons in Investigate This 1.14) within which electron pairs are likely to be found. Note the relationship between the model in Figure 1.14 and the Lewis structure for water, Figure 1.11(b).

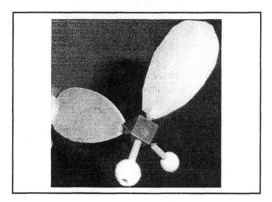

Figure 1.14. Bonding and nonbonding electrons in the molecular model for water.
The sticks represent bonding electron pairs and the paddles nonbonding electron pairs.

1.16. **Consider This**   *How is a Lewis structure translated into a molecular model?*

(a) From their Lewis structures, construct models of water, methane, ammonia, and hydrogen fluoride. Show nonbonding as well as bonding electrons in your models.

(b) How would you describe the *shape* of each molecule you have modeled? First describe the arrangement of the electron pairs around each second-period element. Then, using water as a guide, describe the shape of the molecule. Remember that shape refers to the arrangement of the nuclei with respect to one another. Nonbonding electrons are not considered when determining shape. Discuss your descriptions with your classmates. Develop consistent descriptions that everyone understands and agrees upon.

Models such as the one pictured in Figure 1.14 have limitations that must be taken into account when we use them to visualize molecular properties. For example, while this type of model helps keep track of the number of valence electrons and the shape of the molecule, volumes in space occupied by valence electrons and electron densities within those volumes are not well represented. We will introduce other models for these purposes in the next section.

Section 1.6. Polarity of the Water Molecule

Several kinds of experiments show that the molecules of some substances, **polar molecules**, have their centers of positive and negative charge separated from each other. Polar molecules are **electric dipoles**, that is, they have a positive end and a negative end, as illustrated in Figure 1.15(a). A lower-case Greek delta, δ, is used to symbolize "a small amount." In Figure 1.15(a), this means that there are partial charges, not full + or –, on the ends of the molecule. This charge separation is similar to the induced charge separation shown on the bits of paper in Figure 1.4(c) except that it is a permanent part of the molecule's properties and does not depend on the presence of an external electric charge. In representing molecular polarity, we generally use red to denote parts of molecules or systems that are more negative and blue to represent parts that are more positive. Regions where the positive and negative charges cancel are in green. The text that accompanies Figure 1.17 explains how these regions are determined.

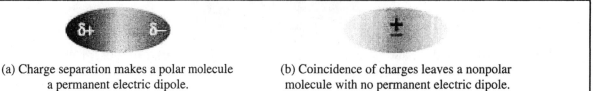

(a) Charge separation makes a polar molecule a permanent electric dipole.

(b) Coincidence of charges leaves a nonpolar molecule with no permanent electric dipole.

Figure 1.15. Schematic illustrations of (a) polar and (b) nonpolar molecules.

Nonpolar molecules are molecules with no electric dipole. In a nonpolar molecule, the centers of positive charge (from the nuclei) and negative charge (from the electrons) coincide, as illustrated in Figure 1.15(b). In a polar molecule, the arrangement of electrons is skewed so that the centers of positive and negative charge do not coincide.

> *1.17.* **Consider This** *How might an electric charge affect polar molecules?*
>
> Imagine that you have a liquid containing polar molecules like those in Figure 1.15(a). If the liquid is sandwiched between two electrically charged metal plates, one negative and the other positive, how would you expect the molecules to be oriented? Draw a sketch to illustrate your answer.

Effect of electric charge on polar molecules

One kind of experiment that is used to test for the polarity of molecules in a liquid is illustrated in Figure 1.16. Two metal plates are placed in the liquid and then one plate is given a negative charge and the other a positive charge. Initially, Figure 1.16(a), when the plates are not charged, the molecules of liquid are randomly oriented. When the plates are charged, Figure 1.16(b), polar molecules orient themselves with their negative ends toward the positive plate and positive ends toward the negative plate. The molecules are continuously jostling each other in the liquid, so they do not line up exactly. The electric circuit that is used to charge the plates responds differently when polar and nonpolar liquids are tested, so polar and nonpolar molecules can be distinguished from one another.

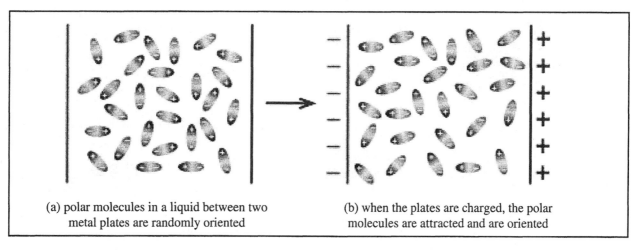

(a) polar molecules in a liquid between two metal plates are randomly oriented

(b) when the plates are charged, the polar molecules are attracted and are oriented

Figure 1.16. Orientation of the molecules of a polar liquid in response to an electric charge.

Water molecules are polar

Experiments such as the one illustrated in Figure 1.16 show that water molecules are polar. A charge-density model of a water molecule is shown in Figure 1.17(a). The surface is drawn to enclose the volume in space where electrons are most likely to be found. These color-coded models are generated from calculations that show net electrical charge on various parts of the molecule. As in Figure 1.15, regions where negative charge from the electrons and positive charge from the nuclei balance one another are colored green. Regions where negative charge from the electrons is not balanced by positive charge (where the electron density is high) have colors at the red end of the spectrum. Regions where positive charge from the nuclei is not balanced by negative charge (where the electron density is low) have colors at the blue end of the spectrum. The extremes, red and blue, respectively, represent the regions with the highest amounts of negative and positive charge.

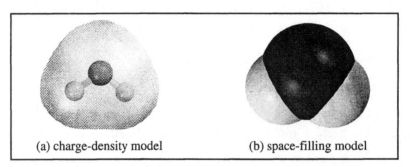

Figure 1.17. Charge-density and space-filling models of the water molecule.
The computed surface in (a) encloses the volume occupied by the electrons in the molecule. The internal ball-and-stick representation is included to show where the nuclei are. Color-coding is explained in the text.

For comparison to the charge-density (computer-generated) model, Figure 1.17(b) shows a space-filling model of water, a physical model molded of plastic with the "atoms" identified by the same colors used for ball-and-stick models. **Space-filling models** are approximate representations of the space (volume) occupied by the electrons around each atomic core in molecules, and you can see that the volume and shape of the two models are similar. Space-filling models often help us visualize the interactions of molecules, as in Figure 1.2. But we still need representations like Figure 1.17(a) to visualize the polarity of molecules.

The significant feature of Figure 1.17(a) is that the side of the molecule where the oxygen atom is located is more negative than the side of the molecule where the hydrogen atoms are located. This unsymmetrical distribution of electric charge makes the molecule an electric dipole. Every polar molecule can be represented using a dipole arrow, as shown in Figure 1.18. The point of the dipole arrow represents the negatively charged end of the molecule, while the "+"

tail of the arrow represents the positively charged end. Quantitative measurements of the effect of electrical charge on polar molecules, as in Figure 1.16, are used to calculate their dipole moments. The **dipole moment** of a molecule is a measure of magnitude of the charge difference between the two ends of the dipole. The larger the dipole moment the larger the effective charge difference.

> The units and numerical values for dipole moments will not concern us except as a relative measure of polarity. For reference, the dipole moment of water is 1.87 Debye (a unit of charge times length).

Figure 1.18. The electric dipole in the water molecule.

Bond polarity

Charge distribution is an important factor in understanding chemical reactions, and is used extensively for explanations throughout the text. Charge-density models, like the one in Figure 1.17(a), are useful for visualizing charge distribution in molecules, but they are not always available, so it is important to be able to estimate charge distributions using simpler models. Unsymmetrical charge distribution results from unequal sharing of valence electrons in covalent bonds. For example, the oxygen atom in water attracts the electrons in the covalent bonds more strongly than do the hydrogen atoms. Thus, the oxygen end of the bonds is slightly negative and the hydrogen ends are slightly positive. The bonds each act like little dipoles, **bond dipoles**, as shown by the dipole arrows in Figure 1.19.

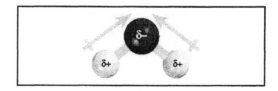

Figure 1.19. Unequal sharing of electron charge and bond dipoles in the water molecule.

Individual bond dipoles are not experimentally measurable, but we can use bond dipoles to figure out the direction of the molecular dipole, Figure 1.18, which can be measured. In Figure

1.19, one bond dipole points up and to the right and the other up and to the left. The two bonds in water are identical, so their bond dipoles should also be the same. The net effect of the dipoles pointing right and left is to cancel each other because they are the same magnitude and point in opposite directions. But both dipoles are pointing up, so they add together (reinforce each other) in the vertical direction and the net effect is the molecular dipole shown in Figure 1.18.

Electronegativity

The positive and negative ends of bond dipoles result from differences in electronegativity (*EN*) of the two atoms. The concept of electronegativity originated with Linus Pauling (American chemist, 1901-1994). The **electronegativity** of an element is a measure of its attraction for the electrons it shares with another element in a covalent bond. *The higher the electronegativity, the greater the attraction for shared electrons.* Electronegativity values for the first 20 elements are shown in Figure 1.20. Electronegativity values increase from element to element across a period. Within groups, the lighter elements are the most electronegative. These trends converge at fluorine, which has the highest electronegativity of all the elements.

	I	II	III	IV	V	VI	VII	VIII
1	hydrogen H 2.2							helium He —
2	lithium Li 1.0	beryllium Be 1.6	boron B 2.0	carbon C 2.6	nitrogen N 3.0	oxygen O 3.4	fluorine F 4.0	neon Ne —
3	sodium Na 0.93	magnesium Mg 1.3	aluminum Al 1.6	silicon Si 1.9	phosphorus P 2.2	sulfur S 2.6	chlorine Cl 3.2	argon Ar —
4	potassium K 0.82	calcium Ca 1.3						

Figure 1.20. Periodic variation of electronegativity for the first 20 elements.
Relative magnitudes for electronegativity values are indicated using the red–green–blue color-coding scheme that was used earlier to indicate charge density.

As a rule of thumb, a significant bond dipole exists for a covalent bond between the atoms of two elements that differ by 0.5 or more *EN* units. For example, a bond between oxygen (*EN* = 3.4) and hydrogen (*EN* = 2.2), which differ by 1.2 *EN* units, has a large bond dipole. In contrast, a bond between carbon (*EN* = 2.6) and hydrogen (*EN* = 2.2), which differ by 0.4 *EN* units, has a small, almost negligible, bond dipole.

> **WEB Chap. 1, Sect. 1.6**
> Use interactive animations to investigate electronegativity and polarity.

Of the elements commonly encountered in living systems, oxygen and nitrogen are the most electronegative, while carbon, hydrogen, phosphorus, and sulfur have intermediate values. Polar bonds formed between N or O and any of these others make a large contribution to molecular dipole moments. Molecules (or parts of molecules) containing these elements are usually polar. Nonpolar bonds formed between any two members of the intermediate electronegativity group make only a small contribution to molecular dipole moments. Molecules composed of hydrogen and carbon, for example, the molecules in oil and gasoline, are nonpolar.

1.18. Check This   *Predicting bond dipoles*

(a) Predict the direction of the bond dipoles in each of the following bonds and indicate whether the bond dipole is relatively small or relatively large. Explain the reasoning for your answers.

   (i) H–F   (ii) N–H   (iii) H–S   (iv) C–Cl   (v) N–F   (vi) Li–H

(b) WEB   Chap 1, Sect 1.6.1-2. Do the charge-density models on these pages support your answers in part (a)? Explain why or why not.

1.19. Worked Example   *Predicting molecular dipoles*

Write the Lewis structure for nitrogen trifluoride, NF_3. On a three-dimensional drawing or model of the molecule, show the direction of the bond dipoles and the resulting molecular dipole.

Necessary information: We need the number of valence electrons in N and F atoms and the electronegativity of these elements. We also need to know that N forms three and F forms one covalent, two-electron bond to other atoms.

Strategy: Write a Lewis structure, as we have done previously. Then make a ball-and-stick model representing the geometry of this structure, determine the direction of the bond dipoles from the relative electronegativities, and use the geometry and direction of the dipoles to predict the direction of the molecular dipole.

Implementation: There are 26 valence electrons in NF_3: 5 from the N and 7 each from the Fs. N bonds with three other atoms and F with only one, so the Fs must all be bonded to the N:

$$\begin{array}{c} F \\ F:N \\ F \end{array}$$

Six electrons are used for the bonds. We can use the remaining 20 electrons to surround all of the atom symbols with eight electrons to give the Lewis structure:

$$\begin{array}{c} :\!\ddot{F}\!: \\ :\!\ddot{F}\!:\!\ddot{N}\!: \\ :\!\ddot{F}\!: \end{array}$$

The tetrahedral array of three covalent bonds between N and F and the nonbonding electron pair on N are shown in this picture of a ball-and-stick model of NF$_3$. The nonbonding electron pairs on the F's are not shown, since they are not relevant to the molecular shape. The molecule can be described as a triangular pyramid with the F's at the corners of the base and the N at the apex.

The electronegativities of N and F are 3.0 and 4.0, respectively, so there is a substantial N–F bond dipole with the negative end toward F, the more electronegative element. Thin arrows in the picture show the bond dipoles. The three bond dipoles point from the apex toward the corners of the base of the pyramid. They are symmetrically arranged around the pyramid, so there is no net dipole in a direction parallel with the base of the pyramid. All three bond dipoles point down from the apex toward the base, so there is a net molecular dipole from the apex perpendicular to the base shown by the thick arrow in the picture.

Does the answer make sense? The Lewis structure obeys our rules and accounts for all the valence electrons, so it seems to be correct. The only way we can test the molecular dipole prediction is to look up NF$_3$ in a reference handbook; the molecule has a dipole moment of 0.24 Debye, which shows that it does have a permanent molecular dipole. The direction of the dipole is harder to determine, but experiments prove that it is in the direction we have shown.

1.20. **Check This**   *Predicting molecular dipoles*

(a) On a three-dimensional drawing or model of the ammonia molecule, NH$_3$, show the direction of the bond dipoles and the resulting molecular dipole. Refer to Check This 1.11 for the Lewis structure.

(b) Do the same for the hydrogen fluoride molecule, HF.

1.21. **Consider This**   *How does molecular geometry affect molecular dipoles?*

(a) Carbon dioxide, CO$_2$, is a linear molecule with the carbon atom midway between the two oxygen atoms. On a Lewis structure of the molecule show partial charge separations, bond dipole arrows, and overall molecular dipole arrow. Would you classify carbon dioxide as polar or nonpolar? What is the basis for your classification?

(b) How is the polarity of the BH$_3$ molecule (WEB Chap 1, Sect 1.6.3-4) similar to CO$_2$? Make a sketch of a likely charge-density model for CO$_2$.

Reflection and projection

The three-dimensional shape of a molecule is determined by the arrangement of valence electron pairs that puts them as close as possible to the center of each atom. In molecules, each second-period element (carbon through fluorine) is surrounded by four pairs of valence electrons (the octet rule) arranged tetrahedrally about the atomic core. However, only the arrangement of the nuclei is considered in describing the shape of a molecule. Nonbonding electron pairs are ignored when describing molecular shape; ball-and-stick molecular models are useful for visualizing molecular shape.

Unequal sharing of electrons between atoms that differ substantially in electronegativity causes small charge separations in covalent bonds between them and produces bond dipoles. If the bond dipoles in a molecule do not cancel one another out, the molecule as a whole has a molecular electric dipole and is said to be polar. One result of this polarity is that polar molecules tend to orient their positive ends toward negative charges and their negative ends toward positive charges. The chemical and physical properties of a substance composed of polar molecules are most unusual when a hydrogen atom is associated with the center of partial positive charge and a highly electronegative atom (nitrogen, oxygen, or fluorine) is associated with the center of partial negative charge. Figures 1.17, 1.18, and 1.19 show that water is one such substance and the consequences are considered in the next sections.

The previous paragraph reminds us that we must not lose sight of why we are spending so much time developing these molecular models. Our primary interest in this and in many of the chapters that follow is the observable properties of water and other substances. We must show how the models answer questions such as: *"Why is water a liquid at ordinary temperatures?" "Why is the solid less dense than the liquid?" "Why is so much energy needed to change the temperature of water?" "Why is so much energy needed to vaporize water?"* And, in later chapters we will be interested in further questions such as: *What kinds of compounds does water dissolve? What chemical reactions occur in aqueous solutions? What chemical reactions involve water itself—either as a reactant or as a product?* Explanations for all of these phenomena are based on models for water and associated molecules and ions. The models we develop in this chapter and extend in later chapters are essential for visualizing the processes and interpreting bulk phenomena in terms of molecular-level explanations. And, as your study of the linkage between macroscopic properties and the molecular model for water progresses, you will see that what you learn about water extends to other substances as well.

Water: A Natural Wonder — Chapter 1

Section 1.7. Why Is Water Liquid at Room Temperature?

In order for a substance to exist as a liquid, the attractions between its molecules have to be strong enough to prevent them from breaking away from one another and going into the gas phase. As energy is added to a liquid, its temperature goes up, the molecules move faster, and finally they are moving fast enough to break away from one another to become a gas. If the pressure exerted by this gas is equal to the pressure of the atmosphere, we say that the liquid is boiling. The boiling point of a liquid is a measure of the strength of the attractions between the molecules in the liquid: the higher the boiling point, the stronger the intermolecular attractions that have to be overcome to produce the gas. We will try to discover how our molecular model accounts for these interactions and leads to the observed behaviors of different compounds.

> "Room temperature" indicates a temperature of about 25 °C.

1.22. **Investigate This** *How does boiling point vary with number of molecular electrons?*

Figure 1.21 shows the boiling points of CH_4 (methane), SiH_4 (silane), GeH_4 (germane), and SnH_4 (stannane) plotted as a function of the total number of electrons in each molecule. These hydrides (compounds of the element with hydrogen) of group IV elements, all have the same symmetrical, tetrahedral structure you found for methane in Consider This 1.16 and all are nonpolar. Describe the relationship between the boiling point and the number of electrons. What conclusions can you draw about the relative attractions between the molecules in these compounds? Discuss in your group and with the class how you might explain any trends you see.

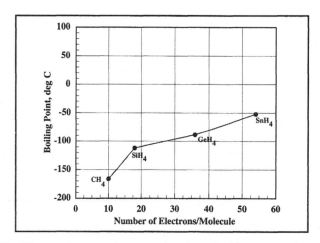

Figure 1.21. Boiling point *vs.* electrons/molecule for group IV hydrides.
The points are connected only to highlight the trend in the data.

Induced-dipole attractions

The boiling points of nonpolar compounds generally increase as the number of electrons in the molecules increases. This is the trend shown for the group IV hydrides in Figure 1.21. Nonpolar molecules have negligible dipole moments. Nonetheless, many of these compounds are liquids or solids at room temperature. Examine the charge-density models of methane, CH_4, ethane, CH_3CH_3, and hexane, $CH_3(CH_2)_4CH_3$, in Figure 1.22. Recall from Section 1.6, that green is used for an electrically neutral surface. Notice that the entire surface of these nonpolar molecules is green.

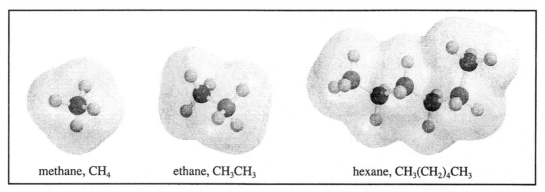

Figure 1.22. Charge-density models of methane, ethane, and hexane.
For comparison, Figure 1.17(a) is a charge-density model of the water molecule computed in the same way.

The electron distributions in Figure 1.22 represent *averages* of the likely locations for finding the electrons. Electrons are continuously in motion and, at any instant, electron distribution can be skewed one way or another. The skewing leads to a fleeting polarity of the molecule. Polarity in one molecule *induces* polarity in a neighboring molecule, as illustrated in Figure 1.23. The two polarized molecules attract one another. These attractions are called **induced-dipole attractions**. Induced-dipole attraction is a molecular-scale effect similar to the observable effect illustrated in Figure 1.4(c), where a charged balloon induces polarity in a neutral object and attracts it. Induced-dipole attractions are short-lived, but as there are many electrons in molecules, there is plenty of opportunity for these attractions to occur. Since the skewing of the electron distribution can occur anywhere in the molecule, these attractions are often called **dispersion forces** or **London dispersion forces** to honor Fritz London (German physicist, 1900-1954) who first proposed these induced-dipole attractions. Dispersion forces exist across the entire surface of a molecule, so dispersion forces among large molecules produce substantial attraction.

> **WEB Chap. 1, Sect. 1.7**
> Use interactive animations to investigate intermolecular attractions.

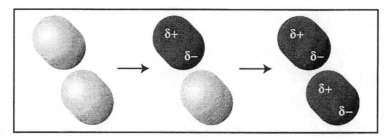

Figure 1.23. Induced-dipole attraction between nonpolar molecules.
The plus sign represents the center of positive charge in the molecule. Displacement of the electron distribution in one molecule creates a transient dipole that induces a dipole in the second molecule.

1.23. **Consider This** *Do boiling points correlate with number of molecular electrons?*

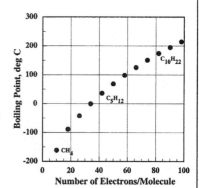

(a) This graph shows boiling points for a related series of nonpolar **hydrocarbons**, compounds containing only carbon and hydrogen atoms. The series starts at methane, CH_4, and adds one "CH_2" after another. The boiling points are plotted as a function of the number of electrons in these molecules. Propose a molecular level interpretation for the observed behavior of the boiling points.

(b) Does your interpretation in part (a) also apply to the data in Figure 1.21? Explain why or why not.

(c) WEB Chap 1, Sect 1.7.1-2. Do these examples follow the same pattern of boiling points as the group IV hydrides in Figure 1.21 and the carbon-hydrogen compounds in part (a)? Does the animation apply to these other cases as well? Explain why or why not.

Attractions among polar molecules

Dispersion forces increase as the size of molecules increase because more electrons are available to form instantaneous dipoles. All molecules have dispersion forces that attract them to one another. Many molecules also have permanent electric dipoles, and we might expect that attractions between the molecular dipoles would increase the attraction of these molecules for one another. In **dipole-dipole attractions**, the positive end of one dipolar molecule and the negative end of another attract one another.

1.24. **Consider This** *How can dipole-dipole attractions be visualized?*

As they jostle about in the liquid, dipolar molecules favor arrangements that maximize their dipole-dipole attractions, such as:

 and

Sketch favorable arrangements for dipole-dipole attractions among *four* dipoles. Compare your arrangements with those made by other students. How many different arrangements did members of the class find?

You found in Consider This 1.24, dipole-dipole attractions do not require one particular alignment of the molecules. There are many orientations for dipole-dipole attractions between a polar molecule and its neighbors in the liquid state. As they are tumbling about in the liquid, polar molecules must often be in position to attract one another. Do the boiling points of polar molecules reflect these additional attractions?

1.25. **Investigate This** *How does polarity affect boiling point?*

Figure 1.24 adds the boiling points of the group V, VI, and VII hydrides to the data for the group IV hydrides from Figure 1.21. The structures of the hydrides in each group are the same:

group	molecules	structure
IV	CH_4, SiH_4, GeH_4, SnH_4	tetrahedral like CH_4
V	NH_3, PH_3, AsH_3, SbH_3	pyramid like NH_3
VI	H_2O, H_2S, H_2Se, H_2Te	bent like H_2O
VII	HF, HCl, HBr, HI	linear like HF

(a) Do the polarities of the hydrides increase or decrease as you go from left to right in Figure 1.24? Explain your reasoning. What trend in boiling points would you expect for this trend in polarities?

(b) Do the dispersion forces (induced dipole attractions) increase or decrease as you go from left to right in Figure 1.24? Explain your reasoning. What trend in boiling points would you expect for this trend in dispersion forces?

(c) What conclusions can you draw about the relative strengths of polarity and dispersion force attractions between the molecules in these compounds? Explain your reasoning. Discuss in your group and with the class how you might explain any deviations from trends you see.

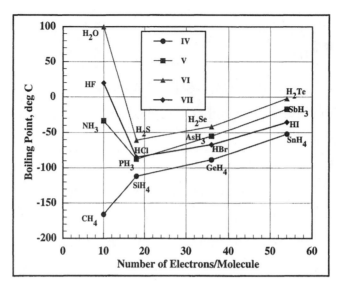

Figure 1.24. Boiling point *vs.* electrons/molecule for groups IV through VII hydrides. The points are connected only to highlight the trends in the data.

There are two striking things to note in Figure 1.24. One is that all the third and higher period (row) hydrides increase in boiling point as the number of electrons/molecule increases. This is the direction we would expect, if dispersion forces are the main contributors to the interactions among these molecules in their liquids. All the valence electron pairs in the group IV hydrides are bonding pairs, while the group V, VI, and VII hydrides have one or more nonbonded electron pairs. These nonbonded electrons are more free to respond to induced dipoles from neighboring molecules, which helps to explain why the third and higher period group V, VI, and VII hydrides have somewhat higher boiling points than the comparable group IV hydrides.

The other feature that stands out in Figure 1.24 is how far out of line the boiling points of ammonia (NH_3), water (H_2O), and hydrogen fluoride (HF) are compared to all the other compounds. The boiling points of these second period compounds are 150 to 250 °C higher than the boiling point of methane (CH_4), even though all four of these molecules have 10 electrons. For comparison, note that the spread in boiling points is only about 50 °C for the hydrides with 18, 36, and 54 electrons/molecule.

Since NH_3, H_2O, and HF are quite polar molecules, perhaps their high boiling points are a result of dipole-dipole attractions. However, HCl and HBr are also quite polar, since the electronegativities of chlorine and bromine are more than 0.5 units higher than the electronegativity of hydrogen. If dipole-dipole attractions are important contributors to intermolecular attractions, we would expect HCl and HBr to have higher boiling points than the hydrides with the same number of electrons. Figure 1.24 shows that HCl has a slightly higher boiling point than is predicted by the trend of the boiling points for the other period three and higher hydrides, but it is not far out of line. It appears that, as molecules are tumbling about in a liquid, dipole-dipole attractions do not contribute much to the intermolecular attractions responsible for holding the molecules together. Therefore, we have to search for another explanation for the anomalous behavior of NH_3, H_2O, and HF.

The hydrogen bond

The model of the water molecule we have developed emphasizes the partial separation of the centers of positive and negative charge that results from the bond dipoles and the bent geometry of the molecule. The partial charge separation produces a molecular dipole and dipolar molecules can attract one another and preferentially orient themselves toward each other with the negative ends of one near the positive ends of another, as you showed in Consider This 1.24. The data shown in Figure 1.24 demonstrate that this attraction is greatly enhanced when it involves hydrogen attached to N, O, or F at the positive end of one molecule and a highly electronegative atom (usually N, O, or F) at the negative end of the other. Models for this interaction between two water molecules are shown in Figure 1.25.

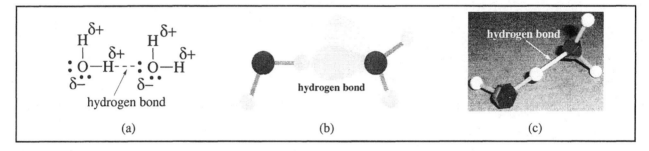

Figure 1.25. Representations of hydrogen bonding between two water molecules.

The bonding between two water molecules represented in Figure 1.25 is called a **hydrogen bond**. Hydrogen bonds are formed when a hydrogen atom covalently bonded to a highly electronegative atom (N, O, and F) is attracted to a nonbonding electron pair on another highly electronegative atom (N, O, or F). The second

> Experiments and calculations both show that the interaction between the hydrogen in one molecule and the lone pair on oxygen in the other has some covalent, electron-sharing, character. The model in Figure 1.25(c) may over-emphasize the amount of covalence.

electronegative atom may be in another molecule, or it may be in the molecule containing the hydrogen atom. The Lewis structure in Figure 1.25(a) highlights the proximity of partial positive and negative charges when the hydrogen atom on one water molecule is near a nonbonding pair of electrons on an adjacent water molecule. The modified ball-and stick model in Figure 1.25(b) shows the same interaction, with the large pink lobe representing the space where the electron pair is most likely to be found. Figure 1.25(c) represents the interaction as two covalent bonds: a shorter (and stronger) bond between the oxygen atom and one of the hydrogen atoms on the left-hand molecule, and a longer (much weaker) bond between the same hydrogen atom and the oxygen atom of the right-hand water molecule.

Figure 1.10 gave the hydrogen–oxygen covalent bond length in a single water molecule as 94 pm. As you can see in Figure 1.26, when water molecules hydrogen bond with one another, this bond stretches a bit to 100 pm, because the hydrogen atomic core is attracted to the nonbonding electron pair in the second oxygen. The length of the hydrogen bond to the second molecule is 180 pm, about twice the length of the hydrogen–oxygen covalent bond. Also note that the hydrogen atom forming the hydrogen bond is located on the line joining the two oxygen atoms. This linear geometry of the three atoms seems to form the strongest hydrogen bonds so hydrogen bonds generally form in situations where the hydrogen can lie roughly on the line between the two electronegative atoms.

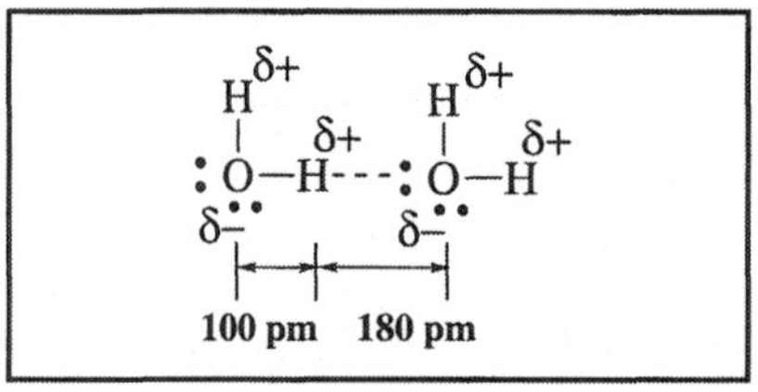

$$H^{\delta+} \quad H^{\delta+}$$
$$| \quad \delta+ \quad | \quad \delta+$$
$$:\!\underset{\delta-}{\overset{..}{O}}\!-\!H\!-\!-\!-:\!\underset{\delta-}{\overset{..}{O}}\!-\!H$$
$$\vdash\!\!\!\!\!\dashv\!\!\!\!\!\vdash\!\!\!\!\!\!\dashv$$
$$100 \text{ pm} \quad 180 \text{ pm}$$

Figure 1.26. Geometry of hydrogen bonding in water.

In water, Figure 1.27 shows that each water molecule can simultaneously form four hydrogen bonds. Two hydrogen bonds [(1) and (2) in the figure] link the central molecule's hydrogen atoms to nonbonding electron pairs on oxygen atoms in other water molecules. The other two hydrogen bonds [(3) and (4) in the figure] use the central molecule's nonbonding electron pairs on oxygen to link to hydrogen atoms on other molecules. A water molecule that forms four hydrogen bonds has a tetrahedral arrangement of bonds around the oxygen: two covalent bonds to hydrogen and two hydrogen bonds to hydrogens on other water molecules.

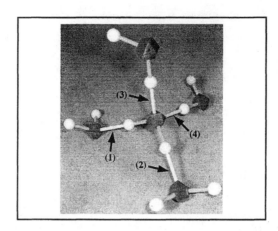

Figure 1.27. A water molecule hydrogen bonded (arrows) to four other water molecules.

In liquid water, hydrogen bonds are continuously broken and reformed as the moving water molecules jostle each other. Although its hydrogen-bonding partners are constantly changing, at any instant, each water molecule in the liquid has hydrogen bonds to about three other water molecules. These hydrogen bonding interactions among water molecules are strong enough to raise the boiling point of water to 100 °C, which is why water is a liquid at room temperature.

1.26. **Consider This**   *Does hydrogen bonding explain the boiling points of NH_3 and HF?*

(a) The boiling points for NH_3 and HF, Figure 1.24, are out of line with other group V and VII compounds. Sketch models to show how hydrogen bonding can account for their relatively high boiling points.

(b) If we have a large number, N, of water molecules, there are $2N$ hydrogen atoms and $2N$ nonbonding pairs that can interact to form hydrogen bonds. Since it takes one hydrogen atom and one nonbonding pair to form each hydrogen bond, the maximum number of hydrogen bonds that can be formed among these molecules is $2N$. If we have N ammonia molecules, how many hydrogen atoms are available for form hydrogen bonds? How many nonbonding pairs? What is the maximum number of hydrogen bonds that can be formed among these N ammonia molecules? Do the same analysis for N hydrogen fluoride molecules.

(c) Do your results in part (b) help to explain the boiling points of NH_3 and HF relative to H_2O in Figure 1.24? Explain why or why not.

1.27. **Check This**   *Predicting relative boiling points*

Lewis structures for *n*-butane, ethyl methyl ether, and 1-propanol are shown below. Predict the relative boiling points of these compounds. Explain your reasoning.

Water: A Natural Wonder Chapter 1

n-butane   ethyl methyl ether   1-propanol

Section 1.8. Further Structural Effects of Hydrogen Bonding in Water

1.28. **Investigate This** *What kinds of hydrogen-bonded networks can you construct?*

Work with a small group to construct about a dozen ball-and-stick models of water molecules. Use the shortest (pink) sticks in your set for the hydrogen–oxygen bonds. Connect one of your water molecules to four others by hydrogen bonds. Use the longer (white) sticks in your set for the hydrogen bonds, as in Figures 1.25(c) and 1.27. This picture shows how to begin such connections. Continue to add water molecules to see how a network of hydrogen-bonded water molecules might form. Can you form a ring of hydrogen-bonded water molecules? If so, how many molecules are required to form a ring without bending any of the sticks? Compare your structure to the representation shown in Figure 1.2(a). What conclusion(s) can you draw?

Ice: a hydrogen-bonded network

Rings of hydrogen-bonded water molecules constructed for Investigate This 1.28 might look something like those in Figure 1.28(a). Each ring contains six oxygen atoms and six hydrogen atoms, with six other hydrogen atoms sticking out from the ring. In this model, covalent H–O bonds alternate with H–O hydrogen bonds in each ring. All bond angles about oxygen are 109.5° tetrahedral angles, so the rings are puckered. Each oxygen atom has one covalent bond to a hydrogen atom and one hydrogen bond that are not part of the ring. Two ring structures can link using three hydrogen bonds, as shown in Figure 1.28(b). This cage-like structure can be extended in all three directions as shown in Figure 1.28(c).

WEB Chap. 1, Sect. 1.8
Try interactive animations to investigate the structure of solid water.

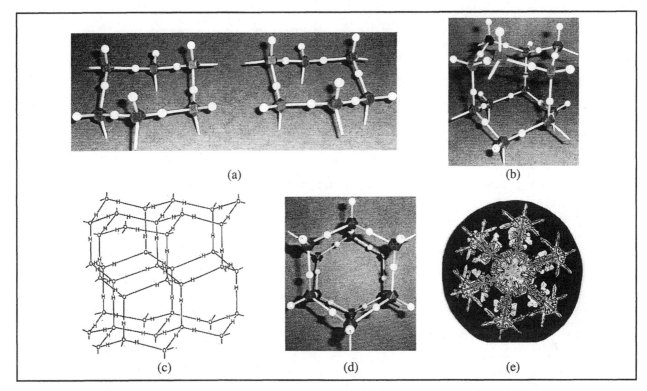

Figure 1.28. Hydrogen-bonded networks of water molecules and the structure of ice.
Panels (a) and (b) show successive stages in building a model of the ice structure. (c) extends the structure to six of the cages in (b) and (d) is a top view of the structure in (b). (e) shows a snowflake for comparison.

The extended regular structure, with every water molecule forming four hydrogen bonds, is ice. The symmetry of the stacked six-member rings can be seen when viewed from the top, as in Figure 1.28(d) and Figure 1.2(a). Snowflakes, Figure 1.28(e), reflect this six-fold symmetry. The extended hydrogen-bonded structure in ice produces more open space between water molecules than in liquid water. Consequently, as we pointed out in Section 1.1, ice is less dense than liquid water. Water is the only common substance that expands when it freezes.

1.29. **Consider This** *How do the structures of liquid and solid water compare?*

Show how breaking one of the hydrogen bonds in the ring of water molecules you made for Investigate This 1.28 allows the structure to "collapse." How would such a collapse affect the ice structures shown in Figure 1.28? Is this collapse related to the relative densities of ice and water? Explain your answer.

1.30. Investigate This *What are the temperatures in an ice-water mixture?*

Do this as a class activity and work in small groups to analyze the results. Fill a tall narrow transparent container with crushed ice and add cold water until it is full. Stir the mixture vigorously for about a minute to get the temperature of the mixture uniform at 0 °C. Place two thermometers or other temperature probes in the mixture with one measuring the temperature near the top and the other near the bottom of the mixture. Allow the system to remain undisturbed while you record the temperature of both thermometers every 5 minutes for 30–60 minutes. Plot the data.

1.31. Consider This *How do you interpret the temperatures in an ice-water mixture?*

Did you expect the results you observed in Investigate This 1.30? Why or why not? At what temperature do you think liquid water is most dense? What is the evidence for your answer?

Variation of liquid water density with temperature

The lower density of solid ice compared to liquid water is unusual, but there is still more to the density story of water. Ice melts when it has enough thermal energy (molecular motion) to begin to break some of the hydrogen bonds. As they break (at 0 °C), the open structure of the solid starts to collapse. Molecules move to take up some of the empty space, producing a liquid that is denser than the solid. As the temperature rises slightly, the increased jostling of the water molecules continues to allow them to find spaces to move into and increases the density even further. This effect reaches its maximum at 3.98 °C. As the temperature continues to increase, above 4 °C, the water molecules move faster, resulting in a larger average distance between them. The density decreases as shown in Figure 1.29. Water at 4 °C is the most dense form of water (under normal pressure) — either as solid or liquid.

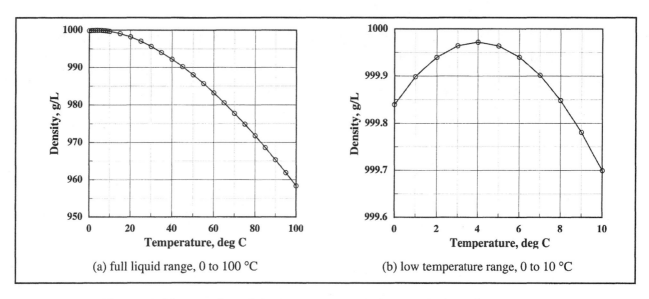

(a) full liquid range, 0 to 100 °C (b) low temperature range, 0 to 10 °C

Figure 1.29. Density of liquid water as a function of temperature.

1.32. **Consider This**   *What are the connections between water density and temperature?*

(a) What is your interpretation of the results from Investigate This 1.30? Do the density properties of water and ice support your interpretation? Why or why not. Try Problems 1.62 and 1.63 to extend your analysis and to examine an important consequence of the density properties of water for aquatic life.

Section 1.9.   Hydrogen Bonds in Biomolecules

Life as we know it would not exist without the hydrogen bond. Hydrogen bonding is central to the structure and function not only of water, but also of proteins, nucleic acids, and many other molecules of biochemical importance. In this section we introduce several biomolecules that are essential constituents of living organisms. Our emphasis is on the role of hydrogen bonding, not the details of their molecular structures.

Proteins

Proteins are high molecular mass compounds formed by linking hundreds or thousands of **amino acid** molecules end-to-end. Figure 1.30 shows how the linking occurs when three amino acids are strung together. The amino acid pictured in the figure is alanine, one of twenty that combine to form proteins. All these amino acids contain the basic structure shown in red in the figure. Different **side groups**, molecular fragments made up of various combinations of H, C, N, O, and S, like the methyl, –CH_3, group on alanine, establish the unique identity of each amino acid. The chain of linked amino acids is called the protein *backbone*. Side groups are spaced along the backbone like charms on a bracelet.

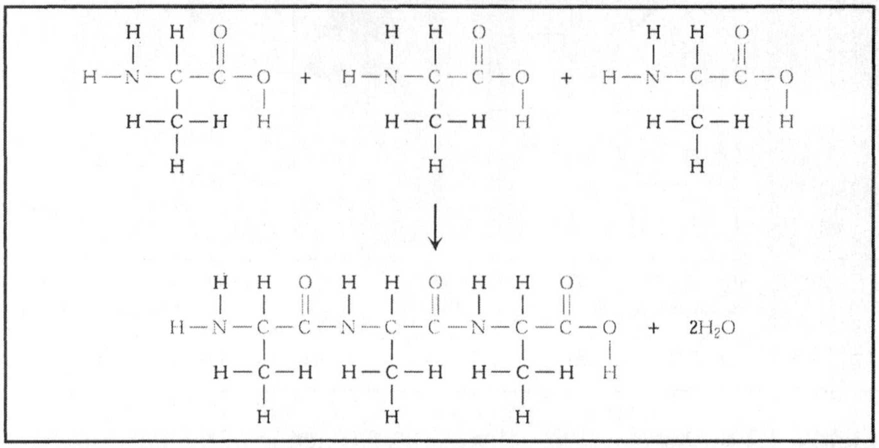

Figure 1.30. Linking amino acids to form a chain.
The atoms shown in red are part of the structure of all 20 amino acids found in proteins. A double line between atomic symbols represents two pairs of shared electrons (see Chapter 5).

Protein folding

For a protein to carry out its tasks in an organism, the chain of amino acids must coil and fold upon itself into what is called its "active form". Many of the interactions that hold proteins in their active forms are hydrogen bonds that form along the protein backbone. The backbone is flexible enough to allow hydrogen bonding between a hydrogen atom covalently bonded to nitrogen and an oxygen atom that is four amino acids away along the chain. A basic structure, the α–helix, present in most proteins is held together by these hydrogen bonds, which are shown as heavy dots in Figure 1.31(a). The chain folds back on itself forming an overall structure such as that shown in Figure 1.31(b). Hydrogen bonds between parts of the backbone that are separated by tens or hundreds of amino acids help to hold this structure together.

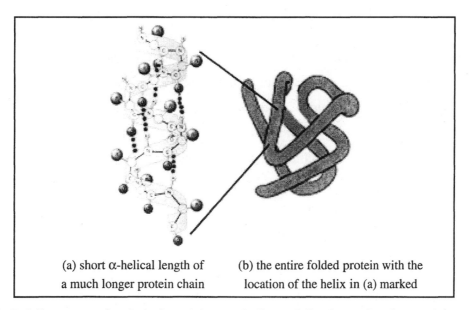

(a) short α-helical length of a much longer protein chain

(b) the entire folded protein with the location of the helix in (a) marked

Figure 1.31. Folding a protein chain into (a) an α-helix and (b) the active form of the protein.

Hydrogen bonds in (a) are shown as heavy dotted lines between the light gray hydrogen atoms and red oxygen atoms. Side groups on the protein chain are represented by the green balls.

When proteins are heated, the higher temperatures increase molecular motions, breaking some of the hydrogen bonds and the protein begins to unfold. Unfolding destroys the molecule's natural structure and activity; the protein becomes **denatured**. Most proteins denature above about 50 to 60 °C. If the solution hasn't been heated too hot or for too long, and is cooled slowly, the chain can sometimes refold and hydrogen bonds can re-form, returning the structure to its original form. Part of the process of cooking protein-containing foods involves heat denaturing of proteins. A familiar example of protein denaturation occurs when you cook an egg, Figure 1.32. As the molecules unfold they become less soluble, their strands begin to tangle with one another, and the more-or-less clear protein solution, Figure 1.32(a), changes to opaque and white, Figure 1.32(b). Unlike "renaturing" following mild heating, the cooking process cannot be reversed; you cannot unfry an egg.

(a) (b)

Figure 1.32. Denaturing egg-white protein changes it from (a) clear to (b) opaque white.

Nucleic acids

The genes of all organisms are composed of **deoxyribonucleic acid, DNA**. DNA is a ladder-like molecule, Figure 1.33, with two parallel strands of alternating sugar and phosphate groups forming the "uprights," and the "rungs" of the ladder formed by hydrogen bonding between pairs of side groups on the chain. Every sugar unit on both strands has an attached side group, or **base**, which may be any one of four molecules — adenine, guanine, cytosine, or thymine, Figure 1.34. The bases pair in two ways: adenine forms two hydrogen bonds with thymine (A–T) and guanine forms three hydrogen bonds with cytosine (G–C). In living cells, the DNA ladder is twisted so that the uprights form a double helix. The order of the bases along the ladder encodes the genetic information stored in DNA.

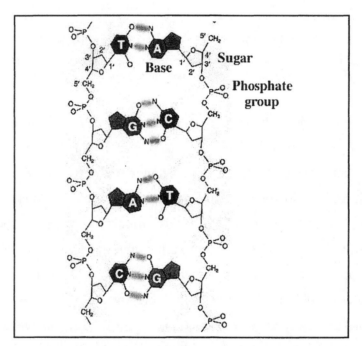

Figure 1.33. Hydrogen bonding in DNA.
A, C, G, and T represent adenine, cytosine, guanine, and thymine. Pink areas represent hydrogen bonds.

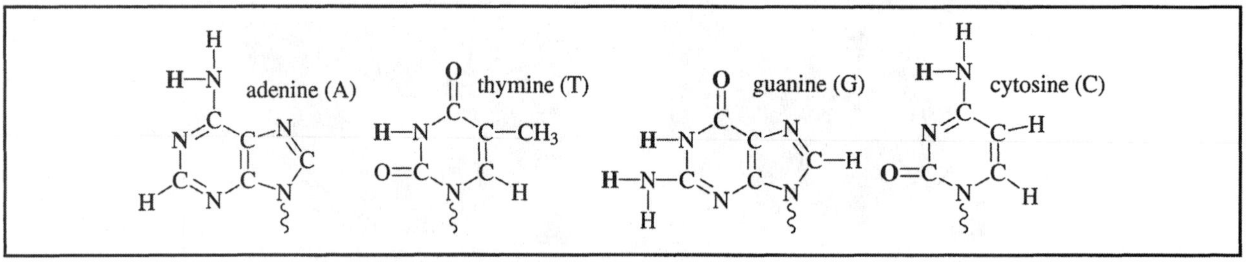

Figure 1.34. Sites for hydrogen bonding in the bases in DNA.
Squiggly lines are the attachment to the sugar-phosphate chain. Atoms used in hydrogen bonding are in blue.

1.33. Investigate This *How do the DNA bases fit together?*

(a) Work in small groups and use your model kits to construct a model of each of the DNA bases — adenine, guanine, cytosine, and thymine, Figure 1.34. Use the light blue polyhedra for the N's, dark gray polyhedra for the C's in the rings, and the standard black and red polyhedra for the other C's and O's. Use light green sticks to connect second-period elements to one another and pink sticks to connect H's. Show how the A–T and G–C pairs can be held together by two and three hydrogen bonds, respectively.

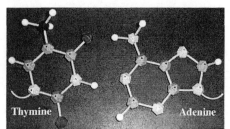

(b) Try making the alternative pairs, A–C and G–T. Can these pairs fit together? Explain why or why not.

At higher temperatures, increased molecular motions of the DNA molecule disrupt the hydrogen bonds, and the two linked strands separate. As with proteins, this process in which DNA loses its double-helical, hydrogen-bonded shape is called denaturation.

1.34. Consider This *How does hydrogen bonding affect the thermal stability of DNA?*

The thermal stability of DNA helices is affected by the relative numbers of **G–C** and **A–T** pairs. Why should changes in the **G–C** to **A–T** ratio affect the temperature at which the helices denature? How do you expect the thermal stability of DNA with a higher **G–C** to **A–T** ratio to compare to the stability of DNA with a lower **G–C** to **A–T** ratio? Explain your reasoning.

Strength of hydrogen bonds

As proteins and nucleic acids carry out their functions in an organism, they must undergo reversible structural changes. Enzymes, for example, are proteins that change their shape slightly as they bind to and catalyze chemical reactions between other biological molecules. And DNA strands have to separate in order to undergo duplication (making new DNA strands) and transcription (making complementary RNA strands). These changes typically involve breaking some hydrogen bonds and forming others without breaking any covalent bonds. Covalent bonds hold the molecules of life together in relatively permanent structures under normal conditions. Though hydrogen bonds also help hold the molecules in their active forms, they allow for changes in shape that can occur with only small energy input. In other words, hydrogen bonds complement stronger covalent bonds in producing the remarkable tapestry of life. Typically, hydrogen bonds are about 5-10% the strength of electron-pair covalent bonds.

Reflection and projection

The hydrogen-bonding model makes it possible to explain a huge range of physical and chemical properties of molecules. The essential features of the model are surprisingly simple: a hydrogen atom bound to a very electronegative atom is attracted to another very electronegative atom elsewhere. The requirements are availability of a nonbonding pair of electrons on the electronegative atom to which the hydrogen is attracted and a geometry that puts the hydrogen roughly on a line between the two electronegative atoms. The examples we have used to illustrate bulk properties of matter where hydrogen bonding is a factor (boiling points, densities of water and ice, denaturation of proteins and nucleic acids) only scratch the surface. All of chemistry is influenced by this remarkable phenomenon in more ways than can be described in any one textbook.

So far, we have not said much about the energy implications of any of the attractions between molecules. From our boiling point comparisons in Section 1.7, we found that, for many compounds, dispersion forces are the most important attractions holding the molecules together. For those polar compounds whose molecules can hydrogen bond with one another, hydrogen-bonding attractions often dominate the interactions among the molecules. And we have said that hydrogen bonds are weaker than covalent bonds, but we have not quantified any of these interactions. In the next sections, we begin a discussion of energy implications when water moves between liquid and gas phases or when it is heated or cooled. Thanks to hydrogen bonding, water is both an enormous reservoir of available energy as well as an enormous sink for waste heat. When we consider the energy associated with phase changes, we will have to deal with the problem of counting molecules, and we will look at techniques for counting by weighing in Section 1.11.

Section 1.10. Phase Changes: Liquid to Gas

1.35. Investigate This *What happens when you sweat?*

For this investigation, you will simulate sweating by wetting a finger. Be sure your hands are clean and dry. Wet one of your fingers and then wave your hand in the air. Do you feel any difference between the wet finger and the others? If so, what is the difference?

> **1.36. Consider This** *What is the purpose of sweating?*
>
> **(a)** What do you think caused your wet finger to feel different when you waved it in the air in Investigate This 1.35? What is required to produce this phase change? What evidence do you have to justify your answer? How could you test your explanation?
>
> **(b)** How is your analysis in part (a) related to sweating?

When a liquid boils to give a gas at atmospheric pressure or evaporates to give the gas at a lower pressure, energy is required to disrupt the attractive interactions between the molecules in the liquid phase. In Section 1.7, we used this idea to correlate boiling points with the strength of these attractions. We did not, however, discuss the source of the required energy. Energy cannot be created or destroyed. If energy enters a liquid to cause boiling or evaporation, the energy has to come from somewhere. When energy is put into one substance, an equivalent amount of energy has to leave another substance.

Since energy can go into or come out of a substance, we need to have a way to indicate the direction of this change. We will use E to symbolize energy and we will use ΔE to symbolize a *change* in energy. The Δ (uppercase Greek letter delta) symbol always means the *final value minus the initial* value of the specified quantity, for example, $\Delta E = E_{final} - E_{initial}$. When energy enters a substance, the final amount of energy in the substance, E_{final}, is larger than the initial amount of energy, $E_{initial}$, that is, $E_{final} > E_{initial}$, so ΔE is positive. Conversely, when energy leaves a substance, $E_{final} < E_{initial}$, and ΔE is negative.

> **1.37. Consider This** *What are the signs of the energy changes for sweating?*
>
> **(a)** What happened to the water on your skin when you waved your wet finger in the air in Investigate This 1.35? Does energy leave the water or enter the water to cause this process? What is the sign of the energy change in the water?
>
> **(b)** Does energy leave your skin or enter your skin in this process? What is the sign of the energy change in your skin? Does your answer help explain the difference between your wet and dry fingers in the investigation?

Energy diagrams

An energy diagram, Figure 1.35, is a graphical way to visualize changes in energy. Energy diagrams appear often in this book, so you need to become familiar with their interpretation. The horizontal lines represent the energy of the substances written directly above them. Energy levels are not always given absolute numeric values, because the diagrams are designed to show energy *differences*, ΔE. The head of an arrow showing the direction of change points to the final energy and the tail is at the initial energy level. An arrow pointing up, as in Figure 1.35, means that the change in energy for the process has a positive sign: $\Delta E > 0$, because the final energy value is greater than the initial energy value. Energy is required to bring about the indicated change. An arrow pointing down means that the change in energy for the process has a negative sign: $\Delta E < 0$, because the final energy value is smaller than the initial energy value. Energy is released when the indicated change occurs.

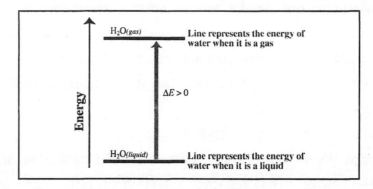

Figure 1.35. An energy diagram for evaporation of a sample of water.

Figure 1.35 is an energy diagram for water evaporating from your skin. Since energy has to enter the water to cause the **evaporation** (vaporization), the energy of the gaseous water is higher than the energy of the liquid water before evaporation. The horizontal line representing the energy of the gas is higher on the graph than the line representing liquid. The upward-pointing red arrow in Figure 1.35 shows you that ΔE for the liquid-to-gas phase change is positive.

1.38. Check This *Interpreting an energy diagram*

(a) The energy diagram shown here is for a change in the temperature of one liter of water. The arrow represents the energy change when the water changes from 50 °C to 25 °C. Does energy enter or leave the water when this change occurs? How is this represented on the diagram? Explain your answers.

(b) Is the $\Delta E > 0$ or is $\Delta E < 0$ for this process? How is this represented on the diagram? Explain your answers.

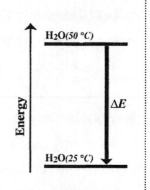

1.39. Consider This *How are energy level diagrams correlated?*

(a) Sketch an energy level diagram that represents the energy change *in your skin* when energy from your skin is used to evaporate some water. Draw an arrow that represents the direction of the change. How is the sign of ΔE for this process related to the direction of the energy-change arrow on your diagram?

(b) Assume that the amount of water that is evaporated by your skin in part (a) is the same as the amount of water represented by Figure 1.35. Should the length of the energy-change, ΔE, arrow on your diagram be longer, shorter, or the same length as the energy-change arrow in Figure 1.35. Explain your reasoning.

1.40. Investigate This *Do all liquids evaporate at the same rate?*

Do this as a class activity and work in small groups to analyze and discuss the results. Use three identical paper towels for this activity. Simultaneously put 10 drops of water in one spot on one towel, 10 drops of methanol on a second towel, and 10 drops of hexane on the third towel. Hold the towels up so everyone can observe the wet spots. *CAUTION:* Methanol and hexane are both flammable. Time how long it takes each of the liquid spots to disappear completely.

1.41. Consider This *What are the relative energies of vaporization of different liquids?*

(a) Is ΔE for the liquids in Investigate This 1.40 positive or negative? Give your reasoning.

(b) For which liquid is the absolute value of ΔE smallest? For which is the absolute value of ΔE largest? Base your reasoning only on our model for vaporization and data from Investigate This 1.40.

(c) Sketch an energy diagram that compares the energy changes, ΔEs, for each liquid evaporating.

In Investigate This 1.40, the liquid films all gained energy from the chalkboard. Each liquid sample gains the same amount of energy per unit time, but they don't all disappear at the same time. Therefore, different quantities of energy must be required for each sample to vaporize. What are these quantities? Our energy diagrams have not shown energy units and all our comparisons have been either directional or in terms of relative size based on experimental evidence. Before we go further, we need to define the units for energy.

Energy units and conversions

The energy unit that is probably most familiar to us is the calorie. A **calorie** was originally defined as the amount of energy required to increase the temperature of exactly 1 g of water by exactly 1 °C. This quantity of energy is called the **specific heat** of water. Scientists now use the **joule** (J) as the standard unit of energy. The joule is defined in terms of an amount of work done rather than an amount of heat transferred. The relationship of joules to calories, is:

> The joule is named to honor James Joule (English scientist, 1818–1889), one of the first scientists to study the quantitative relationships between heat and work.

$$1 \text{ cal} \equiv 4.184 \text{ J} \tag{1.1}$$

(The ≡ symbol means, *"is defined as."*) Most energy values in biological systems are still given in calories. We will often give values in both units. Or, you can work them out for yourself.

1.42. Worked Example *Conversion of energy units*

The energy required to vaporize water is 2.4×10^3 J·g^{-1}. How many calories are required to vaporize one gram of water?

> The unit J·g^{-1} is the same as J/g. The exponential form is less ambiguous and is used throughout the book.

Necessary information: We need the relation between joules and calories in equation (1.1).

Strategy: A *quantity* of something tells you how much of it there is. Quantity has *units* and a *numeric value*. Expressing the quantity in different units requires a different numeric value. If you buy 1 dozen eggs or 12 eggs, for example, you get the same quantity of eggs, but the numeric values for dozen and number are different. We will rewrite equation (1.1) as a unit conversion factor and use it to convert the energy of vaporization in joules to calories.

Implementation: Equation (1.1) can be rewritten as:

$$\frac{1 \text{ cal}}{4.184 \text{ J}} = 1 \tag{1.2}$$

We have been given the quantity of energy required to vaporize one gram of water in units of joules, so we use the ratio given by equation (1.2) to convert to units of calories:

$$\text{energy of vaporization} = 2.4 \times 10^3 \text{ J·g}^{-1} = (2.4 \times 10^3 \text{ J·g}^{-1})\left(\frac{1 \text{ cal}}{4.184 \text{ J}}\right) = 5.7 \times 10^2 \text{ cal·g}^{-1}$$

Does the answer make sense? Equation (1.1) shows that one calorie is a larger amount of energy than one joule. Therefore, a quantity of energy expressed in calories will have a smaller numeric value than the same quantity expressed in joules. This is the result we got.

After finishing any problem involving calculations, check to see if the numeric result is reasonable:

- Check the sign of the result. Here we are converting a positive quantity energy from one unit to another, so the answer must be positive.
- Check the units of the result. If units are always included with numeric values, the units should cancel to give the appropriate units in the result.
- Check the arithmetic. In this case, mentally doing the calculations to one significant figure yields: 2000 ÷ 4 = 500, which is the right order of magnitude.
- Check the number of significant figures. Here, the datum with the greatest uncertainty is the energy in joules, which has an uncertainty of 1 part in 24. We have given the result with two significant figures, which implies an uncertainty of about 1 part in 57. Reporting only one significant figure would imply an uncertainty of 1 part in 6, which would be far too uncertain. Reporting three significant figures would imply an uncertainty of about 1 part in 570, which would not be justified by the original datum.

1.43. Check This   *Conversion of energy units*

(a) The amount of energy required to vaporize hexane is 92 cal·g^{-1}. How many joules are required to vaporize one gram of hexane?

(b) From the data in part (a) and in Worked Example 1.42, what conclusion can you draw about the attractions between hexane molecules relative to those between water molecules? Is your conclusion consistent with the models in Section 1.7? Explain why or why not.

Correlations with molecular models

In Investigate This 1.40, you found that a film of liquid hexane evaporated faster than a similar film of liquid water and could conclude that less energy is required to evaporate hexane than water. The data in Worked Example 1.42 and Check This 1.43 show that the energy required to vaporize a gram of hexane is only about one-sixth the amount required to vaporize a gram of water, so the conclusion you drew from the investigation is confirmed by further data. But we are not ready quite yet to use experiments and data like these to correlate our molecular models of attractions with the observable energies of vaporization.

To compare the energies of vaporization for two compounds, we need to compare the energy required to vaporize the same number of molecules of each compound. A molecule of hexane, C_6H_{14}, has 20 atoms and a molecule of water, H_2O, has three. A water molecule has a smaller mass than a hexane molecule, so one gram of water will contain more molecules than one gram of hexane. An analogy would be the number of grapes and oranges in the same mass of each fruit. A grape has a smaller mass than an orange and Figure 1.36 shows that there are more grapes than oranges in 0.5 kg of each. When one gram of water is vaporized, more molecules change state than when one gram of hexane is vaporized. More interactions have to be broken in the water sample and this might account for the greater quantity of energy required to evaporate it. In order to make comparisons that we can relate to our molecular models of matter, we need to know the amount of energy required to vaporize the same number of molecules of each compound. We need a way to count out the same number of molecules (or know the relative numbers of molecules) for our comparisons.

Figure 1.36. Comparison of equal masses of grapes and oranges.

Section 1.11. Counting Molecules: The Mole

Individual molecules are far too small to count using our usual laboratory apparatus, so we have to devise other ways to count them. A convenient way to measure a quantity of matter is to use a balance to determine its mass. **Mass** is a fundamental property of matter that measures the quantity of matter in a sample. The problem is that the molecules of different compounds have different masses, so the same mass of two compounds will contain different numbers of molecules. In chemistry, *the unit that expresses numbers of molecules is referred to as the amount* of a substance. To measure out the same number of molecules of two different compounds, the ratio of the masses taken must be the same as the ratio of the molecular masses.

> In everyday use, *amount* means just about any measure (mass, volume, number, and so on) that specifies how much of something we have. In chemistry, *amount* specifies number of molecules.

> ***1.44. Consider This*** *How can you count objects by measuring mass?*
>
> In many game arcades, a player gets one or more paper tickets for each game won. Players often bring large batches of tickets to the redemption center to exchange for prizes. The person at the center uses a balance to determine the mass of the tickets and awards prizes based on the mass. If a dozen tickets have a mass of 1.9 g, and a player redeems 24.1 g of tickets, how many dozens of tickets are redeemed? How many tickets are redeemed? Describe your method for solving this problem.

Mole and molar mass

If you know the mass of a dozen identical objects and measure the mass of an unknown number of those same objects, you can determine how many dozens of the objects you have by dividing the measured mass by the mass of a dozen of the objects, as you did in Consider This 1.44. Instead of a dozen, the standard scientific unit for measuring amount (number) is the **mole** (abbreviated "**mol**"). The individual particles that make up the substance being measured can be anything: atoms, molecules, ions, grains of sand, mosquitoes, whatever. Particles the size of atoms and molecules, unlike tickets to be redeemed, are so small that they cannot be counted directly, but we can use a balance to determine the mass of an observable amount of any element or compound. Then, if we know the mass of one mole of the element or compound, we can calculate how many moles are present in the measured mass.

> The *name* mole has an "e" on the end, but the *unit abbreviation* mol does not. Mole is used in text discussions; mol is used with numeric values.

> **WEB Chap. 1, Sect. 1.11**
> Use interactive animations to work through mass, mole, and molar mass interconversions.

1.45. Worked Example *Calculating numbers of moles*

A mole of water has a mass of 18.02 g. If you drink a glass of water, about 225 g of water, how many moles of water do you drink?

Necessary information: The problem statement contains all the needed information.

Strategy: We can use the mass equivalence of a mole of water molecules to write a unit conversion factor to convert the mass of water we drink to number of moles of water.

Implementation: 1.000 mol H_2O = 18.02 g and we rewrite this equivalence as:

$$\frac{1.000 \text{ mol } H_2O}{18.02 \text{ g}} = 1$$

Use this unit factor to convert the mass of our drink to moles:

$$225 \text{ g H}_2\text{O} = (225 \text{ g H}_2\text{O})\left(\frac{1.000 \text{ mol H}_2\text{O}}{18.02 \text{ g}}\right) = 12.5 \text{ mol H}_2\text{O}$$

Does the answer make sense? The mass of water we drink is more than ten times the mass of one mole of water, so our answer, more than ten moles, makes sense. The three significant figures in the result are consistent with the most uncertain value in the original data, the mass of the water.

1.46. **Check This**   *Mass to moles and moles to mass conversions*

(a) Ten drops of water is about 0.5 mL, or about 0.5 g of water. How many moles of water are in 10 drops of water?

(b) The mass of one mole of sucrose (table sugar) is 342 g. A bag of sugar from the market contains 2.26 kg of sugar. How many moles of sugar are in the bag?

(c) The mass of one mole of sodium chloride (table salt) is 58.5 g. A container of salt from the market contains 12.6 mol of salt. How many grams of salt are in the container? You can check your answer at the market.

Recall that the periodic table on the inside front cover of the book gives the *relative atomic masses* of the atoms of each element. The **molar mass** of any element is the mass of the element in grams that is equal to its relative atomic mass. A molar mass of an element contains a mole of atoms of the element. For example, 1.0079 g of hydrogen contains one mole of hydrogen atoms; the molar mass of hydrogen is 1.0079 g. Similarly, the molar mass of carbon is 12.01 g and the molar mass of oxygen is 16.00 g. A molecule contains more than one atom, so the molar mass of a molecule is the sum of the molar masses of all its constituent atoms. For example, oxygen gas consists of O_2 molecules, so a mole of oxygen gas contains two moles of oxygen atoms and the molar mass of oxygen gas is 32.00 g (= 2 × 16.00 g).

This oxygen gas example reminds us to be careful to specify the atomic or molecular unit of interest when we are using molar masses. We often specify the **formula unit**, the chemical formula of the atomic or molecular unit of interest. For example, if we are interested in bromine atoms, we specify the formula unit as Br, the atomic symbol for bromine, and the molar mass of this formula unit is 79.91 g. If we are interested in elemental bromine, which is a liquid of Br_2 molecules, the formula unit is Br_2 and the molar mass in 159.82 g.

Most molecules contain more than one kind of atom and all the atomic masses in their formula unit are combined to get the molar mass. A mole of water, H_2O, for example, contains two moles of hydrogen atoms, (2.0158 g of hydrogen atoms) and one mole of oxygen atoms (16.00 g of oxygen atoms). The molar mass of water used in Worked Example 1.45 and Check This 1.46 is 18.02 g (= 2.0158 g hydrogen + 16.00 g oxygen).

1.47. Worked Example *Calculating molar mass*

What is the molar mass of hexane, C_6H_{14}?

Necessary information: We use the given formula unit and the relative atomic masses of C and H atoms from the inside front cover.

Strategy: The number of atoms of each element in the formula unit multiplied by the molar mass of each elemental atom gives the mass of each element in a mole of formula units. The sum of these masses is the molar mass of the formula unit.

Implementation: The C_6H_{14} formula unit for hexane contains 6 carbon atoms and 14 hydrogen atoms, so in one mole of formula units, we have:

$$6 \text{ mol C} = (6 \text{ mol C})\left(\frac{12.01 \text{ g}}{1 \text{ mol C}}\right) = 72.06 \text{ g} \quad (\text{mass of C in one mole } C_6H_{14})$$

$$14 \text{ mol H} = (14 \text{ mol H})\left(\frac{1.0079 \text{ g}}{1 \text{ mol H}}\right) = 14.11 \text{ g} \quad (\text{mass of H in one mole } C_6H_{14})$$

$$\text{molar mass } C_6H_{14} = (72.06 \text{ g}) + (14.11 \text{ g}) = 86.17 \text{ g}$$

Does the answer make sense? The only way to tell whether our answer makes sense is to check the arithmetic, which you should do.

1.48. Check This *Calculating molar mass*

(a) How many moles of carbon atoms are there in one mole of sucrose, formula unit $C_{12}H_{22}O_{11}$? How many moles of hydrogen atoms? How many moles of oxygen atoms? What is the molar mass of sucrose? Explain how you get your answers.

(b) How many moles of sodium atoms are there in one mole of sodium chloride, formula unit NaCl? How many moles of chlorine atoms? What is the molar mass of sodium chloride? Check your results for parts (a) and (b) with the data in Check This 1.46.

In 1859, a consistent set of relative atomic masses, based on the results of many experiments and the atomic model of matter, was proposed and soon agreed upon by scientists. Since then, many more experiments have been done that refined the original values and produced the relative atomic masses in our modern periodic table. Over these years, the amount of a substance

> More history of atomic masses and the periodic table is discussed in Chapter 4, Section 4.1.

that constitutes a mole has also been redefined and refined. At present, scientists have agreed on one of the forms of carbon, carbon–12 (also written as ^{12}C), as the standard for determining amount (moles). One mole is defined to be the *number* of atoms in *exactly* twelve grams of carbon–12. This number, 6.0221367×10^{23}, which has been determined experimentally, is called **Avogadro's number**. The symbol for Avogadro's number is N_A. The mass of a sample of any pure substance that contains Avogadro's number of particles is one mole of that substance. To summarize:

$$1 \text{ mole} \equiv \text{number of atoms in 12 g of } ^{12}\text{C} = 6.0221367 \times 10^{23} \text{ particles} = N_A \quad (1.3)$$

Although the mole is a number of particles, when we use moles, we rarely need to know the actual number of particles. To compare measurable amounts of substances, we almost always compare number of moles rather than number of molecules (or formula units). Figure 1.37 reminds us that comparing masses of formula units is identical to comparing molar masses of these formula units.

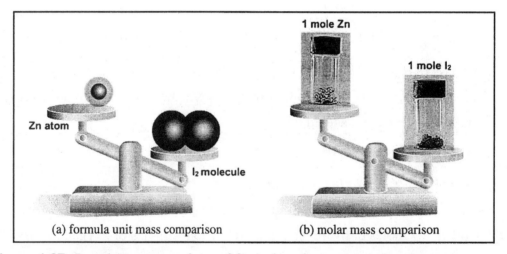

Figure 1.37. Imaginary comparison of formula unit masses and molar masses.
See WEB Chap 1, Sect 1.11 for activities based on these comparisons.

1.49. Check This   *Comparing amounts of compounds*

In Check This 1.46, you calculated the number of moles of water in 10 drops of water, which is the amount of water used in Investigate This 1.40. How many moles of methanol, CH_3OH, and hexane, C_6H_{14}, were used in the investigation? Assume that 10 drops of methanol has a mass of 0.4 g and 10 drops of hexane a mass of 0.3 g. For which compound did the largest number of molecules evaporate? For which did the smallest number evaporate? Explain your answers.

Molar vaporization energies

In Investigate This 1.40, you found that it took the longest time for water to evaporate. Now, your results from Check This 1.46 and 1.49 show that there were more molecules (moles) of water to be evaporated. Was more time required just because there were more molecules to evaporate? The only way to answer this question is to measure the amount of energy that has to be supplied to cause the evaporation of a known amount of liquid. The results of such experiments are shown for five compounds in Table 1.2. The table gives the energies of vaporization *per mole*, the boiling points, and dipole moments (a measure of the relative polarities of the molecules). Figure 1.38 shows ball-and-stick molecular models for the five compounds in Table 1.2

Table 1.2. Vaporization energies, boiling points, and dipole moments for selected compounds.

Compound	Line Formula	Molar Mass, g	Energy of Vaporization $kJ \cdot mol^{-1}$	Boiling Point, °C	Dipole Moment, Debye
water	H_2O	18	44	100	1.87
methanol	CH_3OH	32	39	65	1.70
ethanol	CH_3CH_2OH	46	41	79	1.69
dimethyl ether	CH_3OCH_3	46	23	−25	1.30
hexane	$CH_3(CH_2)_4CH_3$	86	32	69	0

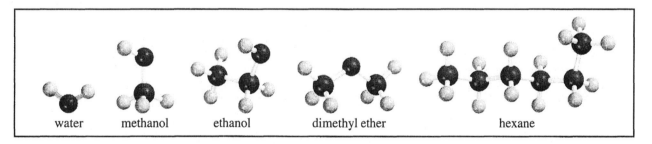

Figure 1.38. Molecular models of the compounds in Table 1.2.

1.50. Consider This *Which compounds in Table 1.2 form hydrogen bonds?*

(a) In addition to water, which compounds in Table 1.2 have molecules that can hydrogen bond to one another? How many hydrogen bonds can each molecule form?

(b) Make molecular models and drawings similar to Figure 1.25(a) to show the hydrogen bonding in the compounds you identified in part (a). Give your reasoning for why the molecules of the remaining compounds can't hydrogen bond among themselves.

Vaporization energies and molecular attractions

We can use molar vaporization energies to compare our molecular models for attractions among molecules, because these energies are for the same number of molecules vaporizing. Let's look first at the polar molecules in Table 1.2. Methanol and ethanol are **alcohols**. Alcohols are related to water in that both have an O–H group. For alcohols, the other oxygen bond is to an **alkyl** group, a group that contains one or more carbon atoms with every unused bond on each carbon atom occupied by hydrogen. Alcohol molecules have an O–H group and can hydrogen bond with one another, but can form a maximum of only about half as many hydrogen bonds as water, that is, about one mole of hydrogen bonds per mole of alcohol (see Consider This 1.26).

If we assume that most of the energy of vaporization of water, 44 kJ·mol^{-1}, is required to break hydrogen bonds, then about half this amount, 22 kJ·mol^{-1}, is required to break the mole of hydrogen bonds in the alcohols. This analysis suggests that breaking hydrogen bonds requires about 22 kJ·mol^{-1}. The remainder of the energy of vaporization for the alcohols, about 20 kJ·mol^{-1}, must be required to overcome the dispersion forces among the molecules.

Like alcohols, **ethers** also contain an oxygen atom, but in this case, both oxygen bonds are to alkyl groups. Ether molecules have no hydrogen atom covalently bonded to the electronegative oxygen, so the molecules of ether cannot hydrogen bond among themselves. The striking differences between the properties of ethanol and dimethyl ether, Table 1.2, are a consequence of hydrogen bonding in the ethanol. These compounds have the same molecular formula, C_2H_6O, and, therefore, the same molar mass, but ethanol is a liquid and dimethyl ether is a gas at room temperature. Their boiling points differ by more than 100 °C. The energy of vaporization for the ether, 23 kJ·mol^{-1}, must be required to overcome the dispersion forces among the molecules. Since ethanol and dimethyl ether have the same molecular formula, the dispersion forces among the molecules in their liquids should be about the same, 23 kJ·mol^{-1}. This value is consistent with our reasoning that gave about 20 kJ·mol^{-1} for the dispersion forces among ethanol molecules.

Finally, consider the nonpolar molecule, hexane, which is a liquid at room temperature and whose molar vaporization energy and boiling point are comparable to those for the alcohols in Table 1.2. The attractions that hold the hexane molecules together are dispersion forces and there are enough electrons in hexane (50) to give liquid hexane some properties that are comparable to smaller hydrogen-bonding molecules. As we will see in Chapter 2, dispersion forces are important in determining how molecules that contain large nonpolar segments as well as a polar group interact with one another and with different molecules.

1.51. Worked Example *Vaporization energies in Investigate This 1.40*

How much energy was required to vaporize the water sample in Investigate This 1.40?

Necessary information: We need the energy of vaporization of water, 44 kJ·mol^{-1}, from Table 1.2 and the number of moles of water in 10 drops, 0.03 mol, from Check This 1.46(a).

Strategy: Use the number of moles of water vaporized and the energy required for each mole to calculate the energy required for the 10-drop sample.

Implementation:

$$\text{energy to vaporize 0.03 mol H}_2\text{O} = (0.03 \text{ mol H}_2\text{O})\left(\frac{44 \text{ kJ}}{1 \text{ mol H}_2\text{O}}\right) = 1 \text{ kJ}$$

Does the answer make sense? Ten drops of water is a good deal less than a mole of water, so it will take a good deal less than the molar energy of vaporization to vaporize it, as we find.

1.52. Check This *Vaporization energies in Investigate This 1.40*

(a) Use the data in Table 1.2 and your results from Check This 1.49 to calculate the energy required to vaporize the methanol and hexane samples in Investigate This 1.40.

(b) Were your observations in the investigation consistent with the energies from part (a) and the energy for water from Worked Example 1.51? Explain why or why not.

Reflection and projection

A central concern of chemistry is to probe matter with whatever tools we have in an attempt to tease out a fundamental understanding of how nature works. In order to explain the bulk properties of matter, scientists invent and refine molecular models that are consistent with the observed properties and help to predict and interpret others as well. Section 1.10 began with consideration of why a wet finger held in the wind feels cooler than a dry finger, and moved on to consider why some liquid films evaporate from a chalkboard faster than others. Our interest in these phenomena was focused on the direction and magnitude of the energy changes. When a liquid is vaporized energy has to be supplied to disrupt the attractions between the molecules and this energy has to come from somewhere.

In order to interpret the data from the liquid films evaporating in molecular terms, we found that we needed a way to assure that we were comparing the same number of molecules (or a known ratio) in each case. This necessity led to our introduction of the scientific unit for amount, the mole. The mole is Avogadro's number of anything, but our usual application is to atoms, molecules, and ions. Molar mass is the mass in grams equal to the sum of the relative atomic masses that make up the formula unit of the element or compound of interest.

When we expressed energy of vaporization as the energy required to vaporize one mole of a compound, we could compare the data and interpret similarities and differences in terms of the molecular models of dispersion forces and hydrogen bonding from our previous discussions. Dispersion forces increase as the size of molecules increase because more electrons are available to form instantaneous dipoles. All molecules have dispersion forces attracting their molecules to each other and some have other interactions that contribute to the attraction. The hydrogen bond is an especially strong polar attraction between molecules in which a hydrogen atom covalently bonded to a very electronegative atom lies approximately on a line between this electronegative atom and another. For small molecules, hydrogen bonding often dominates those physical properties that are related to attractions between molecules. All of these *intermolecular* attractions are weaker than the *intramolecular* bonding (covalent bonds) that hold atoms together in molecules.

We have discussed a few of the properties of water that depend on the extensive hydrogen bonding between water molecules, and there are several more. In the next section, we will continue our discussion of energy and water, and see another way that water is essential to life as we know it.

Section 1.12. Specific Heat of Water: Keeping the Earth's Temperature Stable

On the Fourth of July, 1997, the Mars Pathfinder space probe landed on our planetary neighbor. The next day, a vehicle about the size of a child's wagon, the Sojourner Rover shown in Figure 1.39, left the lander and began moving about, using its various instruments to explore the Martian surface. The design team for the Mars lander had to design instruments and vehicles that could work over a large temperature range. The Martian atmospheric temperature varies about 74 °C, from –80 °C during the Martian night to about –6 °C during the Martian day. In contrast, the night-to-day difference is usually less than 25 °C on Earth, and the average temperature of the Earth is about 16° C, well above the freezing point of water.

Figure 1.39. The Sojourner Rover exploring the Martian surface.

Part of the reason that conditions on the two planets are so different is that more than 70% of the Earth's surface is covered with water. This mass of water acts as a giant energy reservoir. It absorbs energy from the sun and warm air during the day, without changing its temperature very much. The water returns thermal energy to the cooler atmosphere at night. The large mass of water evens out the extremes of temperatures. Mars, without water on its surface, lacks this mechanism for modulating surface temperature changes.

1.53. **Investigate This** *What is observed when liquids are heated at the same rate?*
Do this is a class activity and work in small groups to analyze the results. Add 120–150 g of room temperature water to a 250-mL Styrofoam® cup. Add the same mass of room temperature ethanol to a second identical cup. Clamp identical electrical immersion heaters and thermometers in each cup with the thermometers about one centimeter from the heaters. Read and record the temperature of the liquids every 10 seconds for about one minute. Simultaneously plug both heaters into electrical outlets. Continue reading and recording the temperatures every 10 seconds until one of the liquids reaches about 70 °C. Unplug the heaters.

1.54. **Consider This** *How does temperature change in heated liquids?*

(a) Plot the temperatures of each of the liquids in Investigate This 1.53 as a function of time. How do the plots for the water and ethanol differ? How are they the same?

(b) The same amount of energy has been added to both liquids. How can you account for any differences in the results for the two liquids?

Molar heat capacity

Maintaining the Earth's temperature in a range that supports life is essential for the survival of living things. One of the properties of water that is essential for temperature regulation in organisms and in the environment is its heat capacity. When energy (heat) enters a substance, the molecules begin to move a bit faster. These motions include rotation, vibration, and movement from one location to another. The increased motion is observed as an increase in the temperature of the substance. The **molar heat capacity**, *C*, of a substance is the amount of energy required to raise the temperature of one mole of the substance by one degree Celsius. Molar heat capacities for water and several other compounds are given in Table 1.3.

Table 1.3. Molar heat capacity and specific heat for selected compounds.

Compound	Formula Unit	Molar Mass, g	Molar Heat Capacity, $J \cdot mol^{-1} \cdot {}^\circ C^{-1}$	Specific Heat, $J \cdot g^{-1} \cdot {}^\circ C^{-1}$
water	H_2O	18	75.4	4.18
methanol	CH_4O	32	81.0	2.53
ethanol	C_2H_6O	46	112	2.44
1-propanol	C_3H_8O	60	144	2.39
hexane	C_6H_{14}	86	195	2.27
octane	C_8H_{18}	114	254	2.23
decane	$C_{10}H_{22}$	142	315	2.22
dodecane	$C_{12}H_{26}$	170	375	2.21

You can see from Table 1.3 that the molar heat capacity is larger for higher molar mass, more complex molecules. This makes sense, since it takes more energy to make heavier molecules move about and there are more ways for complex molecules to vibrate and rotate. Although water does not seem to stand out in the listing of molar heat capacities in Table 1.3, Figure 1.40 shows that it does not fit the pattern of the other molecules in the table. The line in Figure 1.40 is the best-fit line through the hydrocarbon data and you see that the alcohols fall pretty well on the line, but that water does not. The molar heat capacity of water is higher than is predicted by the trend in the data for the other liquids.

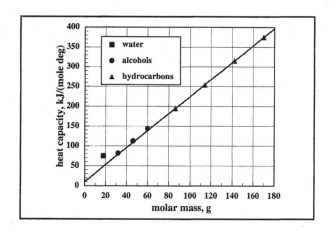

Figure 1.40. Molar heat capacities *vs.* molar mass for the compounds in Table 1.3.

Specific heat

Another way to analyze heat capacity data is to divide the molar heat capacity for a compound by its molar mass. This procedure gives the specific heat, which we introduced briefly in Section 1.10. As we noted there, the **specific heat**, *s*, of a substance is the amount of energy required to increase the temperature of exactly 1 g of the substance by exactly 1 °C. Water has a high specific heat, $s = 4.18$ $J \cdot g^{-1} \cdot °C^{-1}$, which is about ten times that of metals such as copper or iron. If the same amount of heat were added to equal masses of iron and water, the increase in temperature of the iron would be about ten times that of the water. We also see in Table 1.3 that the specific heat of water is almost double that of the other liquids, which are between 2.0 and 2.5 $J \cdot g^{-1} \cdot °C^{-1}$.

1.55. Worked Example *Specific heat and temperature rise*

In Investigate This 1.53, you added the same amount of energy to equal masses of water and ethanol. Use the results of the investigation to show which liquid has the higher specific heat.

Necessary information: We need to know that the ethanol reached a higher temperature than the water when the same amount of energy was added to each in Investigate This 1.53.

Implementation: When energy is added to a liquid that is below its boiling temperature, the temperature of the liquid increases, as we observed for both ethanol and water in Investigate This 1.53. This graph shows the results for such an experiment. The temperature of the ethanol increases more than the temperature of the water, which means that the rate of temperature increase for ethanol is greater than for water. In other words, more energy is

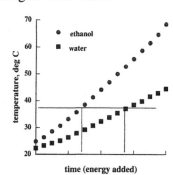

required to increase the temperature of water a given amount than is required to increase the temperature of ethanol by the same amount. [The horizontal line is added to the figure to show that the water gets to a given temperature later (after more energy has been added) than ethanol.] Specific heat is the amount of energy required to bring about a given increase in temperature of a substance. Since more energy is required to bring about a given increase in the temperature of the water, the specific heat of the water is larger than that of ethanol.

Does the answer make sense? The data in Table 1.3 confirm that the specific heat of water is larger than the specific heat of ethanol.

1.56. Check This *Specific heat and temperature rise*

If you carry out an investigation like Investigate This 1.53 with methanol and hexane in the cups, which will reach the higher temperature? Use the data in Table 1.3 and clearly explain the reasoning for your answer.

Temperature

To work with heat capacities and specific heat, it's important to distinguish between temperature and thermal energy. **Temperature** is the scale of hotness or coldness of a substance. It is a measure of the average molecular motions in a substance. We commonly measure temperatures in degrees Celsius, °C, but, in scientific work, we often use the absolute temperature, T, which has units of kelvin, K. The size of the degree is the same on both scales, but their zero points are different. The relationship between absolute and Celsius temperatures is:

No degree symbol, °, is used with, K, the symbol for temperature in kelvin.

$$0 \text{ K} = -273.15 \text{ °C} \quad \text{or} \quad 273.15 \text{ K} = 0 \text{ °C} \quad (1.4)$$

There is a quantitative relationship between the temperature of a substance in kelvin and the motions of its molecules, but all we need here is the qualitative idea that at a higher temperature molecular motion is increased. Temperature is an **intensive property**; it does not depend upon the amount of material present. The temperature of a drop of boiling water is the same as the temperature of a liter of boiling water. The average molecular motions in the two samples are the same.

Thermal energy (heat)

The thermal energies of the drop of boiling water and liter of boiling water are very different, however, since there are so many more molecules in a liter than in a drop. When **thermal** energy, or **heat**, measured in joules or calories, is added to a substance, it increases the motion of the atoms and molecules of the substance. Thermal energy (heat) is an **extensive property**; that is, it depends on the amount of material present. If you spill boiling water on your hand, thermal energy is transferred from the water to your hand. The damage to your hand depends on the amount of water that you spill. The thermal energy, ΔE, transferred to or from a substance can be calculated from the specific heat of the substance, the mass of the substance, and the change in temperature, ΔT ($= T_{final} - T_{initial}$), that the substance undergoes.

1.57. Worked Example *Thermal energy change in water*

Calculate the increase in thermal energy, ΔE, of the water in Investigate This 1.53 if 120. g of water is raised from 21.1 °C to 44.4 °C. [Note that the decimal point in the mass of the water specifies that the zero is significant, that is, the mass of the water is known to three significant figures. Another way to specify this would be to write the mass as 1.20×10^2 g.]

Necessary information: We need the specific heat of water from Table 1.3.

Strategy: The specific heat of water tells us the energy required to raise the temperature of one gram of water one degree Celsius. We multiply the specific heat by the mass of water to which energy has been added to get the energy required to raise its temperature one degree. Then we multiply the result by the temperature change to account for the number of degrees the temperature was raised.

Implementation: We can summarize the strategy in a single equation:

increase in thermal energy = ΔE = (specific heat) × (mass) × (temperature change)

$\Delta E = (4.18 \text{ J·g}^{-1}\text{·°C}^{-1}) \times (120. \text{ g}) \times (44.4 \text{ °C} - 21.1 \text{ °C}) = 11.7 \times 10^3 \text{ J} = 11.7 \text{ kJ}$

Does the answer make sense? We solved the problem by reasoning from the definition and units of the specific heat. If about 4 J are required to raise the temperature of 1 g of water 1 °C, then about 500 J are required to raise the temperature of 120 g of water 1 °C. Each 1 °C increase in temperature requires another 500 J of energy. The temperature change is about 20 °C, so about 10,000 J are required. The answer we got is this same order of magnitude, so it makes sense.

1.58. Check This *Thermal energy change in ethanol*

Calculate the temperature change if 11.7 kJ of heat is added to 120. g of ethanol.

Water: A Natural Wonder   Chapter 1

> **1.59. Consider This** *Are your calculations related to Investigate This 1.53?*
> Compare your answer in Check This 1.58 with the temperature change for the same mass of water from Worked Example 1.57. Is your answer related to the results from Investigate This 1.53? Explain why or why not.

Explaining the high specific heat of water

When thermal energy is added to water, some of the energy is used to break hydrogen bonds. The result is that the temperature rises less than if there were no hydrogen bonding and the average hydrogen bonding is now a bit weaker. Since water is a low molecular mass molecule, a small mass of water is a relatively large number of moles of water and, hence, moles of hydrogen bonds to be broken. A given mass of water can absorb a relatively large amount of thermal energy without as large an increase in temperature as the same mass of another compound.

We have discussed how the high specific heat of water helps to maintain the Earth at a rather constant temperature that is suitable for life. Similarly, because your body is largely water, it is maintained at a rather constant temperature without greatly upsetting your metabolism, even when the temperature of the environment changes substantially. This is yet another way that the bent, polar structure of the water molecule is essential for life.

Section 1.13. Outcomes Review

The next time you plunge your hands in water or watch ice cubes floating in a glass of water, you will be able to interpret the observable properties of water — and some other compounds — in terms of an atomic-molecular model and the interactions of the molecules with one another. In your mind's eye, you'll be able to see the bent, polar water molecules hydrogen bonding to their neighbors with the molecules held together as a liquid or as a solid. And you can relate the properties of different compounds to one another by using our molecular models of matter and comparisons based on counting molecules or moles. You are prepared now to go on, in Chapter 2, to learn how the structure of water molecules affects the interactions of water with other substances and to deepen your understanding of the mole concept.

Before going on, check your understanding of the ideas in this chapter by reviewing these expected outcomes of your study. You should be able to:

•   describe solids, liquids, and gases in terms of their macroscopic properties and write or draw molecular-level descriptions that explain these properties [Section 1.1].

•   make drawings that show how the electrical nature of matter explains the results of electrostatic experiments [Section 1.2].

Chapter 1 — Water: A Natural Wonder

- use the nuclear atomic model, the shell model for electrons, and the periodic table to determine the charge on the atomic core and the number of valence electrons in an atom [Section 1.2].
- use the periodic table and the atomic shell model to predict trends in atomic size and electronegativities [Sections 1.2 and 1.6].
- describe the relationships among different molecular models and the information that each of them provides [Sections 1.3, 1.4, 1.5, and 1.6].
- write Lewis structures for molecules whose molecular formulas contain only first and second period elements (or analogous molecules of higher period elements) [Sections 1.4 and 1.7].
- use drawings, physical models, and words to describe the geometry of the valence electrons and nuclei for molecules whose molecular formulas contain only first and second period elements (or analogous molecules of higher period elements) [Sections 1.5 and 1.7].
- predict the direction and relative magnitude of bond dipoles and the direction of the resultant molecular dipole for simple molecular structures [Section 1.6].
- use drawings, physical models, and words to describe the origin of intermolecular interactions due to London dispersion forces, dipolar attractions, and hydrogen bonding [Section 1.7].
- use intermolecular attractions to predict and/or explain trends in boiling points and energies of vaporization for a series of compounds whose molecular structures you know or can determine [Sections 1.7 and 1.11].
- use drawings, physical models, and words to describe how the structure of the water molecule is responsible for the densities of solid and liquid water, the temperature dependence of the density of liquid water, and the consequences for life on Earth [Section 1.8].
- describe some of the places where hydrogen bonding occurs in biomolecules and explain how hydrogen bonding is important for the functions of these molecules [Section 1.9].
- describe and use energy diagrams to illustrate the direction of energy transfer from one substance to another when phase changes occur [Section 1.10].
- use the relationship among energy change, temperature change, mass, and specific heat to make quantitative comparisons among substances that gain or lose thermal energy [Sections 1.10, 1.11, and 1.12].
- use the molar mass of a compound, determined from the relative atomic masses of its constituent atoms, to calculate the number of moles in a given mass of the compound and *vice versa* [Section 1.11].

- use drawings, physical models, and words to describe the molecular basis for the differences in specific heats among different compounds [Section 1.12].

Section 1.14. EXTENSION — Liquid Viscosity

Viscosity is a measure of the resistance to flow of a liquid. Syrup and motor oil are examples of substances with high viscosity. The higher the viscosity of a liquid, the slower it flows. In this EXTENSION, we are providing an opportunity for you to apply the models of attractions among molecules that we discussed in this chapter to interpret and predict the relative viscosities of different compounds.

1.60. **Investigate This**   *How fast do different liquids flow from a pipet?*

Use two identical glass Pasteur pipets for this investigation. Mark each pipet at the same place about 3 cm from the top. Draw water into one pipet until it is above the mark. Hold the pipet vertical with its tip over a container to catch the water. Time how long it takes the water to drain out of the pipet, starting from the time its top surface passes the mark. Repeat to be sure the time is reproducible to ±1 or 2 seconds. Use the second pipet to carry out the same procedure with hexane. *CAUTION:* Hexane is flammable.

1.61. **Consider This**   *How are viscosities related to molecular attractions?*

(a) How do the outflow times for the water and hexane in Investigate This 1.60 compare? Which liquid has the higher viscosity? Explain the reasoning for your answer.

(b) Consider the effect a network of hydrogen-bonded water molecules has on the ability of individual molecules to move about in the liquid. Could restricted motion at the molecular level affect the observed flow of water from a pipet? If so, how? If not, why not?

(c) Based on our model of attractions among hexane molecules, do you expect the same kind of restricted motion at the molecular level as described for water in part (b)? Why or why not? What might be the affect on the observed flow of hexane from a pipet? Explain your reasoning.

(d) Based on your analyses in parts (b) and (c), what would you predict about the relative viscosities of hexane and water? Is your prediction consistent with your observations in Investigate This 1.60? Explain why or why not.

In Table 1.4, you can see that water has a higher viscosity than several other low-molar-mass compounds. As your analyses have shown, this is because there are many hydrogen bonds between water molecules, so the molecules resist moving past one another.

Table 1.4. Relative viscosities of liquids at 20 °C.

Compound	Line Formula	M, g·mol^{-1}	Relative viscosity*
water	H_2O	18	1.00
methanol	CH_3OH	32	0.59
ethanol	CH_3CH_2OH	46	1.19
acetone	$CH_3C(O)CH_3$	58	0.33
ethylene glycol	$HOCH_2CH_2OH$	62	19.9
diethyl ether	$CH_3CH_2OCH_2CH_3$	74	0.23
hexane	$CH_3(CH_2)_4CH_3$	86	0.33

*Viscosities are all relative to water.

1.62. Consider This *Does hydrogen bonding explain viscosities in other compounds?*

(a) Which of the compounds in Table 1.4 can form hydrogen bonds between molecules in their pure liquids? Is there a correlation between viscosity and the ability to form hydrogen bonds? Explain why or why not. The Lewis structure for acetone is:

$$H-\underset{\underset{H}{|}}{\overset{\overset{H}{|}}{C}}-\overset{\overset{\ddot{O}}{||}}{C}-\underset{\underset{H}{|}}{\overset{\overset{H}{|}}{C}}-H$$

(b) Write a Lewis structure and make a molecular model of ethylene glycol (a compound used in automobile antifreeze). What do you think is responsible for the high viscosity of the glycol compared to the other compounds that contain H–O bonds? Clearly present your reasoning, based on the structure of the molecule.

Figure 1.41 shows the temperature dependence of the viscosity of water. In Section 1.12, we attributed the high molar heat capacity and specific heat of water to hydrogen bond breaking, which uses some of the thermal energy added to the liquid. The result is an average weakening of the hydrogen bonding as energy is added and the temperature increases.

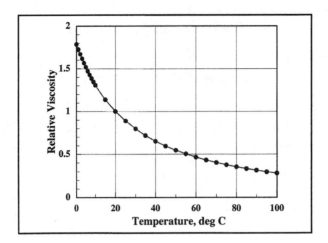

Figure 1.41. Relative viscosity of liquid water as a function of temperature.

1.63. Consider This *Are the models for specific heat and viscosity of water related?*

Do the data in Figure 1.41 provide support and further evidence for our explanation of the high specific heat of water just summarized? Explain why or why not.

Index of Terms

δ, a small amount, 1-28
Δ, change (final − initial), 1-53
ΔE, energy change, 1-53
α–helix, 1-48
alcohol, 1-64
alkyl, 1-64
amino acid, 1-47
amount, 1-58
aqueous, 1-5
atom, 1-13
atomic core, 1-19
atomic number, 1-15
Avogadro's number, 1-62
ball-and-stick model, 1-18
biomolecules, 1-22
boiling point, 1-9, 1-36
bond angle, 1-19
bond dipoles, 1-31
bond length, 1-19
bulk properties, 1-17
C, molar heat capacity, 1-68
calorie, 1-56
compound, 1-18
condensation, 1-9
condensed phase, 1-9
core electron, 1-14
covalent bond, 1-19
denatured, 1-49
density, 1-7
deoxyribonucleic acid, 1-50
dipole moment, 1-31
dipole-dipole attractions, 1-38
dispersion forces, 1-37

DNA, 1-50
electric dipoles, 1-28
electron, 1-13
electron-dot model, 1-18
electronegativity, 1-32
electron-shell model, 1-14
electrostatic attraction, 1-12
element, 1-13
energy, 1-53
energy change, 1-53
ether, 1-64
evaporation, 1-54
extensive property, 1-71
formula unit, 1-60
freezing, 1-9
gas, 1-8
groups, 1-13
heat, 1-71
hydrocarbons, 1-38
hydrogen bond, 1-41
induced-dipole attractions, 1-37
intensive property, 1-70
joule, 1-56
Lewis structure, 1-21
line formula, 1-18
liquid, 1-7
London forces, 1-37
macroscopic, 1-17
mass, 1-58
melting point, 1-9
mol, 1-59
molar heat capacity, 1-68
molar mass, 1-60

mole, 1-59, 1-62
molecular formula, 1-18
molecular model, 1-17
molecular shape, 1-19
molecule, 1-17
neutron, 1-13
nonbonding electrons, 1-21
nonpolar molecule, 1-28
nucleus, 1-13
octet rule, 1-21
periodic table, 1-13
periods, 1-13
phases of matter, 1-6
polar molecules, 1-28
protein, 1-47
proton, 1-13
s, specific heat, 1-69

side group, 1-47
solid, 1-7
space-filling model, 1-30
specific heat, 1-56, 1-69
standard temperature and pressure, 1-8
STP, standard temperature and pressure, 1-8
temperature, 1-70
tetrahedral, 1-26
tetrahedral angle, 1-26
tetrahedron, 1-26
thermal energy, 1-71
unit charge, 1-13
valence electron, 1-14
valence shell, 1-16
vaporization, 1-9
viscosity, 1-74

Chapter 1 Problems

Section 1.1. Phases of Matter

1.1. Identify the following as being either a chemical property or a physical property. Place a *P* by all the physical properties and a *C* by all the chemical properties.

(a) _____ water is clear and colorless

(b) _____ some metals react with water to produce hydrogen gas

(c) _____ water has a density of 1.000 g/cm³ at 4 °C

(d) _____ water boils at 100 °C

(e) _____ water is the product of a reaction between an acid and a base

(f) _____ water is a polar molecule

1.2. Compare solids, liquids, and gases in each given category.

	Solids	Liquids	Gases
Definite volume?			
Definite shape?			
Fixed or changing position of molecules?			
Average distance between molecules small or large?			

1.3. Which phases of matter are referred to as "condensed phases"? What is the justification for the use of this term?

1.4. (a) Name the phase changes between each of the states of matter indicated by the arrows in this diagram.

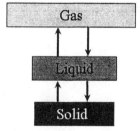

(b) Label each of the four arrows on the diagram to indicate whether energy is released or absorbed in the process.

1.5. Name a substance other than water that commonly exists as a liquid at STP.

1.6. Name a substance that commonly exists as a gas at STP.

Water: A Natural Wonder   **Chapter 1**

1.7. What happens if a closed glass bottle full of water is kept outside while the temperature falls below 0 °C?

1.8. One method for separating a NaCl*(aq)*, sodium chloride solution, into its components is to boil the solution of salt water. In this case, water will evaporate and NaCl*(s)* will be left behind.

    (a) Which property of NaCl*(s)* and water accounts for this separation?

    (b) Design an apparatus that could change the water vapor back into a liquid as well as recover NaCl*(s)*. Name all phase changes that occur.

1.9. Given two liquids that do not dissolve in one another (like oil and water), design an experiment that would allow you to determine which is more dense.

1.10. How could you determine (experimentally) if the solid phase of a particular substance is more dense or less dense than the liquid phase of the same substance?

1.11. Consider Figure 1.3 comparing the densities of solid and liquid phases of *t*-butanol with those of water. What additional information do you need to be able to predict what will happen if a sample of solid *t*-butanol were dropped into liquid water? How would you find that information?

1.12. A block of ice has following dimensions: height = 20 cm, width = 20 cm, and length = 20 cm. The density of liquid water is 1.000 g·mL^{-1} at 0 °C, and the density of ice at the same temperature is 0.917 g·mL^{-1}. Calculate the volume of the puddle of water, at 0 °C, that is left behind when the block of ice melts.

1.13. If an iron bar weighing 100.0 g is heated to a temperature above its melting point (>1535 °C), it will liquefy. What is the mass of the molten (liquid) iron? Is any additional information needed in order to answer this question?

Section 1.2. Atomic Models

1.14. Which of the following are chemical elements? How do you decide?

    (a) water   **(d)** iron oxide

    (b) salt water   **(e)** nitrogen

    (c) iron   **(f)** diamond

1.15. What is the difference between core electrons and valence electrons?

Chapter 1   Water: A Natural Wonder

1.16. How many valence electrons and core electrons do the following elements possess?

    (a) sodium   **(d)** phosphorus

    (b) bromine   **(e)** sulfur

    (c) barium

1.17. Many organisms use these ions in their metabolism: Na^+, K^+, Mg^{2+}, Ca^{2+}, Cl^-, Br^-.

    (a) Complete the following table concerning these ions.

Ion	# of protons	# of electrons	# of valence electrons	Core charge	# of core electrons
Na^+					
K^+					
Mg^{2+}					
Ca^{2+}					
Cl^-					
Br^-					

    (b) What patterns do you observe?

1.18. In terms of electronic structure, what is it that elements in the same period (row) of the periodic table share in common?

1.19. In terms of electronic structure, what is it that elements in the same group (column) of the periodic table share in common?

1.20. In general, do elements from the same period or elements from the same group of the periodic table have similar chemical properties? Justify your answer.

1.21. These metal ions, Mn^{2+}, Fe^{2+}, Fe^{3+}, Cu^{2+}, and Zn^{2+}, are available for uptake by living organisms. How many protons and electrons does each ion have?

Section 1.3. Molecular Models

1.22. Which of these are macroscopic properties? Which are microscopic properties?

    (a) the boiling point of water

    (b) the HOH bond angle in water

    (c) the OH bond length in water

    (d) the ability of water to dissolve salt

    (e) the density of water

    (f) the fact that the oxygen atom of water has two pairs of nonbonding electrons

1.23. Consider these four different models for the ammonia molecule, NH$_3$:

Model 1	Model 2	Model 3	Model 4
NH$_3$	H–N̈–H │ H		

(a) What is the name given to each type of model shown?

(b) What information can be obtained from each type of model?

(c) What other types of models can be used to represent the ammonia molecule?

1.24. Write the molecular formulas (line formulas) of the compounds represented here:

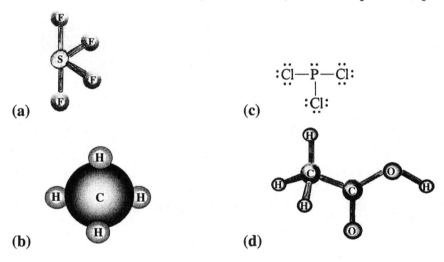

(a)   (b)   (c)   (d)

1.25. Using equations modeled after Figure 1.9(c), show how to form each of the following molecules from their constituent atoms. In each case, count electrons in the products to demonstrate that the octet rule is followed for all second row atoms.

(a) HF (hydrogen fluoride)

(b) NH$_3$ (ammonia)

(c) CH$_3$OH (methanol)

(d) H$_2$O$_2$ (hydrogen peroxide)

1.26. Write the molecular formulas for the following compounds. (If necessary, use a reference handbook and/or other books to find out the structures.)

(a) glucose, a substance known as "blood sugar"

(b) nitrous oxide, a substance used as an anesthetic and as an aerosol propellant. It is commonly called "laughing gas."

(c) methanol, an organic solvent and antifreeze

(d) acetylene, a gas that is used in welding torches

Section 1.4. Valence Electrons in Molecular Models: Lewis Structures

1.27. (a) What information can be obtained from a Lewis structure?

(b) What information cannot be obtained from a Lewis structure?

1.28. Consider this electron-dot model for the ammonium ion, NH_4^+:

$$\left[\begin{array}{c} H \\ H : \overset{..}{\underset{..}{N}} : H \\ H \end{array} \right]^+$$

(a) What information can be obtained from this model?

(b) What information *cannot* be obtained from this model?

(c) Rewrite the ammonium ion using a dash to represent each bonded pair of electrons. Does this change the information found in the model?

(d) How does this electron-dot model for the ammonium ion compare with that for methane, CH_4, given in Worked Example 1.10?

1.29. Examine Table 1.1. Silicon typically makes four covalent bonds and sulfur typically makes two covalent bonds. How many covalent bonds do you expect for each of the following elements? Explain your reasoning in each case.

(a) phosphorus   (c) selenium

(b) chlorine   (d) bromine

1.30. Neon (Ne) is the element to the right of fluorine (F) in the periodic table.

(a) Examine Table 1.1. How many covalent bonds and nonbonding pairs would be expected for neon? Explain the reasoning for your answer.

(b) Ne (along with He, Ar, Kr, Xe, and Rn in the same group of the periodic table) were once known as the "inert gases." Why were they given this name?

1.31. Draw the Lewis structure of ethanol, C_2H_6O. The formula may also be written C_2H_5OH or CH_3CH_2OH to make the connectivity more apparent.

1.32. Draw the Lewis structures for ozone, O_3, sulfur dioxide, SO_2, and nitrite ion, NO_2^-. What do all three of these structures have in common? *Hint:* S is the central atom in SO_2 and N is the central atom in NO_2^-.

Water: A Natural Wonder — Chapter 1

1.33. Each of the following Lewis structures for NCCN has the correct number of electrons. Which is the best Lewis structure for NCCN? Explain your reasoning for rejecting the structures you did not choose. *Hint:* Multiple bonds between atoms will be discussed in Chapter 5. For the purposes of this problem, simply count each stroke (bond) as two electrons shared between the atoms it connects.

(a) :N̈=C=C=N̈:   (d) :N≡C−C≡N:

(b) :N̈−C≡C−N̈:   (e) :N−C̈−C̈−N:

(c) :N̈=C=C=N: (f) :N=C=C−N̈:

1.34. Which of the following Lewis structures are incorrect? In each case, explain why the structure is incorrect. Rewrite each of the incorrect structures so it is correct. *Hint:* See the hint in Problem 1.33.

(a) HOCl   H:Ö:C̈l:

(b) CS_2 :S::C::S:

(c) NH_3 H:N̈:H
              H

(d) $(HO)_2CO$
   :Ö:
   C
H:Ö   Ö:H

(e) H_2Se H:S̈e:H

Section 1.5. Arranging Electron Pairs in Three Dimensions

1.35. Which of the following molecules or ions has a tetrahedral (or close to tetrahedral) orientation of bonding and nonbonding electron pairs?

(a) H_2S (d) NH_2^-

(b) NH_4^+ (e) CH_4

(c) NH_3

1.36. What is the shape of each of the molecules in Problem 1.35? Recall that the shape of a molecule describes the position of the atomic cores with respect to one another.

1.37. Figure 1.12 showed one way to stack four balls so that they are equidistant from a central point. Another way is to arrange them in a square about the point, as shown in this picture. Lines drawn between the centers of adjacent balls form a square of side $2r$, where r is the radius of a ball. The center point of the diagonal of the square is the center of the square. Use the Pythagorean theorem to find the length of the diagonal and the distance from the center of any of the balls to the center of the square.

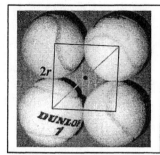

1.38. (a) A cube can be circumscribed about a regular tetrahedron, as shown in this figure. The four dots represent the centers of the four balls in Figure 1.12,. The diagonal of one of the cube faces has a length $2r$, where r is the radius of a ball. The center point of any one of the cube diagonals (one is shown by the dashed line) is the center of the tetrahedron. Use the Pythagorean theorem to find the length of the cube edge and then again to find the length of the cube diagonal. Thus, show that any corner of the cube, that is, the center of any of the balls, is $\left(\frac{\sqrt{6}}{2}\right)r$ from the center of the tetrahedron.

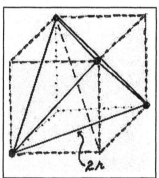

(b) How does the distance you calculated in part (a) compare to the distance of the center of each ball from the center of the square in Problem 1.37? Do your results help justify the statement in the text that "the tetrahedral arrangement puts all the balls as close as possible to the central point"? Explain why or why not.

1.39. SiCl$_4$, silicon tetrachloride, is used for the production of the very pure silicon required in many electronic devices such as transistors.

(a) Draw the Lewis structure for SiCl$_4$.

(b) How does this Lewis structure compare to that of methane, CH$_4$?

(c) Predict the shape of the SiCl$_4$ molecule.

1.40. (a) Draw the Lewis structure for borane, BH$_3$. Compounds like borane are sometimes called "electron deficient." How do you think "electron deficiency" is defined?

(b) What shape do you predict for borane? You might find it useful to try a modified version of Investigate This 1.14 to help make your prediction. Also see WEB Chap 1, Sect 1.6.

Water: A Natural Wonder — Chapter 1

Section 1.6. Polarity of the Water Molecule

1.41. What is electronegativity?

1.42. Answer the following questions on electronegativity.
(a) Which element has the highest electronegativity value?
(b) Where on the periodic table do you find the elements with the lowest electronegativities?
(c) Where on the periodic table do you find the elements with the highest electronegativities?
(d) What is the general trend for electronegativity as you go from the top to the bottom within a group?
(e) What is the general trend for electronegativity as you go from the left to right across a period?
(f) How do the electronegativity trends compare to the atomic size trends? Refer to Figure 1.7 for information about atomic size.

1.43. Without referring to a table of electronegativities, predict which member of each pair has the greater electronegativity. Explain the basis for your prediction in each case.
(a) F *or* S
(b) C *or* H
(c) H *or* O
(d) O *or* C

1.44. You can use the electronegativity difference between two atoms to predict the polarity of the bonds they make. Choose the pair of atoms in each case that you predict to make the more polar bond and explain how you make this prediction.
(a) H-F *or* H-Cl
(b) C-H *or* N-H
(c) K-S *or* Na-S
(d) O-O *or* N-O

1.45. What would be the consequences for life if water were a linear molecule? Consider the effect on the polarity and properties of water if it were linear. You might find it useful to draw pictures of how the linear molecules might interact with one another and compare what you get with the various interactions pictured in this chapter.

1.46. Consider the two molecules H_2O and H_2S.
(a) Compare the Lewis structures of these two molecules.
(b) Compare the molecular shape of these two molecules.
(c) Compare the bond dipoles within each molecule. *Hint:* Use the data in Figure 1.20.
(d) Compare the overall electric dipole of each molecule.

1.47. Draw the Lewis structures for carbon tetrachloride, CCl_4, chloroform, $CHCl_3$, and dichloromethane, CH_2Cl_2. Clearly label the bond dipoles for each molecule. Which molecules are polar and which are non-polar? Explain your answer.

1.48. Each of the following molecules has a dipole moment = 0. In each case, explain why there is no net (molecular) dipole moment. If the molecule has bond dipoles, draw them and explain how they cancel out.

(a) N_2 (molecular nitrogen)

(b) BH_3 (borane) *Hint:* B does not satisfy the octet rule. See Problem 1.40.

(c) $SiCl_4$ (silicon tetrachloride)

(d) BeH_2 (beryllium hydride) *Hint:* Be does not satisfy the octet rule. What shape *must* the molecule have, in order not to have a net dipole moment? You might find it useful to try a modified version of Investigate This 1.14 to help determine this shape.

(e) CH_3CH_3 (ethane) See Check This 1.13.

1.49. Which molecule, ammonia, NH_3, or phosphine, PH_3, has the larger molecular dipole moment? Explain.

Section 1.7. Why Is Water Liquid at Room Temperature?

1.50. Describe each of the following types of intermolecular attractions:

(a) induced-dipole attractions

(b) dipole-dipole attractions

(c) hydrogen bond

1.51. How are an intramolecular covalent bond and an intermolecular hydrogen bond similar? How are they different?

1.52. What type(s) of intermolecular attractions are there between

(a) all molecules?

(b) polar molecules?

(c) a hydrogen atom in a water molecule and a nitrogen atom in ammonia (in a mixture of ammonia and water)?

1.53. Astatine, At, element 85, is radioactive and has a half-life of only 8.3 hours (see Chapter 3). Only minute traces of At have been studied. The hydride, HAt, has been detected but its physical properties are unknown. Based on the data in Figure 1.24, what would you predict for the boiling point of HAt? Explain how you make your prediction.

1.54. Methane, CH_4, and hydrogen sulfide, H_2S, do not form hydrogen bonds. Explain.

1.55. List at least three properties of water that can be attributed to the existence of the hydrogen bond. Briefly describe how each property would be affected if water did not form hydrogen bonds.

1.56. (a) How many hydrogen bonds are possible for one water molecule in a sample of water? Illustrate your answer with a drawing of the structure of water indicating where the hydrogen bonds are possible.

1.57. (a) How many hydrogen bonds are possible for one ammonia molecule in a sample of ammonia? Illustrate your answer with a drawing of the structure of ammonia indicating where the hydrogen bonds are possible.

(b) Can all the ammonia molecules in a sample of liquid ammonia have the maximum number of hydrogen bonds you illustrated in part (a)? If not, what limits the number and how many hydrogen bonds, on average, can each ammonia molecule have?

1.58. Refer to the graph in Consider This 1.23 that relates the boiling points of a series of hydrocarbons to the number of electrons per molecule. What is the smallest hydrocarbon in the series that exists as a liquid at room temperature?

1.59. The boiling points of the noble gases are:

element	He	Ne	Ar	Kr	Xe	Rn
bp, °C	−269	−246	−186	−152	−107	−62

(a) Plot these data on a graph like the one in Figure 1.24. How are these data similar to those for the hydrides plotted in Figure 1.24? How are they different?

(b) Noble gas atoms are spherical and Figure 1.22 shows that tetrahedral hydrides like methane are quite symmetric and almost spherical. What is likely to be the largest factor responsible for the difference in boiling points between a noble gas and the corresponding group IV hydride?

1.60. There are often different compounds having the same formula. As you will learn in Chapter 5, these are known as *isomers*. For C_5H_{12}, there are three isomers. The common names, line formulas, structures, and boiling points of the isomers are given below. All three have the same molecular mass. Why aren't their boiling points closer together? *Hint:* The more compact a molecule, the less surface area it has for its electrons to interact with other molecules. Build models of these molecules to help visualize the surface areas of these molecules.

pentane	isopentane	neopentane
$CH_3CH_2CH_2CH_2CH_3$	$(CH_3)_2CHCH_2CH_3$	$C(CH_3)_4$
bp = 36 °C	bp = 28 °C	bp = 10 °C

Section 1.8. Further Structural Effects of Hydrogen Bonding in Water

1.61. If an ice cube is dropped into liquid water at 85 °C, will it float or sink (before it melts)? Justify your answer.

1.62. In Investigate This 1.30, you found that the temperature at the bottom of the container rises to about 4 °C and remains almost constant as long as there is ice left at the top of the container. Thermal energy (heat) must be entering the container from the warmer room air (the ice does melt). You would expect the water at the bottom to continue to warm up, but the temperature at the bottom stays constant. These seem to be contradictory observations.

(a) What is the special property of water at 4 °C?

(b) If the water at the bottom warmed a bit above 4 °C, how would this property change? What would the water be likely to do? Draw pictures to illustrate what you think would happen.

(c) Would the action of the water shown in your drawings resolve the contradiction suggested above? How could you test your model experimentally?

1.63. The structure of the water molecule is the molecular basis for the survival of plant and animal life in a temperate climate lake. The seasonal "turnover" of such a lake is described in this figure.

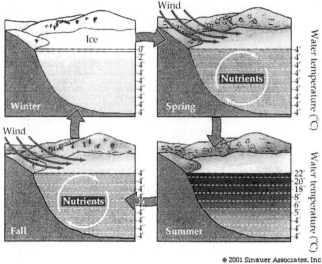

These vertical temperature changes are typical of a lake that freezes in winter. Turnovers, represented by the circling arrows, occur in the spring and fall and mix nutrients and oxygen into the deeper waters. The turnovers are triggered by winds at the surface of the water. How would you relate your observations in Investigate This 1.30 to the changes described in this figure? Explain the connections clearly.

1.64. Consider this representation of covalent bonds and hydrogen bonds in water:

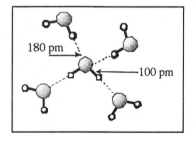

(a) Which bond length is associated with hydrogen bonds and which with covalent bonds?

(b) Offer a reasonable explanation for why there is a difference in bond length between these two bonds.

1.65. The melting points for methane, ammonia, water, and hydrogen fluoride are shown in the table at the right. You can take the melting points as an indication of the relative amount of energy required to disrupt the attractions between molecules in the solids, so they are free to move about as a liquid. Develop an explanation for these data that takes into account the kinds of intermolecular attractions among molecules of each compound. Is water out of line with the rest of the compounds? Why or why not? Give a molecular level interpretation of your answer.

Compound	mp, °C
CH_4	–182
NH_3	–77.7
H_2O	0
HF	–83.1

Section 1.9. Hydrogen Bonds in Biomolecules

1.66. The DNA double helix, held together by the hydrogen bonds shown in Figure 1.33, is quite stiff and resistant to movement through a solution. As shown in this figure, when a solution of DNA is heated, the absorption of ultraviolet light at 260 nm rises sharply over a small temperature range. At the same time the solution suddenly begins to flow more easily (more like water than like syrup). The middle of this range is usually labeled T_m ("melting" temperature). What is happening to the DNA to cause these changes in the solution properties? Explain your response.

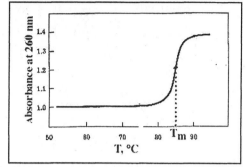

1.67. This figure shows melting temperature, T_m (see Problem 1.66), data for a number of different double-helical DNA's plotted against the fraction of A–T pairs in the DNA's. Why are the melting temperatures a function of the fraction of A–T pairs? Does the direction of the dependence make sense? Clearly explain your reasoning. *Hint:* Review Consider This 1.34.

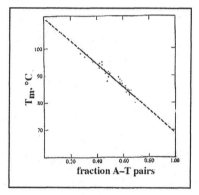

1.68. In Problem 1.66, the "melting" temperature of DNA in solution was defined as the temperature at which sharp changes in ultraviolet light absorption and solution flow occur. Why do you think this is called a "melting" temperature? Is(are) there any analogy(ies) between what happens to DNA in these solutions and what happens when ice melts? Explain your reasoning clearly.

1.69. The proteins in most organisms are denatured at temperatures above about 60 °C. Microorganisms that live in hot springs and organisms that live near deep ocean thermal vents survive at temperatures near or above the boiling point of water, 100 °C. Their proteins are made of the same amino acids as all other organisms. What role do you think hydrogen bonds might play in helping these organisms survive?

1.70. The structure of the DNA of the thermophilic (*therme* = heat + *philos* = loving) organisms discussed in Problem 1.69 also has to be maintained in the high temperature environments where they live. What kind of A–T *versus* G–C composition would you expect to find for the DNA in these organisms? Present your reasoning clearly. *Hint:* See Problem 1.67.

1.71. Cellulose is a long-chain molecule made up of glucose molecules bonded together as shown in this illustration. You will learn more about cellulose later. Many chains like these are hydrogen-bonded to neighboring chains to form the fibers that are used to make paper and cotton and linen cloth. The hydrogen-bonds make cotton cloth soft and flexible because they are easily broken and remade, which allows the fibers to change shape.

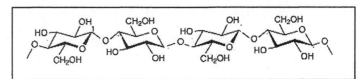

(a) How do you account for the fact that cotton clothing is easily wrinkled? Explain your reasoning clearly.

(b) How does ironing wrinkled cotton clothing restore its "press?" Use diagrams to illustrate your answer.

(c) How might you make "permanent press" cotton cloth? Indicate what you would try to accomplish; don't be concerned about the detailed chemical processes that might be required.

(d) Permanent press cotton clothing is not as soft as regular cotton clothing. Is this the result you might expect from your response to part (c)? Explain why or why not.

Section 1.10. Phase Changes: Liquid to Gas

1.72. How is energy involved for a substance to change from one state to another? Explain.

1.73. Make the following conversions:
(a) 4550 J = _____ kJ
(b) 250. J = _____ calories
(c) 500. Cal = _____ J   [1 Nutritional Calorie (Cal) = 1000 calories]

1.74. A phase diagram such as this one for water, is a common way of representing phase changes. Phase diagrams are pressure-*vs.*-temperature plots that show the pressures at which the phase changes of a substance occur as a function of temperature.

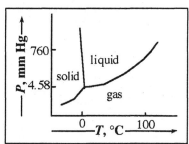

(a) What information does this phase diagram give you about the phases of water at standard atmospheric pressure of 760 mm Hg (= 1 atm)? Explain your reasoning briefly. *Hint:* Consider starting at the pressure axis, moving across the plot (increasing temperature) at a constant 760 mm Hg, and noting when phase changes occur.

(b) For a change from gaseous water to liquid water at 100 °C and 760 mm Hg, will the sign of ΔE be positive or negative? Explain your reasoning.

(c) For a change from liquid water to solid water at 0 °C and 760 mm Hg, will the sign of ΔE be positive or negative? Explain your reasoning.

1.75. Sketch the energy diagrams that describe the following processes. Make sure that you draw an arrow that represents the direction of the change and that you indicate the sign of ΔE for these processes.

(a) A sample of ice melting, $H_2O(s) \rightarrow H_2O(l)$, at 0 °C. $\Delta E > 0$ for the sample.

(b) The combustion of hydrogen gas in oxygen gas is one of many chemical reactions that release considerable quantities of energy. This process is described by the equation:

$$2H_2(g) + O_2(g) \rightarrow 2H_2O(l) + \text{energy}$$

(c) The decomposition of mercury oxide (HgO) occurs at high temperatures. For this process to occur, energy has to be supplied. The process is described by the equation:

$$\text{energy} + 2HgO(s) \rightarrow 2Hg(l) + O_2(g)$$

(d) The burning (or oxidation) of mercury is the reverse of the decomposition process in part (c) and is described by the equation:

$$2Hg(l) + O_2(g) \rightarrow 2HgO(s) + \text{energy}$$

(e) Burning a sample of methane in oxygen gas, for which $\Delta E < 0$. This reaction is described by the equation:

$$CH_4(g) + 2O_2(g) \rightarrow CO_2(g) + 2H_2O(l)$$

1.76. The amount of energy required to vaporize methanol is 1.22 kJ·g^{-1}. How many kcal are required to vaporize one gram of methanol?

Water: A Natural Wonder Chapter 1

Section 1.11. Counting Molecules: The Mole

1.77. Calculate the molar mass of a formula unit of the following substances:
   (a) dimethyl ether, CH_3OCH_3
   (b) ethanol, CH_3CH_2OH

1.78. How many molecules of water are in exactly one mole of water? How many grams of water are in exactly one mole of water?

1.79. How many moles in the following?
   (a) 100.0 g of acetone
   (b) 100.0 g of methanol
   (c) 100.0 g of dimethyl ether
   (d) 100.0 g of sucrose

1.80. How many molecules are there in 1 gram of water, methanol, acetone, ethanol, and dimethyl ether?

1.81. Argon atoms have a diameter of approximately 100 pm. If a mole of argon atoms were lined up one after another, how long, in meters, would the line be? The distance from the Earth to the Sun is 1.5×10^{10} m. What percentage of this distance would the line of Ar atoms reach?

1.82. (a) How many moles of hydrogen bonds are there in a mole of ice? Explain how you get your answer. *Hint:* If each H stopped hydrogen bonding, all the hydrogen bonds would be gone.
   (b) The energy required to melt ice is 6.02 kJ·mol^{-1}. If the model for ice melting presented in the text is correct, how many moles of hydrogen bonds are broken when a mole of ice melts? What percentage of the total hydrogen bonds is this? Clearly explain how you get your answers. Recall that the energy required to break a hydrogen bond between two water molecules is 20–25 kJ·mol^{-1}.

1.83. In which of these compounds can the molecules hydrogen bond to themselves? Draw a diagram of the molecules of each compound hydrogen bonding among themselves.
   (a) water
   (b) methanol
   (c) acetone
   (d) ethanol
   (e) dimethyl ether

1.84. The boiling points of dimethyl ether (CH_3OCH_3) and diethyl ether ($CH_3CH_2OCH_2CH_3$) are –25 °C and 35 °C, respectively. What interaction is mainly responsible for the observed difference?

1.85. Acetone, like dimethyl ether, has no self-hydrogen-bonding capacity, but its dipole moment is more than double that of the ether. Can this high dipole moment explain the properties of acetone compared to the other compounds in Table 1.2? Clearly explain the reasoning for your response and include as many comparisons as possible.

Compound	Line Formula	Molar Mass, g	Energy of Vaporization, $kJ \cdot mol^{-1}$	Boiling Point, °C	Dipole Moment, Debye
acetone	$CH_3C(O)CH_3$	58	32	56	2.88

Section 1.12. Specific Heat of Water: Keeping the Earth's Temperature Stable

1.86. Define specific heat.

1.87. What is the difference between an intensive and an extensive property?

1.88. Are the following properties intensive or extensive?
   (a) the boiling point of water
   (b) the density of water
   (c) the specific heat of water
   (d) the ratio of hydrogen to oxygen atoms in a sample of water
   (e) the (maximum) solubility of salt in water

1.89. Convert 37.0 °C to kelvin.

1.90. How much thermal energy is required to raise the temperature of 1.0 gram of water by
   (a) 10.0 °C?   (c) 25.0 K
   (b) 25.0 °C?

1.91. How much thermal energy is required to raise the temperature of the following by 10.0 °C?
   (a) 10.0 grams of water   (b) 25.0 grams of water

1.92. How much thermal energy is required to raise the temperature of 20.0 grams of acetone by 15.0 °C?

1.93. In dry parts of the world, "air conditioning" is provided by blowing the hot outside air through mats soaked in water before it enters the building. On what scientific principle is this system based? How does it work? What advantages and disadvantages can you see for the people and things in the building?

1.94. Would you expect evaporative cooling of your skin to be more effective on dry days or humid days? Clearly explain the reasoning for your answer.

1.95. Table 1.3 gives the heat capacity of liquid water as 75 $J \cdot mol^{-1} \cdot °C^{-1}$ (18 $cal \cdot mol^{-1} \cdot °C^{-1}$). The heat capacity of solid water (ice) is 38 $J \cdot mol^{-1} \cdot °C^{-1}$ (9.0 $cal \cdot mol^{-1} \cdot °C^{-1}$). Why do you think the heat capacity of the solid is less than that of the liquid? Are our models of liquid and solid water consistent with your explanation?

1.96. The heat capacity of gaseous water (steam) at one atmosphere pressure is 33 $J \cdot mol^{-1} \cdot °C^{-1}$ (9.5 $cal \cdot mol^{-1} \cdot °C^{-1}$). What sort of model do you think would describe gaseous water? Why do you think the heat capacity of the gas is less than that of the liquid (See Table 1.3 or Problem 1.95)? Is your model of the gas and the model of the liquid we have discussed in this chapter consistent with your explanation?

1.97. Table 1.2 shows that 44 $kJ \cdot mol^{-1}$ (10.5 $kcal \cdot mol^{-1}$) are required to change one mole of liquid water to water vapor. This value is for water near 25 °C. The energy of vaporization depends on the temperature of the water, as shown in this figure. Why does the energy required to vaporize water vary with temperature this way? Clearly explain your reasoning.

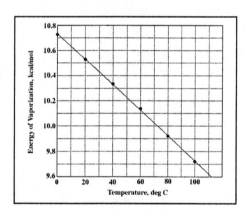

Section 1.13. EXTENSION --Liquid Viscosity

1.98. To which physical property of liquids does the expression "slow as molasses in January" owe its truth? Would molasses be "faster" in June? Why?

1.99. Nature exploits the properties of the hydrogen bond in many ways. Scientists also work to find ways to use this weak bond with its strength in numbers to create materials with interesting and useful properties. One group of researchers has made a compound whose molecules have "sticky" ends; each end of one molecule forms four hydrogen bonds to another to produce long chains, as represented in the figure.

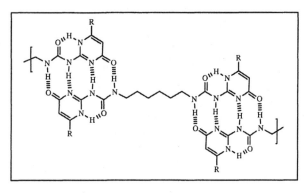

(a) For solutions of this compound (in a non-hydrogen-bonding solvent) that vary in concentration from about 8 to 80 g/L, the viscosity varies as shown in this logarithmic plot. As the concentration changes by a factor of 10, by what factor does the viscosity change? Clearly explain how you might interpret this result. *Hint*: Recall that large molecules can't move rapidly in solution, so their solutions resist flow.

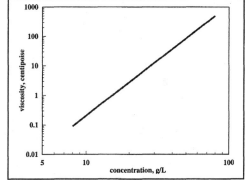

(b) The viscosity of these solutions is temperature dependent. Do you predict an increase or decrease in viscosity as the temperature of a solution is increased? Explain the reasoning for your prediction.

(c) The researchers also made a compound whose molecules are essentially half of one of the molecules shown above. These new molecules have only one "sticky" end. The viscosities of mixtures of 32 g/L of the original compound with small amounts of the new compound are shown in this plot. The horizontal axis shows the decimal fraction of the mixture that is the new compound; 0.01, for example, means that 1 in 100 of the molecules in the solution are the new molecules. Clearly explain how you might interpret what is going on in the solution to produce these results.

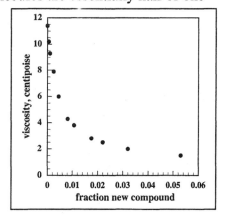

1.100. The viscosity of *n*-heptane, $CH_3(CH_2)_5CH_3$, as a function of temperature is plotted here on the same scale as Figure 1.41, which gives the corresponding data for the viscosity of water. To make the plots comparable, the values here are relative to the viscosity of heptane at 20 °C. What similarities do you observe between these data and those for water? What differences do you observe? What explanation can you provide for the similarities and differences?

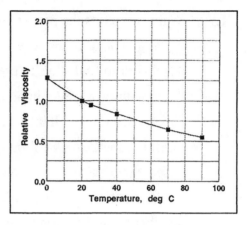

General Problems

1.101. Briefly explain why:

(a) You experience a cooling effect after walking out of the ocean onto a warm, sandy beach, especially on a breezy day.

(b) Liquids can be defined as "disordered" solids. Are there problems with this definition?

(c) Liquids can be defined as "dense" gases. Are there problems with this definition?

(d) Solid water (ice) floats on liquid water.

(e) Steam can badly burn you, if it condenses to water on your skin.

(f) Lakes freeze from the top to the bottom.

(g) Water pipes break, if water freezes in them.

1.102. Chemistry is everywhere. A friend has asked you if the claims here are scientifically accurate: "Keep a bottle of water at your desk and take frequent sips from it. One-third of water is oxygen, so drinking it will help keep you alert." Write a response that will help your friend sort out what is true and what might not be.

1.103. One model of liquid water (an "iceberg" model) is a mixture of molecular-scale, ice-like structures among other less-ordered, less-hydrogen-bonded molecules with the molecules continually exchanging between the two forms. In a sample of liquid water, it is possible to give extra energy of motion only to those molecules that are pointing (oriented) in the same direction. The natural rocking and jiggling of these molecules soon changes their orientation (they become more random). Scientists have measured the time required for the change and find that some of the water molecules make the change in an average of about 0.7×10^{-12} s (0.7 ps). The rest take an average of about 13×10^{-12} s (13 ps). The scientists concluded that liquid water acts like it is made up of two species.

(a) Is this conclusion consistent with the "iceberg" model of liquid water? What might the two species be? Clearly state the reasoning for your answers.

(b) Which of the two species changes orientation rapidly and which more slowly? Explain your reasoning. Use drawings, if they are helpful.

(c) If there are two species, why aren't they apparent to our senses in our everyday contacts with water?

1.104. Gases behave much like liquids in terms of things floating and sinking in them. What can be deduced from each of these facts:

(a) A helium-filled balloon will rise in air.

(b) A hot-air balloon will rise in (colder) air.

(c) A balloon filled with carbon dioxide will sink in air.

1.105. What would happen if an astronaut standing on the Moon let go of a helium-filled balloon she was holding. Recall that the Moon, unlike the Earth, has no atmosphere. *Hint:* See Problem 1.104(a).

1.106. Potassium acid fluoride is a salt composed of the ions, K^+ and $(FHF)^-$. The negative ion, $(FHF)^-$, can be thought of as two fluoride ions hydrogen bonded by a proton with the Lewis structure: $[:\!\ddot{F}\!:\!H\!:\!\ddot{F}\!:]^-$. The H–F bond length in hydrogen fluoride, HF, is 93 pm. Each of the H–F bond lengths in $(FHF)^-$ is 113 pm. Formation of $(FHF)^-$ from HF and F^- releases about 155 kJ·mol^{-1}, making it by far the strongest hydrogen bond known, although much weaker than the two-shared-electron covalent bond in HF which releases about 565 kJ·mol^{-1} when it forms.

(a) What, if anything, is peculiar about the Lewis structure shown for $(FHF)^-$?

(b) Discuss the similarities and differences between this hydrogen bond and the hydrogen bond between two water molecules. Also discuss whether the $(FHF)^-$ example blurs the distinction between covalent bonds and hydrogen bonds.

1.107. (a) About 70% (by mass) of your body is water. How many moles of water does your body contain? How many molecules of water?

(b) The boxed table on page 1-22 gives the elemental composition of your body. Assume that the same number of atoms of oxygen and nitrogen are combined in molecules other than water. How many oxygen atoms (per 100,000 atoms) are combined with hydrogen atoms to make the water in your body?

(c) Use your results from parts (a) and (b) to calculate the total number of atoms in your body. How many of these atoms are nitrogen? carbon? How many moles of nitrogen does your body contain? carbon?

1.108. This table gives the names, structures, and energies of vaporization for most of the hydrocarbons (molecules containing only carbon and hydrogen atoms) whose boiling points are given in the figure in Consider This 1.23.

hydrocarbon	formula	energy of vaporization kJ·mol^{-1}
ethane	CH_3CH_3	15.65
propane	$CH_3CH_2CH_3$	20.13
butane	$CH_3(CH_2)_2CH_3$	24.27
pentane	$CH_3(CH_2)_3CH_3$	27.61
hexane	$CH_3(CH_2)_4CH_3$	31.92
heptane	$CH_3(CH_2)_5CH_3$	35.19
octane	$CH_3(CH_2)_6CH_3$	38.58
nonane	$CH_3(CH_2)_7CH_3$	43.76

(a) Plot these energies of vaporization as a function of the number of electrons in each molecule. Draw the best possible straight line through the points. (If you use a graphing calculator or computer graphing program, it can construct the line for you.) Predict the energy of vaporization for decane, $CH_3(CH_2)_8CH_3$.

(b) Why is there an increase in energy required to vaporize these molecules as -CH_2- groups are added? Use the slope of your line from part (a) to determine how much the energy of vaporization increases for each -CH_2- group added.

(c) Assume that induced dipole attractions (dispersion forces) are directly proportional to the number of electrons in molecules with second row elements connected in a chain, like those in this table. What do you predict for the energy of vaporization of dimethyl ether? How does your prediction compare to the value given in Table 1.2? How do you explain any difference?

(d) The energies of vaporization of diethyl ether, $CH_3CH_2OCH_2CH_3$, and butanol, $CH_3CH_2CH_2CH_2OH$, are 29.1 and 45.9 kJ·mol^{-1}, respectively. Use what you have learned in the previous parts of this problem, plus the data in Table 1.2, to predict these energies and compare them with the experimental values. What attractions among the molecules must you account for in each case?

Chapter 2. Aqueous Solutions and Solubility

Section 2.1. Substances in Solution 2-6
    Solution nomenclature 2-6
    The solution process 2-7
    Favorable and unfavorable factors 2-8

Section 2.2. Solutions of Polar Molecules in Water 2-9
    Hydrogen bonds among unlike molecules 2-10
    Nonpolar solutes 2-11
    Molecular reorganization 2-12
    Intermediate cases 2-12
    Like dissolves like 2-13
    Solutes with multiple polar groups 2-14

Reflection and projection   *2-15*

Section 2.3. Characteristics of Solutions of Ionic Compounds in Water 2-16
    Solution conductivity 2-16
    Ionic solids and solutions 2-17
    Ion-dipole attractions in ionic solutions 2-18

Section 2.4. Formation of Ionic Compounds 2-20
    Names and formulas 2-20
    Why ionic compounds form 2-21
    Chemical reaction equations 2-22
    Formation of ionic crystals 2-24

Section 2.5. Energy Changes When Ionic Compounds Dissolve 2-27
    Solubility: hydration energy and lattice energy 2-27

Reflection and projection   *2-30*

Section 2.6. Precipitation Reactions of Ions in Solution 2-31
    Formation of precipitates 2-32
    Identity of the precipitate 2-33
    Ionic reaction equations 2-33
    Equilibrium 2-35

Aqueous Solutions and Solubility — Chapter 2

Section 2.7. Solubility Rules for Ionic Compounds 2-36
    Solubility rules for ionic compounds 2-38
    Reorganization factors in ionic solubility 2-39
    Exceptions to the rules 2-40

Reflection and projection   *2-40*

Section 2.8. Concentrations and Moles 2-41
    Concentration 2-42
    Moles and molarity 2-43

Section 2.9. Mass-Mole-Volume Calculations 2-46
    Preparing solutions of known molarity 2-48

Reflection and projection   *2-50*

Section 2.10. Reaction Stoichiometry in Solutions 2-50
    Stoichiometric reaction ratios 2-51
    Limiting reactant 2-52
    Amount of product formed 2-53
    Amount of reactant remaining 2-55

Reflection and projection   *2-56*

Section 2.11. Solutions of Gases in Water 2-57
    Water as a reactant 2-60

Section 2.12. The Acid–Base Reaction of Water with Itself 2-61
    The pH scale 2-61

Section 2.13. Acids and Bases in Aqueous Solution 2-64
    Oxyacids 2-64
    Ionic compounds with oxyanions 2-66
    Nucleic acids are oxyanions 2-66
    Carboxylic acids 2-67
    Extent of proton-transfer reactions 2-68
    A Brønsted-Lowry base: ammonia 2-69
    Review of properties of solutions of gases 2-71
    Brønsted-Lowry acid–base pairs 2-71
    Stoichiometry of $H_3O^+(aq)$–$OH^-(aq)$ Reactions 2-73

Reflection and projection   *2-74*

Section 2.14. Outcomes Review 2-75

Section 2.15. EXTENSION — CO$_2$ and Le Chatelier's Principle 2-77
    The carbon cycle 2-77
    Reactions in carbon dioxide-limewater solutions 2-78
    Le Chatelier's principle 2-79

Water that falls as rain and snow drains into streams and rivers that flow to the sea. Along the way, substances from the Earth's solid crust, including the calcium carbonate represented here, dissolve and are carried to the sea. Marine mollusks like these use calcium carbonate to build their solid shells.

Chapter 2. Aqueous Solutions and Solubility

Gutta cavat lapidem
(Dripping water hollows out a stone)

Ovid (43 B.C. – A.D. 17), Epistulae Ex Ponto

2.1. Consider This *What water soluble and water insoluble substances do you know?*

List six substances you know are soluble in water. List six substances you know are not soluble in water. Work in a small group to compare your lists and produce a combined list of all your different soluble and insoluble substances. Discuss your lists in class to find out how similar all your experiences with solubility are and to try to make some generalizations about soluble and insoluble substances.

Life on Earth began in the seas and the chemistry in living cells occurs in aqueous media. If we are to understand the chemistry of living things, we must understand the chemistry of substances that dissolve or do not dissolve in water. For example, a large percentage of the material in eggshells and seashells is calcium carbonate. The mollusks shown on the facing page obtain the calcium and carbonate ions needed for their shells from the sea water around them. The Latin citation above reminds us that water dripping on a stone or flowing in a river can slowly dissolve solid rocks composed of calcium carbonate. So why doesn't the calcium carbonate in seashells simply redissolve and return to the sea? That is one of the questions we will discuss in this chapter.

Chapter 1 focused mainly on the interactions of water molecules with one another in the pure liquid. This chapter extends that discussion to consider interactions of water with other substances. Common experience, as Consider This 2.1 points out, is that some substances, such as salt, sugar, and antifreeze, dissolve readily in water. Other substances, such as chalk, flour, and oil, are not readily soluble—even if the mixture is stirred vigorously or heated. Compounds that dissolve in water and compounds that do not dissolve in water each have characteristic properties. Among the central themes of this chapter are identification of the properties that make some compounds water soluble and others insoluble and understanding the process by which molecules dissolve in water.

Two important ideas from Chapter 1 undergird these discussions: the polar nature of H–O bonds and the V–shaped structure of the water molecule. Together, these properties give the water molecule a permanent dipole moment. The polarity of water enables it to dissolve a variety of molecules and ions, and prevents the dissolution of others. Aqueous solutions have properties that differ from pure water. Properties such as electrical conductivity and acidity or basicity can help us characterize the interactions that occur in these solutions. After a general introduction to the solution process, we will look again at the intermolecular interactions introduced in Chapter 1 — hydrogen bonding, polar attractions, and London dispersion forces — focusing this time on interactions among the molecules in solutions.

Section 2.1. Substances in Solution

Before beginning our discussion of solutions, we need to consider the specific vocabulary that chemists use to describe solutions. Many terms relating to solutions are familiar because they are used in everyday English, but their familiarity can create a problem. In everyday usage a word often has several meanings, but in scientific use, each term has one precise meaning.

Solution nomenclature

A **solution** is a homogeneous mixture of a **solvent**, the substance present in excess (water in the present discussion), and one or more **solutes**, the substances present in smaller amounts that are described as being **dissolved** in the solvent. A **homogeneous mixture** is a *uniform* mixture of two or more substances. The properties of homogeneous mixtures are the same throughout the mixture. A teaspoon of table sugar added to a glass of water and stirred until the solid is no longer visible produces a homogeneous mixture

> Solution, solvent, solute, dissolve, and related terms come from the Latin word *solvere* = to loosen. Solvents "loosen" solutes from their pure form and mix their components with the solvent to give solutions. Homogeneous is derived from Greek: *homo* = the same + *genos* = kind and heterogeneous from *hetero* = different.

of sucrose in water, Figure 2.1(a). The liquid has the same sweetness throughout. Figure 2.1(b), a molecular level representation of the sugar solution, shows the essential idea: *every* sugar molecule is surrounded by water molecules rather than sugar molecules being clumped together. By contrast, **heterogeneous** mixtures are not uniform throughout. Orange juice is a heterogeneous mixture whose composition varies from one part of the mixture to another, the liquids inside and outside of the pulp, for example.

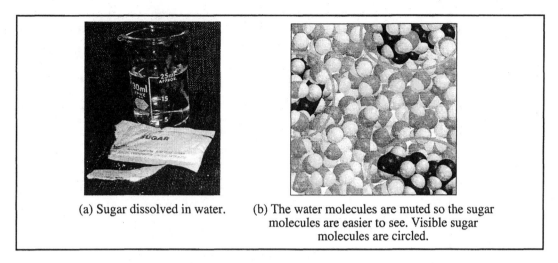

(a) Sugar dissolved in water. (b) The water molecules are muted so the sugar molecules are easier to see. Visible sugar molecules are circled.

Figure 2.1. (a) A homogeneous sugar solution and (b) its molecular level representation.

The solution process

In their pure states, substances that dissolve in water may be solids, liquids, or gases. Understanding the interactions between these solutes and water is the key to understanding the formation and nature of solutions. Without yet going into the details of the interactions, we can begin to think about what has to occur for a substance to dissolve in water (or any other solvent). In any pure solid or liquid, such as sugar or ethanol, molecules attract one another strongly enough to stay in the condensed phase. For solute molecules to become mixed homogeneously among water molecules in a solution, the intermolecular attractions among the solute molecules in their pure solid or liquid state have to be broken. In forming the solution, some attractions between water molecules are broken and new attractions between solute and water molecules are made. Figure 2.2 is an energy diagram we can use to analyze the dissolution.

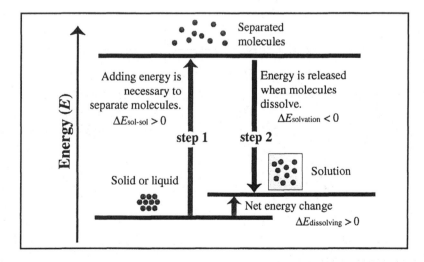

Figure 2.2. Energy changes in the dissolution process.

Aqueous Solutions and Solubility   Chapter 2

To analyze the dissolution process and find its energy, we have broken the process into steps that take us from the solid (or liquid) that dissolves to the solution it forms with water (or other solvent). Step 1 in Figure 2.2 converts the molecules that will be dissolved to their gas phase.

> Note that no covalent bonds are broken in the dissolution process, The molecules stay intact.

This step breaks the solute-solute attractions and requires an input of energy, represented by the long, upward-pointing red arrow, that we have labeled $\Delta E_{sol\text{-}sol}$. In Step 2, these gas-phase solute molecules are mixed with the water molecules to form the solution. In this step, some attractions between water molecules are broken, which requires an input of energy. But new attractions form between the water and solute molecules and these release a good deal of energy. The combined effect of these two interactions is the release of energy represented by the long, downward-facing blue arrow. The figure shows that the process of **solvation**, formation of attractive interactions between separated solute molecules and a solvent, is a downhill process, going from higher to lower energy: $\Delta E_{solvation} < 0$.

If, as is shown in Figure 2.2, the uphill process, breaking solute-solute attractions, $\Delta E_{sol\text{-}sol}$, requires more energy than is released by solvation, $\Delta E_{solvation}$, then the dissolving process requires a net input of energy: $\Delta E_{dissolving} > 0$. A change that requires an input of energy is called **endothermic**. Conversely, if more energy is released by solvation than is required to break the attractions, the overall solution process releases energy. A change that releases energy is called **exothermic**.

2.2.   Check This   *Exothermic energy diagram*

Sketch the energy diagram for an overall solution process that is exothermic.

Favorable and unfavorable factors

We often categorize the factors that affect a process as favorable or unfavorable. A **favorable factor** is one that increases the likelihood that the process will proceed in the direction we are considering. An **unfavorable factor** works against the direction of the process, that is, it favors going in the reverse direction. For example, releasing the energy of solvation lowers the energy of a solute-solvent system, and lower energy favors the formation of the solution. The energy required to separate the solute molecules has to be added to the system, which raises the energy, and is therefore an unfavorable factor. In Chapter 8, we will quantify favorable and unfavorable factors, when we discuss the **entropy** of solutions.

Overall, an exothermic process releases energy and lowers the energy of the system from which the energy comes. The lower final energy favors the process in the exothermic direction.

By this same reasoning, an endothermic process is unfavorable. As we will see, some solutes dissolve in water exothermically and others dissolve endothermically. This observations means that the overall energy change in a solution process must not be the only factor that affects solubility, and we will have to be on the lookout for others.

Section 2.2. Solutions of Polar Molecules in Water

Table 2.1 reminds you of the characteristics of the three intermolecular interactions we discussed in Chapter 1, hydrogen bonding, polar attractions, and London dispersion forces, and illustrates each with representative molecules.

Table 2.1. Comparisons among intermolecular attractions.

	hydrogen bonding	polar attraction	dispersion forces
requires	H atom bonded to O, N, or F	permanent dipolar molecules	all molecules
strength	20-30 kJ·mol^{-1}	2-5 kJ·mol^{-1}	can be quite large — increases with number of electrons per molecule
geometry	H atom approximately on a line between electronegative atoms	positive and negative ends aligned to attract one another	any orientation — best when molecules are side by side
example	water	dimethyl ether	octane

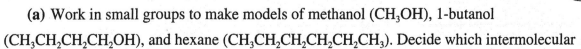

2.3. Investigate This *Which compounds are soluble in water?*

(a) Work in small groups to make models of methanol (CH_3OH), 1-butanol ($CH_3CH_2CH_2CH_2OH$), and hexane ($CH_3CH_2CH_2CH_2CH_2CH_3$). Decide which intermolecular

Aqueous Solutions and Solubility   Chapter 2

attractions, Table 2.1, are present between the molecules in each pure compound. Use these models to predict which molecules will form strong intermolecular attractions with water.

(b) Predict which liquids (methanol, 1-butanol, or hexane) will be soluble in water. Mix one milliliter of each liquid with one milliliter of water; observe which compounds dissolve in water. If one or more of the liquids does not seem to dissolve, try adding two or three drops to one milliliter of water to find out if a small amount will dissolve.

2.4. Consider This *Which compounds are soluble in water?*

(a) Which of your solubility predictions in Investigate This 2.3(b) was correct? What were the reasons for your predictions? If any were incorrect, explain why they were incorrect.

(b) Which of the liquids is most soluble in water? Which is least soluble? Explain how you reach these conclusions.

Hydrogen bonds among unlike molecules

In Chapter 1, we found that hydrogen bonding accounts for many physical properties of water. There, we considered hydrogen bonding between water molecules, but there is nothing to prevent hydrogen bonding between water and other molecules which can hydrogen bond with water, such as alcohols. The maximum number of hydrogen bonds between one methanol molecule and surrounding water molecules is three, as shown in Figure 2.3. The hydrogen covalently bonded to the methanol oxygen takes part in one of these hydrogen bonds. The two nonbonding electron pairs on the methanol oxygen form bonds with hydrogen from water to form the other two. The water molecules have other nonbonding electron pairs and covalently bonded hydrogens, so they can extend the network to other water and methanol molecules.

> WEB Chap 2, Sect 2.2
> Try interactive visualizations of water interacting with polar and nonpolar molecules.

2.5. Check This *Hydrogen-bonded network in water-methanol solution*

WEB Chap 2, Sect 2.2.2. Two of the four water molecules in the top panel look like they are oriented to be able to hydrogen bond with oxygen in methanol, but only one is allowed to replace the "x."

(a) What is the problem with the orientation of the molecule that doesn't "fit"?

(b) How does your answer in part (a) help you understand the network of hydrogen bonds in water around the methanol molecule? Explain.

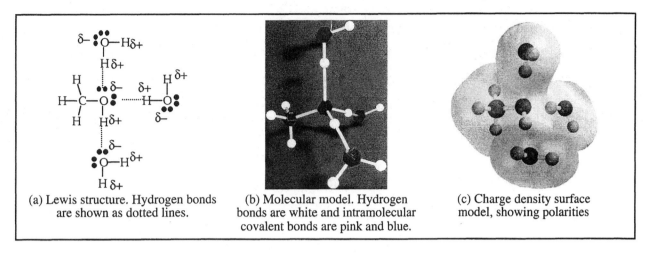

(a) Lewis structure. Hydrogen bonds are shown as dotted lines.

(b) Molecular model. Hydrogen bonds are white and intramolecular covalent bonds are pink and blue.

(c) Charge density surface model, showing polarities

Figure 2.3. Maximum hydrogen bonding between a methanol molecule and water molecules. Molecules are in approximately the same orientation with respect to one another in each representation.

The O–H group in methanol fits into this extensive hydrogen-bonded network and accounts for the high solubility in water that you found in Investigate This 2.3. Other alcohols dissolve in water for exactly the same reason. In fact, methanol, ethanol, CH_3CH_2OH, 1-propanol, $CH_3CH_2CH_2OH$, and 2-propanol, $CH_3CH(OH)CH_3$, are miscible with water. Two liquids are **miscible** if they form homogeneous solutions when they are mixed in any proportion.

Nonpolar solutes

At the other extreme of solubility, you found that hexane, a nonpolar compound that cannot form hydrogen bonds with water, is insoluble in water. A few hexane molecules do dissolve in a volume of water, but there are no specific interactions between them and the surrounding water molecules, Figure 2.4. Rather, there are ever-changing weak attractions, London dispersion forces, between individual water and hexane molecules.

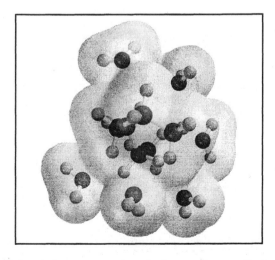

Figure 2.4. A hexane molecule in aqueous solution.

It is difficult to measure the energy change associated with the tiny amount of a hydrocarbon that dissolves in water, but it is close to zero for those that have been measured. This means that the energy of the weak interactions among water and hexane molecules is about the same as the energy required to break up the hexane-hexane attractions and some water-water attractions.

> **2.6. Check This** *Energy diagram for hexane dissolving in water*
> Sketch the energy diagram for the overall solution process of hexane dissolving in water. Assume that the net energy change for the dissolution is zero.

Molecular reorganization

If there is no difference in energy between a solution of hexane in water and the pure liquids, it would be logical for you to question *why* hexane is insoluble in water. The answer lies in the reorganization of the molecules that must occur in going from pure liquids to the solution. The solute molecules go from being all grouped together in the pure liquid to a solution of individual solute molecules scattered among the solvent molecules. The reorganization involved in mixing two kinds of molecules together always favors the mixed state and leads to increased solubility.

On the other hand, Figure 2.4 shows that a molecule of hexane in water takes up space in the liquid without forming strong attractions with the water molecules. The water molecules lose some of their freedom of movement when they have to be reorganized to make way for the hexane molecule. This reorganization, even though it does not require energy, works against the solubility of nonpolar molecules. The solvent reorganization is unfavorable and outweighs the favorable reorganization of mixing. The dominance of the unfavorable solvent reorganization explains why nonpolar compounds are insoluble in water. In Chapter 8, as we have noted before, we will say more about the reasons for this behavior and develop a quantitative model, based on entropy, for these favorable and unfavorable reorganizations. For now, the observation that nonpolar compounds are relatively insoluble in water is enough to allow you to understand and make predictions about solubilities.

Intermediate cases

Your results from Investigate This 2.3 show that 1-butanol is somewhat soluble in water; small amounts do dissolve. A 1-butanol molecule has an alcohol group and a four-carbon alkyl chain that is like a nonpolar hydrocarbon. The alcohol group attractions to water, shown in Figure 2.5, should help 1-butanol dissolve. (We have seen that alcohols with smaller alkyl groups are miscible with water.) On the other hand, the alkyl group is nonpolar and will not help the 1-butanol dissolve. The upshot is limited solubility; alcohols become less and less soluble, as the size of the alkyl group grows.

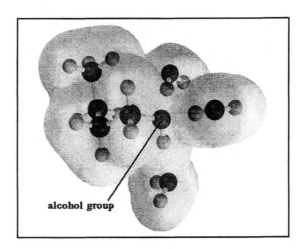

Figure 2.5. A 1-butanol molecule in aqueous solution.
The polar alcohol O-H group in butanol hydrogen bonds with water. The nonpolar end of the butanol, like nonpolar hexane (Figure 2.4), interacts only weakly with water molecules.

Like dissolves like

From what we have seen so far, we can predict that a compound with a polar group that can hydrogen bond with water will be more soluble in water than a similar compound without the polar group: 1-pentanol, $CH_3CH_2CH_2CH_2CH_2OH$, is more soluble than pentane, $CH_3CH_2CH_2CH_2CH_3$. We can also predict that the water solubility of compounds with polar and nonpolar parts will decrease as the nonpolar part gets larger: dimethyl ether, CH_3OCH_3, is more soluble than dipropyl ether, $CH_3CH_2CH_2OCH_2CH_2CH_3$. The generalization that polar molecules that can hydrogen bond with water dissolve in water, a polar hydrogen-bonding solvent, but nonpolar molecules do not, is an example of a "rule" you may have heard before "like dissolves like."

2.7. Check This *Predict relative solubilities*

Write Lewis structures for diethyl ether, $CH_3CH_2OCH_2CH_3$, pentane, $CH_3CH_2CH_2CH_2CH_3$, and 1-butanol, $CH_3CH_2CH_2CH_2OH$.

(a) Which of these molecules can form hydrogen bonds in their pure liquids?

(b) Which can form hydrogen bonds with water?

(c) Which compound is most soluble in water? Which is least soluble? Explain.

(d) Which compound is most soluble in mineral oil, a nonpolar liquid? Which is least soluble? Explain.

Aqueous Solutions and Solubility   Chapter 2

Solutes with multiple polar groups

One of the soluble substances you are likely to have listed in Consider This 2.1 is sugar. You can dissolve about 200 g of sucrose (table sugar), $C_{12}H_{22}O_{11}$, in 100 mL of water. The structure of

> WEB Chap 2, Sect 2.2.3–4
> View and analyze molecular visualizations of sucrose dissolving.

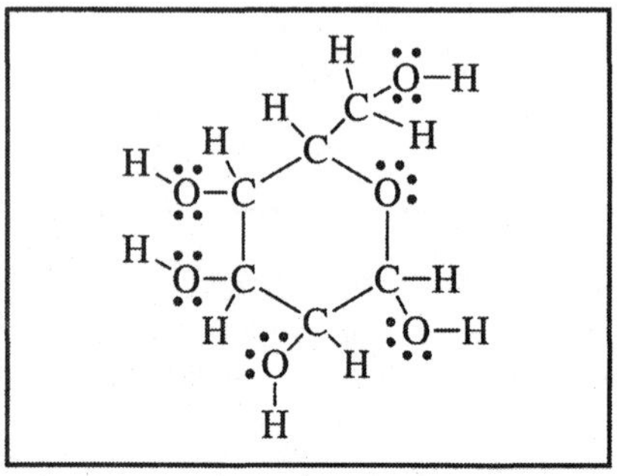

a simpler sugar, glucose, $C_6H_{12}O_6$, that also readily dissolves in water (about 100 g of glucose per 100 mL of water), is shown in Figure 2.6. Note that the glucose molecule contains five alcohol groups as well as the oxygen in the ring. You would expect a glucose molecule to interact and hydrogen bond strongly with water molecules.

Figure 2.6. The structural formula of glucose.

2.8. Consider This *What are the interactions of glucose with water?*

Use the structural formula in Figure 2.6 to determine the maximum number of water molecules that can be hydrogen bonded to a glucose molecule. Compare the maximum number of hydrogen bonds between water and glucose (a six-carbon compound) to the maximum number between water and methanol (a one-carbon compound). Does this comparison help to explain the high solubility of glucose? If so, how does it help?

Your comparison in Consider This 2.8 shows that, carbon-for-carbon, glucose and methanol have about the same number of attractive, hydrogen-bonding interactions with water. A complex compound whose molecules contain several polar groups that can hydrogen bond with water can be relatively soluble in water. Another comparison is between the solubility of glucose and cyclohexanol. Cyclohexanol, Figure 2.7, has only one alcohol group on the six-carbon ring and has a solubility of about 3.6 g per 100 mL of water. The much higher solubility of glucose is due to its many polar, hydrogen-bonding groups.

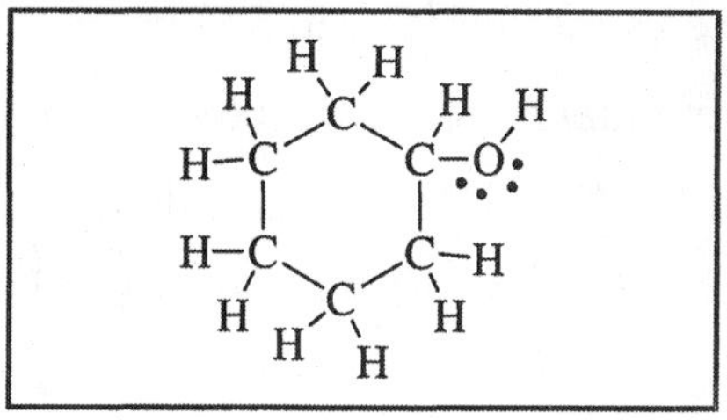

Figure 2.7. The structural formula of cyclohexanol.

2.9. Check This   *Predict relative solubilities*

(a) Write out the Lewis structures for these compounds with multiple polar groups.

$H(OCH_2CH_2)_4OH$ $HO(CH_2CH_2)_5OH$

polyethylene glycol (PEG 200)   1, 10-decanediol

(b) Which compound is more soluble in water? Explain.

Reflection and projection

Solutions are mixtures in which solute molecules separate from each other and disperse uniformly throughout the solution. Dissolving a solute in a solvent requires energy to break the attractions among the pure solute molecules (and also some of the attractions among the solvent molecules). Some or all of this energy is supplied by the attractions between the solute and solvent molecules in the solution. Many polar molecules dissolve in water (and in one another) because of hydrogen bonding between the solute and water molecules. Polar compounds that contain no hydrogen atom attached to an oxygen or nitrogen atom, cannot hydrogen bond among themselves. They can, however, hydrogen bond with water, which makes them more water soluble than comparable nonpolar compounds.

Nonpolar compounds are generally insoluble in water, mainly because the water molecules must reorganize to accommodate them. This reorganization is an unfavorable process that outweighs the favorable reorganization of mixing the solute and solvent. Compounds whose molecules contain both polar parts that can hydrogen bond with water and large nonpolar parts are less soluble than comparable compounds with smaller nonpolar parts. Compounds with several polar groups are more soluble than comparable compounds with fewer polar groups. Many complex compounds with several polar groups, including many biomolecules like sugars, dissolve readily in water.

The solutes we have discussed so far have been carbon-containing polar compounds that dissolve to give solutions of individual molecules mixed among the water molecules. Among the soluble substances that you listed in Consider This 2.1, there must have been ones like table salt that are not carbon-containing compounds. We will now examine how solutions of these substances are similar to and different from the solutions we have been discussing.

Section 2.3. Characteristics of Solutions of Ionic Compounds in Water

2.10. Investigate This   *Which solutions conduct an electric current?*

Use an electrical conductivity tester to determine whether or not solutions conduct an electric current. The tester consists of a source of electrical potential, two wires that dip into the test solution, and a meter or a signal light to signal electrical current flow. Place 2–3 mL of distilled water in a small test tube or well plate. Test the conductivity of pure water. Dissolve a small amount of glucose in water in another test tube or well and test the conductivity of this sugar solution. Repeat the conductivity test with a solution of table salt in another test tube or well.

2.11. Consider This   *Why do some solutions conduct an electric current?*

(a) Which of the liquids in Investigate This 2.10 conduct an electrical current? How do you tell?

(b) Can you distinguish among pure water, an aqueous glucose solution, and an aqueous sodium chloride solution by their conductivities? Use sketches like these of the conductivity experiment to show what you *observed* for each solution. What molecular-level picture can you suggest to explain your answer? Illustrate your explanation on your sketches.

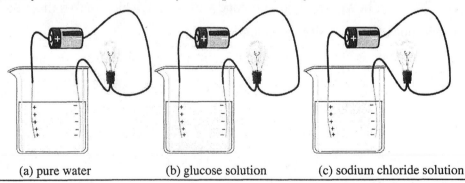

(a) pure water   (b) glucose solution   (c) sodium chloride solution

Solution conductivity

For a solution to give a positive electrical conductivity test (bulb lights up), electric charge has to be transported from one test wire to the other. Solutions that conduct a current must contain mobile charged particles. When an electrical potential is applied to the solution, negative particles move toward the positive wire and positive particles move toward the negative wire. Water molecules are electrically neutral and cannot transport electrical charge; this case is illustrated in Figure 2.8(a). In the previous section, we pictured a solution of glucose as electrically neutral glucose molecules mixed with and hydrogen bonding with water molecules. This picture is confirmed by the electrical conductivity results: a solution of glucose, like pure water, does not conduct electricity, as illustrated in Figure 2.8(b).

WEB Chap 2, Sect 2.3
Use molecular visualizations to check your understanding of solution conductivity.

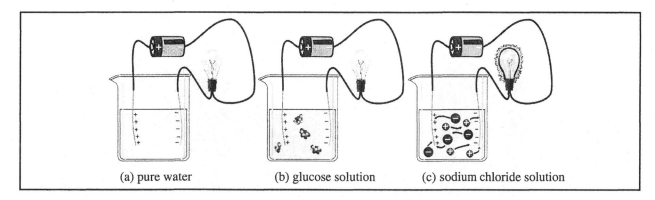

(a) pure water   (b) glucose solution   (c) sodium chloride solution

Figure 2.8. Observed results and molecular-level interpretation of solution conductivity.

Many solids, like sodium chloride, dissolve in water to form solutions that *do* conduct an electric current, as illustrated by the bulb lighting up in Figure 2.8(c). These solutions must contain charged particles that move in response to the electric field between the wires, as illustrated in the figure. There must be equal numbers of positive and negative charges in the solution, since neither the water nor the solid had an electrical charge before they were mixed to make the solution.

We have said nothing about what happens when the charged particles reach the wires. In Investigate This 2.10, you may have noticed that gas bubbles formed on the wires in the conducting solution. Formation of the gas indicates that a chemical reaction is occurring to produce the gas. For our purposes in this chapter, we are interested only in whether solutions do or do not conduct an electric current. We will return in Chapter 10 to discuss the chemical reactions that result from the flow of electric charge through conducting solutions.

Ionic solids and solutions

Figure 2.9 shows a molecular-level model for the dissolving of solid compounds whose solutions conduct electricity. The solid, Figure 2.9(a), is made up of **ions**, species that have too few or too many electrons to balance the positive charge of their nuclei. A **cation** (**positive ion**)

is an atom, or a covalently bonded group of atoms, that has lost one or more electrons. The

> Cation is pronounced "CAT-ion" and anion is pronounced "AN-ion."

positive charge results from there being too few electrons to balance the positive charges of nuclear protons. An **anion** (**negative ion**) is an atom, or a covalently bonded group of atoms, that has gained one or more extra electrons. The negative charge results from there being more electrons than are needed to balance the positive charges of the nuclear protons. The ions are held in place by strong attractive forces to their nearest neighbors, forming an orderly arrangement called an **ionic crystal**.

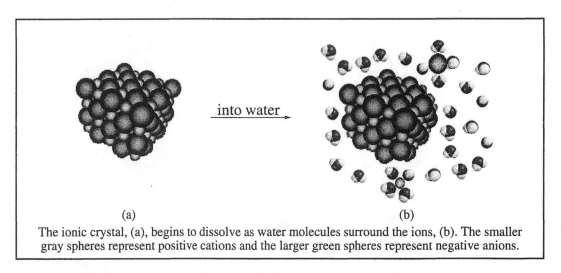

(a)   (b)
The ionic crystal, (a), begins to dissolve as water molecules surround the ions, (b). The smaller gray spheres represent positive cations and the larger green spheres represent negative anions.

Figure 2.9. Model of an ionic crystal dissolving in water.

2.12. **Check This** *Properties of crystals*

(a) If each cation has a +1 charge and each anion a −1 charge, what is the net charge on the crystal in Figure 2.9(a)? What is the net charge on the remaining crystal in Figure 2.9(b)? Explain how you get your answers.

(b) WEB Chap 2, Sect 2.4.1. Are there properties of crystals shown in the movie that are not shown in Figure 2.9(a)? If so, what are they and why aren't they shown in the figure?

Ion-dipole attractions in ionic solutions

When a soluble ionic solid dissolves in water, the resulting solution, is called an **ionic solution**. Figure 2.9(b) shows how the ions that enter the solution are surrounded by the polar

> WEB Chap 2, Sect 2.5.1–2
> Use interactive visualizations to examine the ionic dissolving process.

water molecules that are attracted by the electrically charged ions. Once they have left the crystal and are surrounded by water molecules, the ions are free to move about. When conductivity tester wires are placed in the solution, positive ions and negative ions carry current through the solution, as shown in Figure 2.8(c).

Look closely at the model for dissolved ions in Figure 2.9(b). The figure shows that water molecules orient themselves differently around positive and negative ions. (For clarity, Figure 2.9(b) shows only a few of the water molecules that surround each ion.) Cations attract the negative (oxygen) ends of water molecules. Anions attract the positive (hydrogen) ends of water molecules. Attractions between ions and water molecules are called **ion–dipole attractions**, and they are relatively strong. Ions at the corners and along the edges of a crystal dissolve more readily than those in the crystal faces, because more water molecules can gather around the ions while they are still part of the crystal.

As an ion leaves the crystal, water molecules surround it completely, forming a **hydration layer** [*hydr*o = water]. Dissolved ions are often called **hydrated ions**. Although individual water molecules are constantly being exchanged in hydration layers, the hydration layer helps keep oppositely charged ions from getting close to one another. Positive and negative ions are usually separated by at least two layers of water molecules. Consequently, each dissolved ion moves almost independently of all the others.

2.13. **Check This**   *Ionic hydration layers and crystal dissolution*

(a) About how many water molecules can surround an ion to form its first hydration layer? Explain how you arrive at your answer.

(b) WEB   Chap 2, Sect 2.5.2. Is your answer in part (a) consistent with the molecular level representations in these animations? Explain why or why not. Crystal dissolution is represented in the movie. What interactions favor dissolving? What interactions favor the crystal remaining intact? Explain clearly how the competition among these interactions is represented.

Aqueous ionic solutions are everywhere around you and in you. Seas and oceans contain many dissolved ionic compounds; that's why they taste salty. Even freshwater streams, such as the one shown in the chapter opening, contain some dissolved ionic compounds. Almost every liquid you drink, including tap water, is an ionic solution and all biological fluids are ionic solutions. Yet, many solid ionic compounds, for example, seashells and your teeth, do not dissolve in water. In the next two sections, we will consider ionic compounds in more detail and find out how to predict which are soluble and which are not.

Section 2.4. Formation of Ionic Compounds

Names and formulas

Before we examine how ionic compounds form, let's consider how we name these compounds and write their formulas. Names and formulas for a few common **monatomic** (single atom) ions and **polyatomic** (many atom) ions are given in Table 2.2. Ionic compounds do not contain individual molecules, so a chemical formula such as NaCl does not represent a molecule. The formula represents a ratio of ions. For any ionic compound, the ratio is determined by the numbers needed to cancel positive and negative charges. The smallest whole-number ratio of positive ions to negative ions that is electrically neutral is used as the chemical formula. Calcium chloride crystals, for example, contain Ca^{2+} and Cl^- ions. Two chloride ions are required to balance the charge on one calcium ion, and the chemical formula is written as $CaCl_2$. Ionic charges are almost never indicated in the formula. Note that the cation is always named first in the names of ionic compounds.

Table 2.2. Names and formulas of a few common ions.

Cations				Anions			
Monatomic		Polyatomic		Monatomic		Polyatomic	
calcium	Ca^{2+}	ammonium	NH_4^+	bromide	Br^-	carbonate	CO_3^{2-}
magnesium	Mg^{2+}	hydronium	H_3O^+	chloride	Cl^-	hydroxide	OH^-
potassium	K^+			fluoride	F^-	nitrate	NO_3^-
sodium	Na^+			oxide	O^{2-}	phosphate	PO_4^{3-}
				sulfide	S^{2-}	sulfate	SO_4^{2-}

2.14. Worked Example *Chemical formulas for ionic compounds*

Write the chemical formulas for the ionic compounds **(a)** ammonium carbonate and **(b)** calcium phosphate.

Necessary information: We need to know, from the names, that compound (a) contains the ammonium and carbonate ions and (b) contains the calcium and phosphate ions. Formulas and charges for the ions are given in Table 2.2.

Strategy: The numbers of cations and anions must be such that the total positive charge cancels the total negative charge. The easiest ways to find the numbers are by simply inspecting the charges on the ions or by trial and error.

Chapter 2   Aqueous Solutions and Solubility

> *Implementation:* (a) The ammonium ion is +1 and the carbonate ion is –2. There must be two +1 ammonium ions to balance each –2 carbonate. The formula is $(NH_4)_2CO_3$. The parentheses around the ammonium ion are necessary to distinguish the number of ammonium ions (2) from the number of hydrogen atoms in the ammonium ion (4).
>
> (b) The calcium ion is +2 and the phosphate ion is –3. We'll need more than one of each ion. Try two phosphate ions to give a –6 charge. We will need three calcium ions to cancel the –6. The formula is $Ca_3(PO_4)_2$. As before, parentheses around the phosphate are required to avoid subscript confusion.
>
> *Does the Answer Make Sense:* Multiply the charge times the subscript for both the positive and negative ion to be sure that the total plus and minus charge does cancel out. The easiest mistake to make in writing formulas is to switch the subscripts; this check helps avoid that error.

2.15. Check This   *Formulas for ionic compounds*

Write the chemical formulas for the ionic compounds (a) ammonium chloride, (b) sodium sulfide, and (c) calcium carbonate.

Why ionic compounds form

Before considering what happens to ions in solution, we will consider why ionic compounds form in the first place. Table 2.2 shows that atoms of elements in Group I (**alkali metals**) and Group II (**alkaline earths**) lose electrons to form monatomic positive ions. Atoms of elements in Groups VI (oxygen and sulfur) and Group VII (**halogens**) gain electrons to form monatomic negative ions. The keys to this behavior lie in values for the electronegativities of these elements, which are given in Figure 2.10, and in the attraction between oppositely charged ions.

	I	II	III	IV	V	VI	VII	VIII
1	hydrogen H 2.2							helium He —
2	lithium Li 1.0	beryllium Be 1.6	boron B 2.0	carbon C 2.6	nitrogen N 3.0	oxygen O 3.4	fluorine F 4.0	neon Ne —
3	sodium Na 0.93	magnesium Mg 1.3	aluminum Al 1.6	silicon Si 1.9	phosphorus P 2.2	sulfur S 2.6	chlorine Cl 3.2	argon Ar —
4	potassium K 0.82	calcium Ca 1.3						

Figure 2.10. Electronegativities for the first 20 elements.

On the left-hand side of the periodic table, the elements in Groups I and II have low electronegativities. These elements have a relatively weak attraction for electrons. On the other side of the periodic table, elements in Groups VI and VII have high electronegativities. These elements have a strong attraction for electrons. Imagine a gas phase reaction between a sodium atom losing a loosely held electron and a chlorine atom attracting that electron to yield a sodium cation and chloride anion, as represented in Figure 2.11.

WEB Chap 2, Sect 2.4
Study interactive visualizations of ionic crystals and crystal lattice energy.

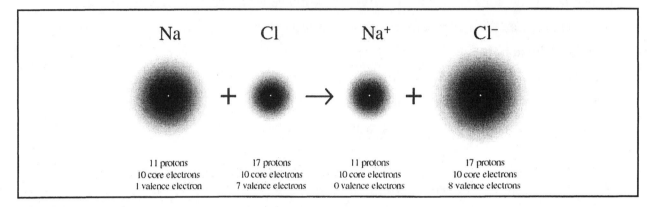

Figure 2.11. Molecular (atomic) level representation of an electron transfer reaction.

In Figure 2.11, we are trying to represent the overall changes that occur at the atomic level in this reaction. An electron is transferred from atomic sodium to atomic chlorine. The transfer gives an 11+ sodium nucleus surrounded by only 10 electrons; this is a sodium cation. The transfer also gives a 17+ chlorine nucleus surrounded by 18 electrons; this is a chloride anion. The sizes of the atoms and ions in expression (2.1) are drawn to scale. The sodium cation (which has fewer electrons than the atom) is smaller than the chloride anion (which has more electrons than the atom). This is a general result: *almost all monatomic anions are larger than any monatomic cation.*

Chemical reaction equations

Another way to write the reaction between a gaseous sodium atom and a gaseous chlorine atom to yield a gaseous sodium cation and a gaseous chloride anion is this symbolic representation:

$$Na(g) + Cl(g) \rightarrow Na^+(g) + Cl^-(g) \qquad (2.1)$$

Chemical equations, like equation (2.1), are valuable shorthand representations describing chemical change. The physical state of each substance is often indicated in parentheses following the chemical formula. Gases are represented using *(g)*, liquids using *(l)*, solids using *(s)*, and aqueous solutions using *(aq)*. The right-facing arrow means "yields." Substances to the left of the arrow are called **reactants** and substances to the left are called the **products** of the reaction.

When there are multiple reactants or products, they are separated by a plus sign, which is read as "and." Chemical equations are **balanced** when *equal numbers of each atom* appear on each side of the arrow and the *net electrical charge is the same* on each side of the arrow. When we want to refer to "the chemical reaction represented by chemical equation (xx)," we will usually shorten the reference to "reaction (xx)." For example, we would refer to the "reaction of Na(g) and Cl(g) represented by chemical equation (2.1)" as "reaction (2.1)."

2.16. **Worked Example** *Writing a balanced chemical reaction equation*
Write a balanced gas phase chemical reaction equation that represents the reaction between calcium atoms and fluorine atoms to yield calcium cations and fluoride anions.

Necessary information: Table 2.2 shows us that the calcium cation has a 2+ charge and the fluoride anion a 1– charge.

Strategy: Since a calcium atom has to lose two electrons to form a calcium cation, it must transfer these electrons to two fluorine atoms, each of which gains an electron to form a fluoride ion.

Implementation: The chemical reaction representing this transfer is:
$$Ca(g) + 2F(g) \rightarrow Ca^{2+}(g) + 2F^-(g)$$
Note that the numbers of atoms or ions taking part in the reaction are specified by coefficients written in front of the symbol for the species.

Does the answer make sense? The same number of calcium atoms (or ions) are represented on each side of the yields arrow and the same is true for number of fluorine atoms (and ions). The net electric charge on the reactants is zero; none of the reactants has an electric charge. The net electric charge on the products is also zero; the 2+ charge on the calcium cation and the 1– charges on each of *two* fluoride anions cancel out. Equal numbers of atoms of each kind appear in the reactants and products and the net electrical charge is the same for both reactants and products. The chemical reaction equation is balanced and makes sense.

2.17. **Check This** *Writing a balanced chemical reaction equation*
Write a balanced gas phase chemical reaction equation that represents the reaction between magnesium atoms and oxygen atoms to yield magnesium cations and oxide anions.

Formation of ionic crystals

Figure 2.12 shows an energy diagram for reaction (2.1), the formation of positive and negative gaseous ions from gaseous atoms. You see that the reaction is endothermic; 145 kJ·mol^{-1} are *required* for the reaction. Energetically, the electron transfer is an uphill process. The energy change, $\Delta E_{\text{ion form}}$, is unfavorable for the formation of ions from atoms.

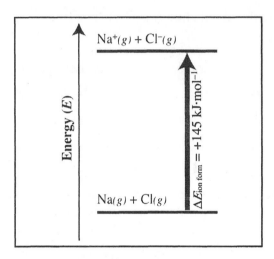

Figure 2.12. Energy change for formation of Na$^+$(g) + Cl$^-$(g) from Na(g) + Cl(g).

Experimentally, however, an enormous number of compounds of metal cations and nonmetal anions *do* exist. All such compounds of Group I and II metal cations with Group VI and VII anions are white ionic crystals that resemble ordinary table salt. The source of energy that makes formation of the compounds possible is the attraction of the cations and anions for one another. They combine to form extended three-dimensional **lattices**, regular repeating patterns of positive and negative ions, the ionic crystals shown in Figure 2.9. The attractions between cations and anions are represented in a two-dimensional lattice in Figure 2.13.

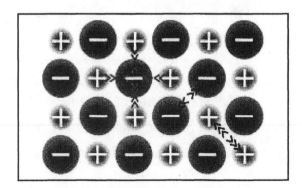

Figure 2.13. Attractions and repulsions in an ionic lattice.
Red arrowheads facing *toward* one another represent Coulombic attractions and blue arrowheads facing *away* from one another represent Coulombic repulsions.

The energy of attraction between unlike charges (or repulsion between like charges) is given by **Coulomb's law**:

$$E \propto \frac{Q_1 Q_2}{d} \tag{2.2}$$

In words, Coulomb's law states that the energy (E) of attraction or repulsion between two electric charges is directly proportional to the magnitude of each charge (Q) and inversely proportional to the distance (d) separating the charges. If the charges have opposite signs, the energy has a negative value, that is, it is less than zero. Attraction of opposite charges lowers the energy of the lattice (crystal) in Figure 2.13. Conversely, the energy is positive if the charges are the same; repulsion of like charges raises the energy of the lattice (crystal).

2.18. Consider This *Are attractions or repulsions stronger in a crystal?*

In Figure 2.13, is the magnitude of the attractive energy, represented by red arrowheads, larger or smaller than the magnitude of the repulsive energy, represented by blue arrowheads? Explain the reasoning for your answer.

In an ionic crystal, represented by the lattice in Figure 2.13, oppositely charged ions are closest together. Ions with the same charges are separated by greater distances. Positive–negative attractions between adjacent ions are strong and outweigh the repulsions between more distant like-charged ions. An ionic crystal is lower in energy than its separated ions. The **lattice energy** of an ionic crystal is the energy that would have to be supplied to convert the solid crystal to separated gaseous cations and gaseous anions:

$$Na^+Cl^-(s) \rightarrow Na^+(g) + Cl^-(g) \tag{2.3}$$

For NaCl, the lattice energy, $\Delta E_{lattice}$, is 787 kJ·mol^{-1}. (In equation (2.3), the charges are shown in Na^+Cl^- to emphasize that it is an ionic crystal.)

Figure 2.14 brings together the energies we have been discussing, in order to determine the energy change, $\Delta E_{xtal\ form}$, when Na^+Cl^- ionic crystals (abbreviated "xtal") are formed directly from gaseous Na and Cl atoms:

$$Na(g) + Cl(g) \rightarrow Na^+Cl^-(s) \tag{2.4}$$

Reaction (2.4) is represented by the left-hand, downward-pointing blue arrow in Figure 2.14. Experimental difficulties make it impossible to measure the energy change for this reaction directly, but $\Delta E_{xtal\ form}$ can be calculated by combining the energy changes for reactions (2.1) and (2.3).

Aqueous Solutions and Solubility — Chapter 2

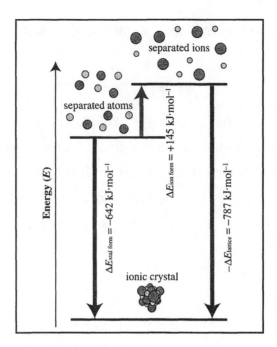

Figure 2.14. Energy diagram for the formation of one mole of ionic crystals of NaCl.

The energy change for reaction (2.1), $\Delta E_{\text{ion form}}$, is represented by the upward-pointing, red arrow in Figure 2.14. The energy change for the formation of the NaCl crystals from the ions, the reverse of reaction (2.3), is $-\Delta E_{\text{lattice}} = -787$ kJ·mol^{-1}, and is represented by the right-hand, downward-pointing blue arrow on the energy diagram. Whenever a reaction is reversed, the energy change for the reverse reaction is equal in magnitude, but opposite in sign to the energy change for the forward reaction. The energy change for reaction (2.4), formation of Na$^+$Cl$^-$ ionic crystals from gaseous Na and Cl atoms, is the sum of the energy changes for the steps that bring about this same change:

$$\Delta E_{\text{xtal form}} = \Delta E_{\text{ion form}} + (-\Delta E_{\text{lattice}})$$

$$= (145 \text{ kJ·mol}^{-1}) + (-787 \text{ kJ·mol}^{-1}) = -642 \text{ kJ·mol}^{-1}$$

Reaction (2.4) forming Na$^+$Cl$^-$ ionic crystals from gaseous Na and Cl atoms is exothermic.

2.19. Check This   *Energy diagram for formation of one mole of CaCl$_2$*

The lattice energy for calcium chloride, CaCl$_2$, crystals is 2260 kJ·mol^{-1}. The gaseous reaction forming the ions from the atoms,

$$\text{Ca}(g) + 2\text{Cl}(g) \rightarrow \text{Ca}^{2+}(g) + 2\text{Cl}^-(g),$$

requires 1039 kJ·mol^{-1}. Draw an energy diagram, analogous to Figure 2.14, and use it to find $\Delta E_{\text{xtal form}}$ for the formation of ionic crystals of CaCl$_2$ from the gaseous atoms.

Chapter 2 — Aqueous Solutions and Solubility

Section 2.5.   Energy Changes When Ionic Compounds Dissolve

2.20. Investigate This *What temperature changes occur when solids dissolve?*

Work in small groups on this investigation. You will use three capped, numbered vials, each containing about 1 g of a white solid. The solids are (1) ammonium chloride, NH_4Cl, (2) calcium chloride, $CaCl_2$, and (3) sodium chloride, $NaCl$. Open one of the vials, hold the vial near the bottom, and have a partner pour about 5 mL of room temperature water into the vial. Feel the temperature of the vial and watch what happens inside. Let your partners also feel the vial. Cap the vial so you don't spill the contents. Gently shake its contents while continuing to observe any additional changes. Continue recording your observations until no further change is evident. Repeat this procedure with the other two vials containing white solids. Uncap each vial and use an electrical conductivity tester to determine whether the solutions conduct electricity.

Before and after photos of $CaCl_2$ added to water with a digital thermometer showing the temperatures and a conductivity meter in the liquid showing before and after conductivity.
Photos to come.

2.21. Consider This *What energy changes occur when solids dissolve?*

   (a) In Investigate This 2.20, were the three solids you dissolved ionic? What is the evidence for your answers?

   (b) When a process releases energy, the energy usually warms surrounding objects, such as the container. When a process requires energy, the energy usually comes from surrounding objects, which get cooler. In Investigate This 2.20, were the dissolving processes for the three solids exothermic or endothermic? What is the evidence for your answers?

   (c) Did any of your results surprise you? Why or why not? What might be happening when these white solids are mixed with water?

Solubility: hydration energy and lattice energy

You found in Investigate This 2.20 that sodium chloride, ammonium chloride, and calcium chloride all dissolve readily in water and that their solutions are good conductors of electric current, so the solutions must contain ions. You also found that the dissolving process was exothermic for some compounds (the solution got warm) and endothermic for others (the solution got cool). These energy changes are the subject of this section.

Attractions between ions and water in the solution, which we discussed in Section 2.3, is a major factor that favors dissolving of ionic solids. We express this attraction as the **hydration energy**, for the ions. Hydration energy is just a special name for solvation energy when the solvent is water. The hydration energy is the energy change, $\Delta E_{hydration}$, when the gaseous ions dissolve in water. For NaCl, the process is:

$$Na^+(g) + Cl^-(g) \rightarrow Na^+(aq) + Cl^-(aq) \tag{2.5}$$

The hydration energy for NaCl is -784 kJ·mol^{-1}. Attractions between positive and negative ions in the solid, the lattice energy, is the major factor that favors keeping the crystal intact. The balance between the hydration and lattice energies helps determine whether an ionic solid is soluble. Figure 2.15 is an energy diagram showing these two energies and the resulting net energy change, $\Delta E_{dissolving} = 3$ kJ·mol^{-1}, for dissolving solid NaCl. (Figure 2.15 is another specific example of the general energy diagram for the solution process shown in Figure 2.2.)

> WEB Chap 2, Sect 2.5
> Correlate energies and molecular level visualizations of dissolving.

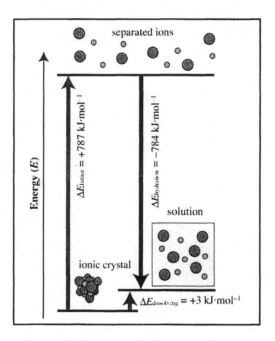

Figure 2.15. Energy changes for NaCl dissolving in water.
Dissolving NaCl(s) in water is a slightly endothermic process, $\Delta E_{dissolving} = 3$ kJ·mol^{-1}.

2.22. Consider This *How does Figure 2.15 correlate with Investigate This 2.20?*
The net energy change, $\Delta E_{dissolving}$, for dissolving solid NaCl, Figure 2.15, is quite small. Is this consistent with your observations in Investigate This 2.20? Explain why or why not.

For some ionic compounds, as you observed with $CaCl_2$ in Investigate This 2.20, the hydration energy is much greater than the lattice energy, and the energy released during dissolving, $\Delta E_{dissolving} < 0$, warms the solution quite substantially. For other ionic compounds, such as ammonium chloride, the lattice energy is greater than the hydration energies of the ions, and the energy taken in during dissolving, $\Delta E_{dissolving} > 0$, removes energy from the solution and makes it cooler. Some of the hot packs and cold packs you might have used to ease minor injuries employ dissolution reactions like these to produce their thermal effects. The lattice and hydration energies for a few ionic solids are given in Table 2.3.

Table 2.3. Lattice and hydration energies (in kJ·mol⁻¹) for some ionic compounds. The entries where the cation row and anion columns intersect are $\Delta E_{lattice}$ and $\Delta E_{hydration}$ values for the compound. For example, 2440 kJ·mol⁻¹ is $\Delta E_{lattice}$ for $MgBr_2$.

	Anion					
	Cl^-		Br^-		I^-	
Cation	$\Delta E_{lattice}$	$\Delta E_{hydration}$	$\Delta E_{lattice}$	$\Delta E_{hydration}$	$\Delta E_{lattice}$	$\Delta E_{hydration}$
Li^+	861	−898	818	−867	759	−854
Na^+	787	−784	751	−753	700	−740
K^+	717	−701	689	−670	645	−657
Ag^+	916	−850	903	−819	887	−806
Mg^{2+}	2524	−2679	2440	−2626	2327	−2540
Ca^{2+}	2260	−2337	2176	−2285	2074	−2194
Sr^{2+}	2153	−2205	2075	−2142	1963	−2053
	S^{2-}		CO_3^{2-}			
Mg^{2+}	3406	−3480	3122	−3148		
Ca^{2+}	3119	−3140	2804	−2817		
Sr^{2+}	2974	−3030	2720	−2725		

2.23. Worked Example *Energy change for dissolving $CaCl_2$ in water*

Use the data in Table 2.3 to draw an energy diagram, analogous to Figure 2.15, for dissolving solid calcium chloride, $CaCl_2$, in water and find $\Delta E_{dissolving}$.

Necessary Information: All the data we need are in Table 2.3.

Strategy: We use the values for $\Delta E_{lattice}$ and $\Delta E_{hydration}$ from Table 2.3 to determine the energy levels for the ionic solid crystal and the ionic solution relative to the energy of the separated ions. The energy difference between these two levels is $\Delta E_{dissolving}$.

Implementation: The energy level diagram is:

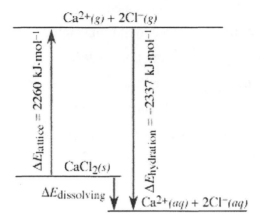

$\Delta E_{dissolving}$ is the sum of the energy changes for the two steps that give the equivalent change:

$$\Delta E_{dissolving} = \Delta E_{lattice} + \Delta E_{hydration} = (2260 \text{ kJ·mol}^{-1}) + (-2337 \text{ kJ·mol}^{-1}) = -77 \text{ kJ·mol}^{-1}$$

Both the energy level diagram and this calculation show that dissolving $CaCl_2(c)$ in water is an exothermic process.

Does the Answer Make Sense? In Investigate This 2.20, you found that the solution got warm when solid calcium chloride was dissolved. The dissolution reaction is exothermic, as our energy level diagram predicts.

2.24. Check This   *Energy change for dissolving LiI in water*

Lithium iodide, LiI, is very soluble in water. Draw an energy level diagram, analogous to Figure 2.15 and the one in Worked Example 2.23, for dissolving solid LiI in water. Does the solution get warmer or cooler as the solid dissolves? Explain your answer.

Reflection and projection

The polar model for water developed in Chapter 1 provides the basis for understanding how water dissolves ionic compounds by hydrating (solvating) the cations and anions and keeping them apart from one another. Solutions of ions conduct an electric current because individual positive and negative ions, being almost independent of each other and surrounded by water molecules, can move and carry electric charges through the solution. Sizes and charges of ions reflect their electron arrangements. Ionic compounds do not exist as molecules, so formulas are written as the smallest whole-number ratio of positive and negative ions that cancel charge to give an overall electrically neutral formula.

Forming an ionic compound from a metal and a nonmetal requires that metal atoms form cations by giving up electrons. Nonmetal atoms, in turn, accept those electrons to form anions. More energy is required to remove the electrons than is returned when anions form. Ionic compounds form because their lattice energies, the energies required to break the net attractions among ions in crystal lattices, are large. The energy associated with dissolving an ionic compound in water, $\Delta E_{dissolving}$, depends on the relative magnitudes of the lattice energy, $\Delta E_{lattice}$, and hydration energy, $\Delta E_{hydration}$, for that compound.

You observed that some soluble ionic compounds dissolve exothermically and others dissolve endothermically. This result means that you cannot use the energy of the solution process to predict solubilities. Therefore, we will investigate solubilities experimentally to see if we can discover some rules or generalizations that will make it possible for you to predict which ionic compounds will be soluble and which will be insoluble.

Section 2.6. Precipitation Reactions of Ions in Solution

2.25. Investigate This *What reactions of ions in solution can you observe?*

Use an electrical conductivity tester to test the conductivity of aqueous solutions of $CaCl_2$, Na_2SO_4, and $NaNO_3$. Then add about 1 mL of the $CaCl_2$ solution to each of two empty vials. Add about 1 mL of the Na_2SO_4 solution to the first vial. Screw on the lid and gently swirl the contents. Record any changes you observe. Add 1 mL of $NaNO_3$ solution to the second vial, screw on the lid, gently swirl the contents, and record any changes you observe. Test the conductivity of each of the two mixtures you made.

Before and after photos of $CaCl_2$ solution mixed with Na_2SO_4 solution with conductivity testers showing the conductivity in the original solutions and in the one with $CaSO_4$ precipitate. Photos to come.

2.26. Consider This *How do you explain the reactions of ions in solution?*

How are your observations of the two mixtures in Investigate This 2.25 the same? How are they different? How do you explain any changes that occurred when you mixed the solutions to make the mixtures? Are the conductivities consistent with your explanation? Explain, why or why not.

Ions in solution in living cells, including Na^+, K^+, Ca^{2+}, Fe^{2+}, Cl^-, HPO_4^{2-}, $H_2PO_4^-$, HCO_3^-, and CO_3^{2-}, are essential in transport, construction, metabolism, and energy transfer. In the previous section, we looked at the energy associated with bringing ions into solution when an ionic compound dissolves, but we did not consider the reverse of that process. Ionic compounds can come out of solution, or **precipitate** when their solubility is exceeded. You observed the formation of a precipitate, a solid substance separating from the liquid, when you mixed $CaCl_2$ and Na_2SO_4 solutions in Investigate This 2.25.

> "Precipitate" and "precipitation" generally have to do with falling. Rain falls from the sky. An insoluble solid falls from solution. In chemistry, "precipitate" is used as both a verb (the action of precipitating) and as a noun (the solid formed by a precipitation reaction).

People who suffer from kidney stones have first-hand experience with precipitation reactions from solutions in their body. Kidney stones, Figure 2.16(a), are primarily calcium phosphate, $Ca_3(PO_4)_2$ and calcium oxalate, CaC_2O_4. Other precipitates, such as the $CaCO_3$ crystals in the balance organ of the inner ear, Figure 2.16(b), and calcium phosphate in our bones and teeth are essential to our well being. And shelled animals like the mollusks in the chapter opening depend on precipitation of $CaCO_3$ to form their shells. Why do some substances, such as calcium phosphate, precipitate, while others, such as sodium phosphate, remain in solution? And how do we establish the identity of a precipitate that forms when ionic solutions are mixed?

(a) kidney stones, $Ca_3(PO_4)_3$ and CaC_2O_4 (b) balance organ crystals, $CaCO_3$

Figure 2.16. Two precipitates that form in humans.
Note the great difference in size of these solids. The kidney stones are 1000 times larger than the inner ear balance organ crystals.

Formation of precipitates

Let's analyze the formation of the precipitate you observed when you mixed solutions of calcium chloride and sodium sulfate in Investigate This 2.25. When the calcium chloride solution was in a container by itself, both the calcium ions and the chloride ions remained in solution. The mixture of sodium ions and sulfate ions and the mixture of sodium ions and nitrate ions in their separate containers also remained in solution. The solid precipitate must have resulted from interactions of the ions in the mixed solution. Table 2.4 is a review of the starting conditions, the ions together in each mixture, and the result for each mixture.

> **WEB Chap 2, Sect 2.6**
> Interactive animation and visualization of a precipitation reaction and ionic equations.

Table 2.4. Results of mixing $CaCl_2$ with Na_2SO_4 and $NaNO_3$.

	Before Mixing			After Mixing	
	$CaCl_2$ solution	Na_2SO_4 solution	$NaNO_3$ solution	$CaCl_2$ and Na_2SO_4	$CaCl_2$ and $NaNO_3$
Positive ion(s)	$Ca^{2+}(aq)$	$Na^+(aq)$	$Na^+(aq)$	$Ca^{2+}(aq) + Na^+(aq)$	$Ca^{2+}(aq) + Na^+(aq)$
Negative ion(s)	$Cl^-(aq)$	$SO_4^{2-}(aq)$	$NO_3^-(aq)$	$Cl^-(aq) + SO_4^{2-}(aq)$	$Cl^-(aq) + NO_3^-(aq)$
Conductivity?	yes	yes	yes	yes	yes
Precipitate?				yes (white solid)	no

Identity of the precipitate

When an ionic solid dissolves, water molecules solvate ions from the crystal lattice and take them into solution, as we have represented in Figure 2.9(b) and the WEB Chap 2, Sect 2.5. Precipitation is the reverse of dissolving. If an ionic compound, such as ordinary table salt, NaCl, dissolves readily, then we know that the ions — $Na^+(aq)$ and $Cl^-(aq)$ in this case — will probably not form a precipitate when they are mixed together. Initially, as shown in Table 2.4, only one kind of cation and one kind of anion is present in each container. When the contents of the calcium chloride and sodium sulfate containers are mixed, all four ions are present together. If a precipitate forms, at least one kind of cation must react with at least one kind of anion.

The four possible ionic combinations are Na_2SO_4, $CaCl_2$, NaCl, and $CaSO_4$. The first two possibilities can be eliminated because they were in solution together before mixing. If the third combination, NaCl, were insoluble, it would have precipitated when solutions of $CaCl_2$ and $NaNO_3$ were mixed, but no precipitate formed. $CaSO_4(s)$, calcium sulfate, is the only remaining possibility, and it is the precipitate formed when solutions of $CaCl_2$ and Na_2SO_4 are mixed. It is widely distributed as a solid in nature, and geologists call it gypsum. The construction industry uses enormous quantities of gypsum as "sheet rock" or "dry wall" in homes and offices. You may have used it as a casting material called "plaster of Paris" and some chalk is $CaSO_4(s)$.

Ionic reaction equations

One way to write the chemical equation that describes the precipitation reaction is:

$$Ca^{2+}(aq) + 2Cl^-(aq) + 2Na^+(aq) + SO_4^{2-}(aq) \rightarrow CaSO_4(s) + 2Na^+(aq) + 2Cl^-(aq) \quad (2.6)$$

The reaction represented symbolically by reaction equation (2.6) is represented at the molecular level in Figure 2.17.

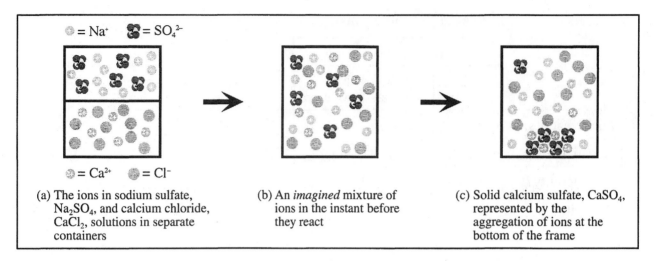

Figure 2.17. Molecular level representation of the precipitation of calcium sulfate.

2.27. Check This *Electrical conductivity in Investigate This 2.25*

Do equation (2.6) and Figure 2.17 explain the electrical conductivities you observed in Investigate This 2.25. Explain why or why not.

An important observation from equation (2.6) and Figure 2.17 is that the sodium ions and the chloride ions play no part in the overall reaction. The ions must be present in the solutions to balance electrical charge, but they appear as independent aqueous ions on both sides of equation (2.6). The ions are shown in Figure 2.17 as independent ions before and after the reaction. Ions that are present, but do not participate in a chemical reaction are called **spectator ions.** They are often omitted when writing chemical equations:

$$Ca^{2+}(aq) + SO_4^{2-}(aq) \rightarrow CaSO_4(s) \qquad (2.7)$$

Equation (2.7) is an example of a **net ionic equation**, which shows only the ions that react and the product of their reaction.

2.28. Worked Example *Product of a precipitation reaction and net ionic reaction*

When clear, colorless solutions of cadmium nitrate, $Cd(NO_3)_2$, and sodium sulfide, Na_2S, are mixed, a beautiful orange precipitate is formed. What is the ionic compound that precipitates? Write the net ionic equation for the precipitation reaction.

Necessary information: We need to recall the solutions we used in Investigate This 2.25.

Strategy: List all the combinations of a cation and an anion (accounting for charge balance) that can be formed in the mixture and eliminate those that are soluble.

Chapter 2 — Aqueous Solutions and Solubility

Implementation: The cations and anions present in the mixture are $Cd^{2+}(aq)$, $Na^+(aq)$, $NO_3^{2-}(aq)$, and $S^{2-}(aq)$. The four possible combinations are $Cd(NO_3)_2$, Na_2S, CdS, and $NaNO_3$. The first two are in the solutions we started with, so they must be soluble. The last $NaNO_3$, sodium nitrate, was one of the solutions we used in Investigate This 2.25, so it must be soluble. The only possibility remaining is CdS, cadmium sulfide, which is the orange solid. The complete ionic equation is:

$$Cd^{2+}(aq) + 2NO_3^{2-}(aq) + 2Na^+(aq) + S^{2-}(aq) \rightarrow CdS(s) + 2Na^+(aq) + 2NO_3^{2-}(aq)$$

The net ionic reaction equation is:

$$Cd^{2+}(aq) + S^{2-}(aq) \rightarrow CdS(s)$$

Does the answer make sense? We eliminated all the other possibilities, so cadmium sulfide must be the solid that is formed. This solid is used in oil paints to make colors that range from cadmium yellow to cadmium orange. Note that the complete ionic equation and the net ionic equation are balanced in both atoms and charge, as they must be.

2.29. Check This *Product of a precipitation reaction and net ionic reaction*

When aqueous solutions of silver nitrate, $AgNO_3$, and sodium chloride, NaCl, are mixed, a white precipitate is formed. What is the ionic compound that precipitates? Write the net ionic equation for the precipitation reaction. The WEB Chap 2, Sect 2.5 provides lab, symbolic, and molecular (interactive animations) level representations of this reaction system.

Equilibrium

Another important observation, shown in Figure 2.17, but not indicated by either equation (2.6) or (2.7), is that tiny amounts of $Ca^{2+}(aq)$ and $SO_4^{2-}(aq)$ remain in solution after the solutions are mixed. A few ions are always left behind when any ionic solid precipitates from aqueous solution. After a time, when an ionic solid is left in contact with an aqueous solution of its ions, a balance is reached between dissolving and precipitation and there is no further *net* change in the amount of solid or of ions in the solution. At the balance point, $Ca^{2+}(aq)$ and $SO_4^{2-}(aq)$ continue to react to form $CaSO_4(s)$, but elsewhere in the mixture $CaSO_4(s)$ is dissolving:

$$CaSO_4(s) \rightarrow Ca^{2+}(aq) + SO_4^{2-}(aq) \tag{2.8}$$

Dissolving, reaction (2.8), and precipitation, reaction (2.7), are still going on but at exactly the same rate, so no net change is observed. When this balance has been achieved, the solution is said to be in **equilibrium** with the solid. (We will continue to use the concept of equilibrium qualitatively, but delay a quantitative discussion of equilibrium until Chapter 9.)

Aqueous Solutions and Solubility Chapter 2

Equilibrium can be represented by an equation like this, for the case of calcium sulfate:

$$Ca^{2+}(aq) + SO_4^{2-}(aq) \rightleftharpoons CaSO_4(s) \qquad (2.9)$$

The two oppositely pointing arrows in equation (2.9) indicate that the reaction is **reversible** — it is, as we said, going in the forward and reverse directions at the same time. A solution that contains its equilibrium amount of ions is called a **saturated solution**. The amount of an ionic compound that will dissolve to give a solution at equilibrium (saturation) is determined by the identity of the compound and the temperature of the solution. The amounts of solutes in solutions are expressed as concentrations. We will return to concentrations in Section 2.8, but first we'll finish what we set out to do, figure out which ionic compounds are soluble and which are not.

Section 2.7. Solubility Rules for Ionic Compounds

2.30. **Investigate This** *Which mixtures of ionic compounds yield precipitates?*

Work together in a small group to do this investigation and interpret its results. You will use a 24-well microtiter plate and aqueous solutions of several ionic compounds. Place the well plate on a white sheet of paper and label the rows and columns as shown in the drawing. Add about 1 mL (equivalent to a well about one-third full) of each one of the specified pair of solutions to each well. Well A1, for example, should have 1 mL of NaCl and 1 mL of KNO_3 added to it. Observe and record which solution pairs produce a precipitate.

	1 KNO_3	2 KBr	3 K_2SO_4	4 K_2CO_3	5 $K_2C_2O_4$	6 K_3PO_4
NaCl (A)	○	○	○	○	○	○
$CaCl_2$ (B)	○	○	○	○	○	○
LiCl (C)	○	○	○	○	○	○
$BaCl_2$ (D)	○	○	○	○	○	○

2.31. **Consider This** *Which ionic compounds are soluble and insoluble?*

(a) Several of the mixtures in Investigate This 2.30 *did not* form precipitates. What do you learn from these results about the solubility of the ionic compounds that can be formed from each mixture? What compounds *do not* form insoluble precipitates?

(b) Several of the mixtures *did* form precipitates. What are the solids formed in these mixtures? How do you know? Write a net ionic equation [like equation (2.7)] to describe the formation of each precipitate.

Although the data you have to work with are limited, you can use your results from Investigate This 2.30 to develop some generalizations about which ionic compounds are soluble and which are not. We'll begin the reasoning with a Worked Example and then you can finish it.

2.32. Worked Example *Solubility trends for alkali metal ions, halides, and nitrate*

What generalizations can we make about the solubility of ionic compounds with alkali metal cations (Li^+, Na^+, and K^+)? What generalizations can we make about the solubility of ionic compounds with halide anions (Cl^- and Br^-) and nitrate anion (NO_3^-)?

Necessary Information: This problem uses only the observations from Investigate This 2.30.

Reasoning: The anion test solutions all had K^+ as the cation, so potassium-ion compounds with these anions are soluble. No precipitates formed in any of the Na^+ cation (A-row) wells, so sodium-ion compounds with these anions are soluble. The behavior of Li^+ cation (C-row wells), another alkali metal, is more complicated. No precipitates are formed by Li^+ with mononegative anions (Cl^-, NO_3^-, and Br^-), but Li^+ does form a precipitate with phosphate.

The cation test solutions all had Cl^- as the anion, so chloride-ion compounds with these cations are soluble, and we already know from above that KCl is soluble. No precipitates are formed with the Br^- anion in the 2-column wells, so bromide-ion compounds of all the cations are soluble. No precipitates are formed with the NO_3^- anion in the 1-column wells, so nitrate-ion compounds of all the cations are soluble.

Conclusions: These tentative generalizations about water solubilities are justified by the data:

• Ionic compounds of the alkali metal cations are soluble, except for some Li^+ compounds with multiply-charged anions.

• Ionic compounds of the halide anions (Cl^- and Br^-) are soluble.

• Ionic compounds of the nitrate anion are soluble.

Determining the solubilities of ionic compounds requires extensive data collection under a range of conditions, but observations like those in Investigate This 2.30 are useful in establishing trends that allow us to make predictions.

Aqueous Solutions and Solubility Chapter 2

> **2.33. Consider This** *What are solubilities for multiply-charged cations and anions?*
>
> **(a)** What generalizations can you make about the solubility of ionic compounds with alkaline earth cations (Ca^{2+} and Ba^{2+})?
>
> **(b)** What generalizations can you make about the solubility of ionic compounds with multiply-charged anions (SO_4^{2-}, CO_3^{2-}, $C_2O_4^{2-}$, and PO_4^{3-})?
>
> **(c)** Can you make any generalizations about the probable solubility of an ionic compound, if both the cation and anion are multiply charged? Explain your reasoning.

Solubility rules for ionic compounds

We can combine much of what we have learned so far into a few simple solubility rules that work for a wide variety of ionic compounds:

- Ionic compounds with an alkali metal cation or a halide or nitrate anion are likely to be soluble.
- Ionic compounds of a monopositive cation and mononegative anion are likely to be soluble.
- Ionic compounds of a multiply-charged cation and a multiply-charged anion are likely to be insoluble.

There are, of course, exceptions to these broad generalizations and they do not address the intermediate case of ionic compounds in which only one of the ions is multiply charged. You can, however, use the rules to make predictions that will be correct in many cases you are likely to meet.

> **2.34. Check This** *Predicting solubilities*
>
> **(a)** Cesium oxalate, $Cs_2C_2O_4$, and strontium hydrogen carbonate, $Sr(HCO_3)_2$, are both soluble in water. Is this what you would predict? Explain why or why not.
>
> **(b)** Consider the possible outcomes when solutions of the two ionic compounds in part (a) are mixed. Would a precipitate be likely from either (or both) new cation–anion pairing? Explain your reasoning. Write a net ionic equation for the reaction(s) you predict to occur. Use a reference handbook to find solubility data to check your predictions.

Although the rules above give us predictive power, they do not provide an underlying molecular basis for understanding solubility. The energetics of the solution process are one factor in determining solubility, but we have seen that the energetics for soluble compounds can be favorable in some cases and unfavorable in others. Compare the data in Table 2.3 for the alkali metal halides, all of which are quite soluble, and you will find both exothermic and endothermic overall energies for dissolving. The data in the table show that the solution process for the alkaline earth sulfides and carbonates is exothermic, that is, energetically favorable, but these compounds are all insoluble.

Reorganization factors in ionic solubility

In Section 2.2, we found that nonpolar molecules and molecules with large nonpolar parts are not soluble in water, even though mixing of the solute with water is favorable process. For these cases. the reorganization of the water molecules required to accommodate the solute is unfavorable and limits the solubility. Similar factors are at work in ionic solutions. The reorganization of the solute from ions fixed in a rigid lattice to ions mixed with water molecules and free to move about in solution is favorable. The problems arise because of the reorganization of the water molecules required to hydrate the ions. The orientation of the water molecules about the ions reduces the water molecules' freedom of movement, which makes this reorganization unfavorable.

Multiply-charged ions attract water molecules more strongly and have a larger orientation effect on the water molecules than do singly-charged ions. Consequently, the unfavorable reorganization that multiply-charged ions impose on the water molecules is large enough to lower the solubility of these ions. The solubility of an ionic compound composed of a cation and anion that are both multiply-charged is doubly unfavored and, as we observe, most of these compounds are insoluble. Ionic compounds composed of singly-charged cations and anions are usually soluble. From this observation, we can infer that the favorable mixing of singly-charged solute ions with water molecules outweighs their unfavorable reorganization of hydration. We will discuss these reorganization factors further in Chapter 8.

2.35. Consider This *What might cause some exceptions to the solubility rules?*

(a) The silver halides, AgCl, AgBr, and AgI, are exceptions to our solubility rules. All these ionic halides are insoluble. Ag^+ and K^+ cations are almost the same size. How would you compare their reorganization effects on water molecules? Based on your comparison, what would you predict about the solubilities of their halides? Explain

Photos of precipitates formed by adding $AgNO_3$ solution to NaCl, NaBr, and NaI solutions.
   Photos to come.

(b) Do the data in Table 2.3 help explain the difference in solubility of the potassium and silver halides? If so, show how. If not, explain why not.

Exceptions to the rules

Ag^+ and K^+ are about the same size, so we would expect that, when dissolved, both ions would have about the same effect on the reorganization of water molecules. Any difference between the solubility of silver and potassium ionic solids is likely to be a result of a difference in their dissolution energies, $\Delta E_{dissolving}$. The data in Table 2.3 show that the K^+ halides (except KI(s)) dissolve endothermically. This unfavorable effect for KCl(s) and KBr(s) is relatively small, less than 20 kJ·mol^{-1}, and the potassium halides are quite soluble. The silver halides also dissolve endothermically, but the unfavorable energy effect is a good deal larger, 66 to 84 kJ·mol^{-1}, and the silver halides are insoluble. This substantial energy requirement apparently greatly limits the solubility of the silver compounds. Despite exceptions like these, the simple rules stated above are useful and will lead to correct predictions in most cases.

Reflection and projection

We accomplished what we set out to do and have a short set of simple rules you can use to predict the solubility of many ionic compounds. Coupled with the general rules for predicting the solubility of polar and nonpolar molecular compounds, these are powerful tools for understanding phenomena that involve substances dissolving in or precipitating from aqueous solutions. Along the way, we have learned about the interactions that govern the energetics of solution processes. We have also learned about another important factor that helps to determine solubilities, the effect of reorganization of solute and solvent molecules when a solution is formed. The reorganization of solute molecules mixing with solvent molecules always favors solution. The reorganization of solvent molecules to orient around solute molecules is an unfavorable factor for solution and favors the separate solute and solvent. The balance among all these factors leads to the observed solubilities and rules.

We use chemical equations as a shorthand way to represent chemical reactions. When forward and reverse reactions are balanced, which we show, by the use of double arrows, $\rightleftharpoons$, we say that a reaction is at equilibrium. A solution in equilibrium with pure solute, known as a saturated solution, contains an unchanging amount of solute. The amount of solute dissolved, expressed as its concentration, is a quantitative measure of solubility. Using concentrations enables us to make meaningful comparisons among solubilities and to study and characterize many other kinds of reactions as well. Next we will turn our attention to determining concentrations.

Chapter 2 — Aqueous Solutions and Solubility

Section 2.8. Concentrations and Moles

2.36. Investigate This *What determines the amount of precipitate formed?*

Mixture 1 immediately after mixing and after sitting for several minutes.

Do this as a class investigation, but work in small groups to discuss and analyze the results. The reagents for this investigation are 5% and 10% by weight aqueous solutions of lead nitrate, $Pb(NO_3)_2$, and 5% and 10% by weight aqueous solutions of sodium iodide, NaI. (A 5% by weight aqueous solution contains 5 g of the solute dissolved in 95 g of water.) Prepare the mixtures shown in this table in three separate test tubes.

	Mixture 1	Mixture 2	Mixture 3
5% $Pb(NO_3)_2$	10 mL	10 mL	
10% $Pb(NO_3)_2$			10 mL
5% NaI	10 mL		10 mL
10% NaI		10 mL	

Record your observations on the appearance of the original solutions, what occurs during mixing, and the final appearance of the mixtures. Note the relative amounts of precipitate in each test tube.

2.37. Consider This *How are concentration and amount of precipitate related?*

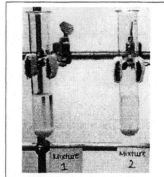
Mixtures 1 and 2 after sitting for several minutes.

(a) What is the precipitate formed in the mixtures in Investigate This 2.35? How do you know? Do our solubility rules predict this precipitate? Why or why not? Write the net ionic equation for the precipitation reaction.

(b) Were the amounts of precipitate in the three mixtures the same or different? If they were different, which one had the most precipitate? Reasoning on the basis of your net ionic equation, is this the mixture you would predict to have the most precipitate? Why or why not?

Aqueous Solutions and Solubility — Chapter 2

Concentration

The **concentration** of a solution is a measure of the amount of solute present in a given quantity, often a volume, of the solution. The higher the concentration, the more solute is present in the same volume of solution, as shown in Figure 2.18. We will use this concept to try to understand the observations in Investigate This 2.36.

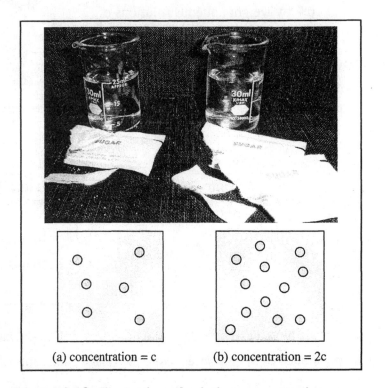

Figure 2.18. Illustration of solution concentration.
When twice as much sugar is dissolved, the molecular level representation shows that there are twice as many solute particles (sugar molecules) in the same volume of solution; the concentration is doubled.

Your results from Investigate This 2.36 may seem puzzling. When you compare the contents of Mixture 1 to Mixture 2, you might expect that adding a higher concentration of NaI solution in Mixture 2, 10% instead of 5%, would produce more precipitate. There is more iodide to react and the net ionic reaction equation requires two iodide anions for each lead cation:

$$Pb^{2+}(aq) + 2I^-(aq) \rightarrow PbI_2(s) \tag{2.10}$$

You observe, however, that the amounts of precipitate formed in the two mixtures are about the same. You also observe that the amount of precipitate formed in Mixture 3, with twice as much added lead cation, is about the same as the amount in Mixture 1.

One explanation for these results is that Mixture 1 contains about the right amounts of the reactants to react completely with each other. Then adding more of either reactant gives no further precipitate, because the other ion has already been used up. If this explanation is correct, a 5% by weight sodium iodide solution contains enough iodide to react with the lead cations in an equal volume of the same concentration, 5% by weight, lead nitrate solution. Solutions with concentrations in weight percent are easy to prepare, but interpreting the results of reactions between substances in these solutions can lead to puzzles like this one.

Moles and molarity

Since chemical equations, like equation (2.10), relate numbers of ions and molecules, we need to express solution concentrations in a way that tells us about the numbers of ions and molecules in a given amount of solution. The appropriate way to express the amount of solute is in numbers of moles. Recall, from Chapter 1, Section 1.11, that the mole is the chemist's counting unit, 6.02×10^{23} particles of anything. You seldom need to use the actual number of particles, but you do need to remember that the mole is for *counting* particles. A common way of expressing the amount of solute in a volume of solution is as the number of moles present per liter of solution. The number of moles of solute per liter of solution is defined as the **molarity** (M) of the solution. A solution containing 1.00 mol of a solute in 1.00 L of solution has a concentration of 1.00 mol·L^{-1}, which is the same as 1.00 M.

One **molar mass** of a compound is the mass of the compound *in grams* that has the same numeric value as the relative mass of the compound. A molar mass of a compound is a mole of the compound and contains 6.02×10^{23} molecules of the compound. To calculate the molar mass of a compound, you *multiply the relative mass of each atom* (from the inside front cover of this book) *by the number of that atom in a molecule of the compound and then sum the products to get the number of grams in a molar mass of the compound*. Ionic compounds do not contain discrete molecules, but the mass of one **formula unit**, the formula we use to specify the atom ratio in the ionic compound, is treated as if it were a molecule. For example, lead iodide, formula unit PbI$_2$, contains one lead ion for every two iodide ions in the ionic lattice. The molar mass of PbI$_2$ is:

> **WEB Chap 1, Sect 1.11**
> Review moles and molecules with interactive questions and visualizations.

$$\text{molar mass PbI}_2 = \left(1 \text{ mol Pb} \times \frac{207.2 \text{ g Pb}}{1 \text{ mol Pb}}\right) + \left(2 \text{ mol I} \times \frac{126.9 \text{ g I}}{1 \text{ mol I}}\right) \quad (2.11)$$

$$\text{molar mass PbI}_2 = \left(1 \times 207.2 \text{ g Pb}\right) + \left(2 \times 126.9 \text{ g I}\right) = 461.0 \text{ g} \quad (2.12)$$

2.38. Check This *The molar mass of lead nitrate*

How many atoms of each element are in a formula unit of lead nitrate, Pb(NO$_3$)$_2$? What is the molar mass of lead nitrate, Pb(NO$_3$)$_2$?

Aqueous Solutions and Solubility   Chapter 2

To understand more about the results from Investigate This 2.36, we need to know the molar concentration of the solutions used. Then we can compare the reacting numbers of moles, which are proportional to the reacting number of ions. We will calculate the reacting numbers of moles in a series of Worked Examples and Check This examples you can use to check your understanding. The steps are: find the number of moles in a mass of a reactant, find the molarity of the stock solution containing this mass, and find the number of moles in each mixture.

2.39. Worked Example   *Mass to mole conversion*

How many moles of lead nitrate, $Pb(NO_3)_2$, are in a 5.0 g sample of the ionic crystals?

Necessary Information: We need the molar mass of $Pb(NO_3)_2$ from Check This 2.38.

Strategy: Use the molar mass to write a unit factor for mass-to-mole conversion.

Implementation:

$$1 \text{ mol } Pb(NO_3)_2 = 331.2 \text{ g } Pb(NO_3)_2 \text{ [from molar mass]}$$

Mass to moles conversion:

$$5.0 \text{ g } Pb(NO_3)_2 = 5.0 \text{ g } Pb(NO_3)_2 \left(\frac{1 \text{ mol } Pb(NO_3)_2}{331.2 \text{ g } Pb(NO_3)_2} \right) = 0.015 \text{ mol } Pb(NO_3)_2$$

Does the Answer Make Sense: 5.0 g is much less than the molar mass, 331.2 g, so the sample must contain much less than one mole of lead nitrate. Our answer is much less than one mole, so it is in the right direction.

2.40. Check This   *Mass to mole conversion*

How many moles of sodium iodide, NaI, are in a 5.0 g sample of the ionic crystals? Note that you first have to find the molar mass of NaI.

2.41. Worked Example   *Molarity of a solution*

What is the molarity of a 5.0% by weight aqueous solution of lead nitrate, $Pb(NO_3)_2$? The solution contains 5.0 g of $Pb(NO_3)_2$ in 95 g of water. 95 g of water is 95 mL; assume that the final solution volume is also 95 mL (0.095 L).

Necessary Information: We need the moles of $Pb(NO_3)_2$ in 5.0 g of the solid, 0.015 mol, from Worked Example 2.39.

Strategy: Divide the number of moles in the solution by the volume of solution to get the $mol \cdot L^{-1} = M$.

Chapter 2 Aqueous Solutions and Solubility

Implementation:

$$\text{molarity of Pb(NO}_3)_2 \text{ solution} = \left(\frac{0.015 \text{ mol Pb(NO}_3)_2}{0.095 \text{ L}}\right) = 0.16 \text{ M}$$

Does the Answer Make Sense: The solution we started with had 0.015 mol of solute in about one-tenth of a liter of solution. In a liter of solution of the same concentration, there would be about ten times as much solute. The calculated number of moles per liter, 0.16, is about ten times as much solute.

2.42. Check This *Molarity of a solution*

What is the molarity of a 5.0% by weight aqueous solution of sodium iodide, NaI? Make the same assumptions we made in Worked Example 2.41.

2.43. Worked Example *Volume to moles conversion*

How many moles of lead nitrate, $Pb(NO_3)_2$, are present in Mixture 1 in Investigate This 2.36?

Necessary Information: We need the molarity of the $Pb(NO_3)_2$ solution, 0.16 M, from Worked Example 2.41 and the volume of this solution, 10 mL, used in the investigation.

Strategy: Use the molarity to write a unit factor for volume-to-mole conversion.

Implementation:

1 L $Pb(NO_3)_2$ solution = 0.16 mol $Pb(NO_3)_2$ [from molarity]

We used 10 mL (0.010 L) of the 0.16 M $Pb(NO_3)_2$ solution in Mixture 1. The volume to moles conversion is:

$$0.010 \text{ L Pb(NO}_3)_2 \text{ solution} = 0.010 \text{ L} \left(\frac{0.16 \text{ mol Pb(NO}_3)_2}{1 \text{ L solution}}\right) = 0.0016 \text{ mol Pb(NO}_3)_2$$

Does the Answer Make Sense: We used $1/100$ of a liter of solution, so we expect the number of moles in the mixture to be $1/100$ as many as in a liter. This is what we found.

2.44. Check This *Volume to moles conversion*

How many moles of sodium iodide, NaI, are present in Mixture 1 in Investigate This 2.36?

The results from Worked Example 2.43 and Check This 2.44, show you that Mixture 1 contains a little more than twice as many moles of NaI as moles of $Pb(NO_3)_2$. Although the mixture contains the same *mass* of each solute, their molar masses are so different that the number of *moles* of each is quite different. Table 2.5 summarizes the results of our previous calculations and gives the results for the other Pb(NO$_3$)$_2$ and NaI solutions and mixtures. We now have the information necessary to understand how the reactions going on in the mixtures produce the observed results. Before going on to this analysis in Section 2.10, let's review and generalize the sequence of calculations we have just done and practice a bit more with mass-mole-volume calculations in Section 2.9.

Table 2.5. Molar composition of the solutions and mixtures from Investigate This 2.36.

	molarity, mol·L^{-1}	moles Pb(NO$_3$)$_2$	moles NaI
5% Pb(NO$_3$)$_2$ solution	0.16		
10% Pb(NO$_3$)$_2$ solution	0.34		
5% NaI solution	0.35		
10% NaI solution	0.74		
Mixture 1		0.0016	0.0035
Mixture 2		0.0016	0.0074
Mixture 3		0.0034	0.0035

Section 2.9. Mass-Mole-Volume Calculations

We can summarize the process we followed in the previous calculations with a "route map," Figure 2.19(a). The blue arrow from mass (grams) to moles represents the calculations in Worked Example 2.39 and Check This 2.40. The unit conversion factor, mol/g, beside the arrow is a reminder that this is the way the molar mass is used to convert mass to moles. The red arrow from volume of solution to moles represents the calculations in Worked Example 2.43 and Check This 2.44. The unit conversion factor, mol/L, beside the arrow is a reminder that this is the way the molarity is used to convert volume to moles.

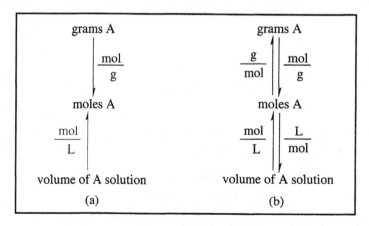

Figure 2.19. Route map for mass-mole-volume calculations.

We can generalize the map to show that these conversions can go the other way as well, moles to mass and moles to volume, Figure 2.19(b). Again, the unit conversion factors beside the arrows remind you how the molar mass and molarity are used to make the conversions. We will often use this route map to illustrate the calculations we make. If you find it a helpful tool to direct your thinking about other problems like these, please use it there also.

2.45. Worked Example   *Moles to mass conversion*

We need 0.045 mol of glucose, $C_6H_{12}O_6$, for an experiment. What mass of glucose do we have to weigh out to get this amount?

Necessary Information: We need relative atomic masses, so we can determine the molar mass of glucose.

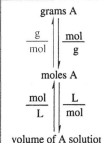

Strategy: Calculate the molar mass of glucose and use it to convert moles to mass. The conversion is illustrated by the red arrow and unit factor on this route map.

Implementation:

Molar mass of glucose:

  C: 6×12.0 g   = 72.0 g
  H: 12×1.008 g = 12.1 g
  O: 6×16.0 g   = 96.0 g

  molar mass   = 180.1 g glucose

Mole to mass conversion:

$$0.045 \text{ mol glucose} = 0.045 \text{ mol glucose} \left(\frac{180.1 \text{ g glucose}}{1 \text{ mol glucose}} \right) = 8.1 \text{ g glucose}$$

Does This Answer Make Sense? We need about $1/20$ of a mole of glucose and 8 g is about $1/20$ of 180 g, so our answer makes sense.

2.46. Check This   *Moles to mass conversion*

What mass of water do you have to weigh out to obtain 12.5 mol of water?

2.47. Worked Example   *Moles to volume conversion*

Suppose the glucose we need in Worked Example 2.45, 0.045 mol, is in an aqueous solution that is 0.10 M in glucose. What volume of the solution do we need to use?

Necessary Information: All the information we need is in the statement of the problem.

Strategy: Use the molarity of the solution to convert moles to volume of solution. The conversion is illustrated in red on this route map.

Implementation:

$$0.045 \text{ mol glucose} = 0.045 \text{ mol glucose}\left(\frac{1 \text{ L solution}}{0.10 \text{ mol glucose}}\right)$$

$$= 0.45 \text{ L solution}$$

Does This Answer Make Sense? We need about $1/20$ of a mole of glucose and a liter of solution contains $1/10$ of a mole, so about half a liter is required, as we found.

2.48. Check This   *Moles to volume conversion*

What volume of a 0.050 M solution of potassium phosphate contains 1.00×10^{-3} mol of potassium phosphate?

Preparing solutions of known molarity

So far, our calculations have taken only single steps on the route map in Figure 2.19(b). You will, however, encounter problems that require multiple steps. A common example that arises many times in chemistry and biology experiments is the need to prepare a volume of aqueous solution that contains a solute at a known molarity. The procedure for making the solution is to dissolve the appropriate mass of the solute in a small volume of water and then add water to obtain the desired volume of solution. You need to calculate the appropriate mass of solute to weigh out.

2.49. Worked Example   *Preparing a solution of known molarity*

How many grams of the amino acid glycine, H_2NCH_2COOH, do you need to weigh out to prepare 250.0 mL (0.2500 L) of aqueous 0.150 M glycine solution?

Necessary Information: We need the relative atomic masses, so we can determine the molar mass of glycine.

Strategy: We know the desired volume of solution and its concentration, so we need to work back through moles to the mass, as shown by the red arrows on this route map.

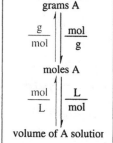

Implementation:

Molar mass of glycine:

C:	2 × 12.01 g	= 24.02 g
H:	5 × 1.008 g	= 5.04 g
O:	2 × 16.00 g	= 32.00 g
N:	1 × 14.01 g	= 14.01 g
	molar mass	= 75.07 g glycine

Volume to moles conversion:

$$0.2500 \text{ L solution} = 0.2500 \text{ L solution}\left(\frac{0.150 \text{ mol glycine}}{1 \text{ L solution}}\right) = 0.0375 \text{ mol glycine}$$

Moles to mass conversion:

$$0.0375 \text{ mol glycine} = 0.0375 \text{ mol glycine}\left(\frac{75.07 \text{ g glycine}}{1 \text{ mol glycine}}\right) = 2.82 \text{ g glycine}$$

Does the Answer Make Sense? The solution contains less than one mole of solute per liter and we only want one-quarter of a liter, so the required mass should be a good deal less than one molar mass, as it is.

2.50. Check This   *Preparing a solution of known molarity*

How many grams of glucose, $C_6H_{12}O_6$, do you need to weigh out to prepare 50.0 mL (0.0500 L) of an aqueous 2.50 M glucose solution?

2.51. Consider This   *Do the solubility rules explain the composition of seawater?*

(a) Rivers on Earth, like the one in the chapter opener, have been flowing to the sea for billions of years. Everything the river waters dissolve eventually ends up in the sea. This table shows the ions that are present in highest concentration in modern seawater. Do our solubility rules help explain this composition? Why or why not?

Ionic composition of seawater

Ion	Molarity
Cl^-	0.545
Na^+	0.468
SO_4^{2-}	0.028
Mg^{2+}	0.054
Ca^{2+}	0.010
K^+	0.010
HCO_3^-	0.002

(b) Show that the number of moles of positive charge and of negative charge per liter of seawater are almost the same, that is, that seawater is an electrically neutral ionic solution.

Aqueous Solutions and Solubility — Chapter 2

Reflection and projection

Chemists, biologists, and other scientists who work with solutions are often faced with the need to make mass-mole-volume interconversions like those we have discussed in these previous two sections. We introduced the mass-mole-volume route map as a guide that you might find helpful in directing your approach to these kinds of problems. If it is helpful, please continue to use it, but if you are more comfortable with another approach, by all means use that. Fundamental to all these conversions is the concept of the mole and, as you see on the route map, moles are central in all the conversions. Applying the mole concept to chemical reactions is our next task.

Section 2.10. Reaction Stoichiometry in Solutions

2.52. Consider This *Are concentration and amount of precipitate related?*

The quantitative calculations in Section 2.8 were concerned with the molarity of the stock solutions of reactants, $Pb(NO_3)_2$ and NaI, and the number of moles of each in Mixture 1 from Investigate This 2.36. Return to Consider This 2.37(b) and answer the questions again. Are your answers different now? Why or why not? Explain how the molar concentrations and numbers of moles in the three mixtures, from Table 2.5, are consistent with the observed results.

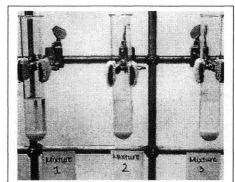

Mixtures 1, 2, and 3 after sitting for several minutes.

All of the reactions and calculations in this section refer to the results from Investigate This 2.36, which are shown in Consider This 2.52. We have said that the net ionic reaction equation is:

$$Pb^{2+}(aq) + 2I^-(aq) \rightarrow PbI_2(s) \tag{2.10}$$

To analyze the mixtures in which this reaction occurs, we need to know how many moles of each ion are present in the mixtures. So far, however, we have only referred to the concentrations and numbers of moles of $Pb(NO_3)_2$ and NaI in the stock solutions and mixtures. We need to account for the formation of ions when these ionic compounds dissolve, and then determine the number of moles (and concentrations) of the reactants, $Pb^{2+}(aq)$ and $I^-(aq)$, in the mixtures.

Chapter 2 — Aqueous Solutions and Solubility

2.53. Worked Example *Moles of ions in an ionic compound solution*

$Pb(NO_3)_2(s)$ dissolves in water to form $Pb^{2+}(aq)$ and $NO_3^-(aq)$ ions:

$$Pb(NO_3)_2(s) \rightarrow Pb^{2+}(aq) + 2NO_3^-(aq) \qquad (2.13)$$

How many moles of $Pb^{2+}(aq)$ are present in Mixture 1?

Necessary Information: We need the balanced reaction equation (2.13) and the number of moles of $Pb(NO_3)_2(s)$ in the mixture, from Table 2.5 (repeated in this marginal box).

| Molar composition of the mixtures |||
Mixture	$Pb(NO_3)_2$	NaI
1	0.0016	0.0035
2	0.0016	0.0074
3	0.0034	0.0035

Strategy: The reaction equation shows us that each mole of $Pb(NO_3)_2(s)$ that goes into solution yields one mole of $Pb^{2+}(aq)$ in solution. Use this relation to write a unit factor.

Implementation:

1 mol $Pb(NO_3)_2(s)$ = ["yields" written as an equality] 1 mol $Pb^{2+}(aq)$ [from reaction ratio]

$$0.0016 \text{ mol } Pb(NO_3)_2(s) \text{ dissolved} = 0.0016 \text{ mol } Pb(NO_3)_2(s) \left(\frac{1 \text{ mol } Pb^{2+}(aq)}{1 \text{ mol } Pb(NO_3)_2(s)} \right)$$

$$= 0.0016 \text{ mol } Pb^{2+}(aq) \text{ in solution}$$

Does the Answer Make Sense? Equation (2.13) indicates that all the ionic compound is completely ionized in solution. The one-to-one relationship in equation (2.13) gives the same number of moles of dissolved lead ions as number of moles of lead nitrate that dissolve.

2.54. Check This *Moles of ions in an ionic compound solution*

Write the balanced chemical reaction equation for dissolving sodium iodide, NaI, an ionic compound, in water. How many moles of $I^-(aq)$ are present in Mixture 1?

Stoichiometric reaction ratios

In equation (2.13), or the equation you wrote in Check This 2.54, there is a simple one-to-one relation between the ion in its compound and the ion in solution, so you might think that writing a unit factor is a waste of effort. However, these relationships can be more complicated and we need to be sure to account correctly for them in reactions. **Stoichiometry** is the quantitative relationship of moles of reactants and products in a chemical equation. To remind you how to account for the stoichiometry of reactions, we will add a step to our route map for doing mass-mole-

> Stoichiometry is from Greek *stoicheon* = element + *metry* = measurement and refers to quantitative relationships among atoms, molecules or moles. We will use it to refer to reactant-product ratios in balanced reactions.

volume calculations, as shown in Figure 2.20. If A represents $Pb(NO_3)_2(s)$ and B represents $Pb^{2+}(aq)$, the red arrow on the map represents the conversion we made in Worked Example 2.53.

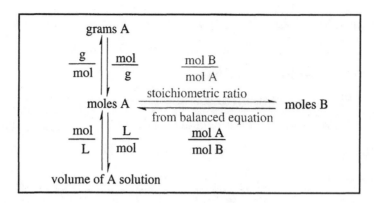

Figure 2.20. Addition of reaction stoichiometry to the mass-mole-volume route map.

For the mixtures in Investigate This 2.36, we can use the stoichiometric ratios from reaction equation (2.10) to calculate how much of each reactant, $Pb^{2+}(aq)$ and $I^-(aq)$, is used up in the reaction and how much product, $PbI_2(s)$, could have been formed. Reaction equation (2.10) shows that one mole of $Pb^{2+}(aq)$ reacts with two moles of $I^-(aq)$ to yield one mole of $PbI_2(s)$. Let's use this stoichiometric ratio of reactants to find out how many moles of $I^-(aq)$ are required to react with all the $Pb^{2+}(aq)$ in Mixture 1.

2.55. Worked Example *Moles of $I^-(aq)$ required for complete reaction with $Pb^{2+}(aq)$*

How many moles of $I^-(aq)$ are required to react with 0.0016 mol $Pb^{2+}(aq)$ (Mixture 1) by reaction equation (2.10)?

Necessary Information: We need the reactant mole ratio from reaction equation (2.10).

Implementation:

1 mol $Pb^{2+}(aq)$ = 2 mol $I^-(aq)$ [from reaction equation (2.10)]

$$0.0016 \text{ mol } Pb^{2+}(aq) = 0.0016 \text{ mol } Pb^{2+}(aq)\left(\frac{2 \text{ mol } I^-(aq)}{1 \text{ mol } Pb^{2+}(aq)}\right) = 0.0032 \text{ mol } I^-(aq)$$

Does the Answer Make Sense? The two-to-one mole ratio is exactly what we expect.

2.56. Check This *Moles of $Pb^{2+}(aq)$ required for complete reaction with $I^-(aq)$*

How many moles of $Pb^{2+}(aq)$ are required to react with 0.0035 mol $I^-(aq)$ (the amount in Table 2.5 that you calculated for Mixture 1 in Check This 2.44) by reaction equation (2.10)?

Limiting reactant

The result from Worked Example 2.55 shows that 0.0032 mol of $I^-(aq)$ are required to react with all the $Pb^{2+}(aq)$ in Mixture 1. The mixture contains 0.0035 mol $I^-(aq)$, enough to react with all the $Pb^{2+}(aq)$. As we would then expect, your result from Check This 2.56 shows that there is not enough $Pb^{2+}(aq)$ to react with all the $I^-(aq)$; about 0.0018 mol of $Pb^{2+}(aq)$ are required and the mixture contains only 0.0016 mol. The amount of precipitate that can be formed by reaction (2.10) is *limited* by the amount of $Pb^{2+}(aq)$ present in the mixture. The reactant that limits the amount of reaction that can occur in a particular reaction mixture is called the **limiting reactant**.

To determine the limiting reactant in a reaction mixture, you do a set of calculations like those in Worked Example 2.55 and Check This 2.56. Then you compare the results with the composition of the mixture to see which reactant is present in excess; the other reactant is the limiting reactant. Sometimes, if you know something about similar mixtures, you can reason out which is the limiting reactant in a mixture without going through all the calculations. For example, Mixture 2 has about twice as much $I^-(aq)$ as Mixture 1. You already know that there is more than enough $I^-(aq)$ in Mixture 1 to react with all the $Pb^{2+}(aq)$ present in the mixture. Adding more $I^-(aq)$ cannot produce more reaction, because

Molar composition of the mixtures		
Mixture	$Pb(NO_3)_2$	NaI
1	0.0016	0.0035
2	0.0016	0.0074
3	0.0034	0.0035

$Pb^{2+}(aq)$ is the limiting reactant. This explains why these two mixtures produce the same amount of precipitate.

2.57. Check This   Limiting reactant

What is the limiting reactant in Mixture 3? Reason out the answer and then do the least amount of calculation necessary to check your reasoning.

Amount of product formed

What mass of precipitate is produced by the reaction in Mixtures 1, 2, and 3? To show how to answer this question, we will complete our mass-mole-volume route map, which we will now call the stoichiometric route map, Figure 2.21.

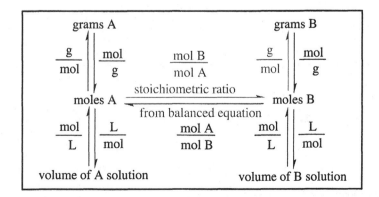

Figure 2.21. Addition of product stoichiometry to the stoichiometric route map.

Aqueous Solutions and Solubility   **Chapter 2**

Note that any conversion you carry out involves moles. There are two parts to the calculation of the mass of product from the known number of moles of $Pb^{2+}(aq)$ that limit the amount of reaction; these are shown by the red arrows in Figure 2.21. First we have to do the mole-mole conversion to determine the number of moles of product, $PbI_2(s)$, that can be formed and then a mole-mass conversion, as we have done in the previous sections, to get the mass of $PbI_2(s)$.

2.58. Worked Example   *Mass of product formed in an ionic precipitation reaction*

What mass of $PbI_2(s)$ is produced in Mixtures 1 and 2 from Investigate This 2.36?

Necessary Information: We need the stoichiometry of the reaction from reaction equation (2.10), the number of moles of limiting reactant in each case, and the molar mass of $PbI_2(s)$.

Strategy: The approach is presented above and illustrated in Figure 2.21.

Implementation:

1 mol $Pb^{2+}(aq)$ = 1 mol $PbI_2(s)$ [from reaction equation (2.10)]

$$0.0016 \text{ mol } Pb^{2+}(aq) = 0.0016 \text{ mol } Pb^{2+}(aq) \left(\frac{1 \text{ mol } PbI_2(aq)}{1 \text{ mol } Pb^{2+}(aq)} \right) = 0.0016 \text{ mol } PbI_2(s)$$

1 mol $PbI_2(s)$ = 461.0 g $PbI_2(s)$ [from molar mass]

$$0.0016 \text{ mol } PbI_2(s) = 0.0016 \text{ mol } PbI_2(s) \left(\frac{461.0 \text{ g } PbI_2(s)}{1 \text{ mol } PbI_2(s)} \right) = 0.74 \text{ g } PbI_2(s)$$

Does the Answer Make Sense? There is somewhat more than $1/1000$ of a mole of $Pb^{2+}(aq)$ to react; we should get somewhat more than $1/1000$ of a mole of $PbI_2(s)$, which is what we find.

2.59. Check This   *Mass of product formed in an ionic precipitation reaction*

(a) In reaction equation (2.10), what is the stoichiometric ratio of $PbI_2(s)$ formed to $I^-(aq)$ that reacts?

(b) What mass of $PbI_2(s)$ is produced in Mixture 3? Use your results from Check This 2.44, Check This 2.57, and the stoichiometric ratio in part (a). How does this mass compare with the amount from Mixtures 1 and 2?

The mass of $PbI_2(s)$ you calculated in Check This 2.59 is only about ten percent greater than the mass we calculated in Worked Example 2.58. Such a small difference is hard to detect by looking at the amounts of solid formed in the three mixtures in Investigate This 2.36. This is why all the mixtures appeared to produce about the same amount of solid and finally explains all the precipitation results from the investigation.

Amount of reactant remaining

There is another question we can ask about the reaction mixtures: What is left in the solutions? The complete balanced ionic reaction equation for the $PbI_2(s)$ precipitation reaction is:

$$Pb^{2+}(aq) + 2NO_3^-(aq) + 2Na^+(aq) + 2I^-(aq) \rightarrow PbI_2(s) + 2Na^+(aq) + 2NO_3^-(aq) \quad (2.14)$$

Reaction equation (2.14) only partially represents the actual state of affairs in our mixtures. We know that the ratio of $Pb^{2+}(aq)$ to $I^-(aq)$ in the mixtures is not exactly one to two. In Mixtures 1 and 2, $Pb^{2+}(aq)$ is the limiting reactant, so some $I^-(aq)$ is left over in the solution, together with the spectator ions, $Na^+(aq)$ and $NO_3^-(aq)$. The left over $I^-(aq)$ is not represented in equation (2.14). A good way to represent such cases is with an illustration like Figure 2.22.

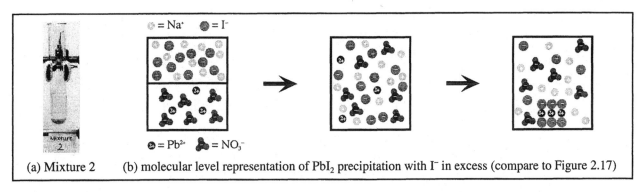

(a) Mixture 2   (b) molecular level representation of PbI_2 precipitation with I^- in excess (compare to Figure 2.17)

Figure 2.22. Representation of the reaction between $Pb(NO_3)_2$ and NaI with NaI in excess.

2.60. Check This   *Illustrations of reactant mixtures*
(a) Show (by counting) that all the solutions and the solid represented in Figure 2.21 are electrically neutral.
(b) Make a sketch, modeled after Figure 2.22, that represents the reaction in Mixture 3.

Figure 2.22 gives you a qualitative picture of what is happening at the molecular level in the mixture and shows that some $I^-(aq)$ is left over in the solution. To find the actual amount that is left over in Mixture 1, we subtract the number of moles that react with $Pb^{2+}(aq)$ (0.0032 mol) from the original number of moles present (0.0035 mol). There are 0.0003 mol of $I^-(aq)$ remaining in the solution.

Molar composition of the mixtures		
Mixture	$Pb(NO_3)_2$	NaI
1	0.0016	0.0035
2	0.0016	0.0074
3	0.0034	0.0035

Aqueous Solutions and Solubility   Chapter 2

2.61. Worked Example   *Concentrations of ions in a reactant mixture*

The volume of Mixture 1 is 20 mL (0.020 L). What is the molar concentration of $I^-(aq)$ in this solution?

Necessary information: We need the number of moles of $I^-(aq)$ left in this solution, 0.0003 mol, from the preceding paragraph.

Strategy: Divide the number of moles by the volume of solution to get molarity.

Implementation:

$$\text{molarity of } I^-(aq) = [I^-(aq)] = \frac{0.0003 \text{ mol}}{0.020 \text{ L}} = 0.02 \text{ M [one significant figure]}$$

Does the answer make sense? The $I^-(aq)$ is in a small volume of solution, so we expect the molarity, mol·L^{-1}, to be larger than the number of moles, as it is.

2.62. Check This   *Concentrations of ions in a reactant mixture*

(a) How many moles of $I^-(aq)$ are left unreacted in Mixture 2? Show how you get your answer.

(b) The volume of Mixture 2 is 20 mL (0.020 L). What are the molar concentrations of $I^-(aq)$, $Na^+(aq)$, and $NO_3^-(aq)$ in this solution? Use the results in Table 2.5 (or the marginal table) to get the moles of $Na^+(aq)$ and $NO_3^-(aq$. What part(s) of the stoichiometric route map could you use for each ion?

All the calculations we have made on the number of moles and the molarities in the reaction mixtures have neglected the reverse reaction:

$$PbI_2(s) \rightarrow Pb^{2+}(aq) + 2I^-(aq) \tag{2.15}$$

In Section 2.6, we said that no ionic precipitation reaction uses up all the ions. A few ions are always left in the solution and the concentrations of ions in solution ultimately reach equilibrium with the ions in the solid. For the kind of precipitation reactions we have considered here, there is only a tiny concentration of the limiting reactant ion left in solution. The assumption that the reactions go to completion is a good one for many, but not all, cases.

Reflection and projection

This section dealt with the application of the mole concept to ionic reactions in solution. As a specific example to guide the discussion, we worked through an analysis of the reaction mixtures and observations made in Investigate This 2.36. In order to do this, we expanded our mass-mole-volume route map to include the stoichiometry of reactions. The centrality of moles to stoichiometry is shown clearly on the expanded map where every conversion involves moles. A balanced chemical equation provides stoichiometric mole ratios among the reactants and products of a reaction which permit you to do a variety of conversions and comparisons. We found, for example, that the amount of precipitate formed in each mixture was controlled by the number of moles of a limiting reactant. And, knowing the amount of each reactant used up, we could calculate the composition of the final solution (assuming complete precipitation of the ionic solid).

So far, we have discussed the aqueous solubility of polar and nonpolar liquids and solids and of ionic solids. What about the solubility of gases? A discussion of gas solubility leads to another fundamental characteristic of aqueous solutions: their acid-base properties. These are the topics of the next sections.

Section 2.11. Solutions of Gases in Water

2.63. **Investigate This**   *What are the properties of aqueous solutions of gases?*

Do this as a class investigation and work in small groups to discuss and analyze the results.

(a) Use a electrical conductivity tester to test these solutions:

• distilled or de-ionized water (dissolved nitrogen, $N_2(g)$, oxygen, $O_2(g)$, and a tiny bit of carbon dioxide, $CO_2(g)$, from the air)

• carbon dioxide, $CO_2(g)$, dissolved in water

• ammonia, $NH_3(g)$, dissolved in water

• hydrogen chloride, $HCl(g)$, dissolved in water

• glucose, $C_6H_{12}O_6(s)$, dissolved in water (provides a reference to previous discussions)

Which solutions conduct an electric current? Which do not?

> Aqueous solutions of $CO_2(g)$, $NH_3(g)$, and $HCl(g)$, are available at the supermarket as seltzer water, household ammonia, and some liquid drain cleaners, respectively.

(b) Use a pH meter or pH paper to measure the pH of each of the solutions. Record the results and save them for analysis in Consider This 2.69, Section 2.12.

2.64. **Consider This**   *Why do some aqueous solutions of gas conduct electrical current?*

(a) What conclusion(s) can you draw about the composition of the solutions of gases that conduct an electrical current in Investigate This 2.63?

(b) Are all the conductivities the same?

(c) Can you use your answer in part (a) to explain your answer in Part (b)? Why or why not?

Aqueous Solutions and Solubility   Chapter 2

Water dissolves gases just as it does solids and liquids. This is fortunate, because dissolved carbon dioxide and oxygen are essential for the plant and animal life in rivers and seas. The solubilities of the gases whose solutions you investigated are given in Table 2.6. Some gases are much more soluble than others, just as we found for the solubilities of other molecular and ionic solutes.

Table 2.6. Aqueous solubility of gases at 25 °C.
Solubilities are in g·kg^{-1} (grams of gas per kilogram of water) at a total pressure of 101 kPa (1 atmosphere).

gas	N_2	O_2	CO_2	NH_3	HCl
solubility	0.018	0.039	1.45	470	695

2.65. Worked Example   *Molarity of saturated aqueous solutions of gases*

Table 2.6 gives the maximum amount of each gas (at atmospheric pressure) that will dissolve in 1 kg (≈ 1.00 L) of water. Assuming that N_2 does not change the volume of the water it dissolves in, what is the molarity of a saturated solution of N_2?

Necessary information: The problem statement and Table 2.6 contain all we need.

Strategy: Convert grams of dissolved gas to moles and divide by the volume to get molarity.

Implementation:

$$0.018 \text{ g } N_2 = (0.018 \text{ g } N_2)\left(\frac{1 \text{ mol } N_2}{28.0 \text{ g } N_2}\right) = 6.4 \times 10^{-4} \text{ mol}$$

$$\text{molarity of } N_2 = \frac{6.4 \times 10^{-1} \text{ mol}}{1.00 \text{ L}} = 6.4 \times 10^{-4} \text{ M}$$

Does the answer make sense? Much less than one mole of nitrogen gas dissolves in a liter of water, so the low concentration makes sense.

2.66. Check This   *Molarity of saturated aqueous solutions of gases*

What are the molarities of saturated solutions of O_2, and CO_2 gases? Use the same approximations as in Worked Example 2.65 and show how you get your answers.

2.67. Worked Example   *Molarity of a saturated aqueous solution of ammonia*

What is the molarity of the ammonia solution in Table 2.6?

Necessary Information: When this much ammonia dissolves in 1 L of water, the volume of the solution will not be 1 L, so we need the density of the solution, 0.89 kg·L^{-1}, as well as the molar mass of ammonia.

Strategy: Use the density to convert the mass of the solution to volume. Use the molar mass to convert the mass of ammonia to moles and then apply the definition of molarity.

Implementation: 0.470 kg of ammonia is dissolved in exactly 1 kg of water, so the mass of the ammonia solution is 1.470 kg. The volume of the solution is:

$$1.470 \text{ kg solution} = 1.470 \text{ kg solution}\left(\frac{1 \text{ L solution}}{0.89 \text{ kg solution}}\right) = 1.65 \text{ L solution}$$

$$1 \text{ mol NH}_3 = 17.0 \text{ g} \quad [\text{from molar mass}]$$

$$470 \text{ g NH}_3 = 470 \text{ g NH}_3 \left(\frac{1 \text{ mol NH}_3}{17.0 \text{ g NH}_3}\right) = 27.6 \text{ mol NH}_3$$

$$\text{NH}_3 \text{ molarity} = \left(\frac{27.6 \text{ mol NH}_3}{1.65 \text{ L}}\right) = 16.8 \text{ M NH}_3$$

Does the Answer Make Sense: More than 27 moles of ammonia dissolve in a liter of water, so this high molarity is expected and makes sense.

2.68. Check This *Molarity of a saturated aqueous solution of hydrogen chloride*

What is the molarity of the hydrogen chloride solution in Table 2.6? The density of the solution is 1.20 kg·L^{-1}.

Molecular models of the gases are shown in Figure 2.23. The molecules of nitrogen and oxygen are nonpolar. There should be little attraction between them and water molecules and, indeed, Table 2.6 and the calculations in Worked Example 2.65 and Check This 2.66 show that these gases are not very soluble. The gas molecules in solution can readily escape into the gas phase when they are near the surface of the solution. The explanation(s) for the larger solubilities of the other three gases must lie in their interactions with water molecules.

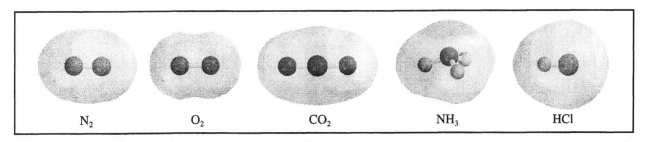

Figure 2.23. Models showing the polarity of five gaseous molecules.

Aqueous Solutions and Solubility — Chapter 2

Water as a reactant

Your results in Investigate This 2.63 provide a clue to the explanation. Solutions of $CO_2(g)$, $NH_3(g)$, and $HCl(g)$ in water conduct an electric current; therefore, ions must be present in these solutions. Solutions of glucose or $O_2(g)$ and $N_2(g)$ do not conduct an electric current; ions are not present in these solutions. Our model for the solubility of glucose involves multiple hydrogen bonds, Consider This 2.8, but no ions. Similarly, no ions are formed in solutions of N_2 and/or O_2. How are ions formed when molecules of CO_2, NH_3, and HCl dissolve in water?

There are differences among the three conducting solutions. The solution of dissolved HCl is a better conductor of electric current than solutions of either CO_2 or NH_3. The higher conductivity must mean that more ions are present in the HCl solution. Aqueous solutions of hydrogen chloride, $HCl(aq)$, are called hydrochloric acid. Acidic aqueous solutions contain substantial concentrations of **hydronium ion**, $H_3O^+(aq)$, water molecules to which a hydrogen cation (H^+) has been added. The addition of H^+ to a molecule is referred to as **protonation** because the added H^+ is simply a proton. In $HCl(aq)$, the HCl molecules are the source of protons:

> WEB, Chap 1, Sect 2.11.1
> View and analyze molecular level visualizations of HCl reacting with H_2O.

$$HCl(aq) + H_2O(l) \rightarrow Cl^-(aq) + H_3O^+(aq) \qquad (2.16)$$

$$:\!\ddot{Cl}\!:\!H\,(aq) + :\!\ddot{O}\!:\!H\,(l) \longrightarrow :\!\ddot{Cl}\!:^-(aq) + H\!:\!\ddot{O}\!:\!H\,^+(aq) \qquad (2.17)$$
$$HH$$

Reaction (2.16) produces equal numbers of positive hydronium ions and negative chloride ions, $Cl^-(aq)$. These ions are responsible for the conductivity of the solution.

In reaction (2.16), water is a reactant in the chemical reaction, rather than just an inert solvent for the reaction. Reactions between water and CO_2, and between water and NH_3, must also produce ions that enable their solutions to conduct electric current. Explaining the behavior of substances such as carbon dioxide or ammonia in water requires the concepts and terminology of acids and bases introduced in the next two sections. We'll also use this knowledge of acids and bases to find out how high-molecular-mass biomolecules can dissolve in water.

2.69. Check This *Compare solutions of NaCl and HCl*

WEB Chap 2, Sect 2.5.2–4 and 2.11.1. Use the movies to compare molecular level representations of solutions of NaCl and HCl, including the way they are formed. Make a list of the similarities and differences and briefly discuss each one.

Chapter 2 — Aqueous Solutions and Solubility

Section 2.12. The Acid–Base Reaction of Water with Itself

2.70. Consider This *Which aqueous solutions of gas are acidic? which basic?*

Refer to your record of pH measurements on the solutions of gases in Investigate This 2.63. At room temperature, aqueous solutions with a pH < 7 are defined as being **acidic**. Those solutions with a pH > 7 are defined as being **basic**. Which of the solutions are acidic? Which basic? Which have a pH of 7? Is there any correlation between the pH and the electrical conductivity of these solutions?

> Aqueous solutions in contact with air contain a small amount of $CO_2(aq)$ dissolved from the air. Does this explain the pH you observe for distilled water? How?

The pH scale

Hydronium ion concentrations in water vary from greater than 1 M in highly acidic solutions to less than 10^{-14} M in highly basic solutions. The pH scale was invented to describe the wide range of hydronium concentrations in water without using exponential numbers. pH is defined mathematically as a logarithmic function of the hydronium ion concentration:

$$pH \equiv -\log_{10}[H_3O^+] \tag{2.18}$$

pH is the negative of the base-10 logarithm of the hydronium ion concentration. Square brackets are used to mean the molar concentration of whatever is inside the brackets.

Another way to think about the pH scale is in terms of the exponent of ten when hydronium ion concentration is expressed as molarity. Figure 2.24 shows graphically how the pH and hydronium ion concentration are related; as $[H_3O^+]$ gets smaller (larger negative exponent) pH increases. Keep in mind that the pH is a logarithmic function. When the pH of a solution decreases from 4 to 2, for example, the concentration of hydronium ion, $[H_3O^+]$, has increased by a factor of 100, from 0.0001 M to 0.01 M.

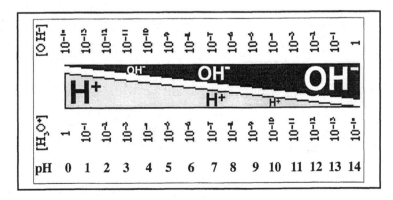

Figure 2.24. The pH scale and corresponding concentrations of $H_3O^+(aq)$ and $HO^-(aq)$.
In addition to the numerical concentration values, the wedges and symbols in the wedges are reminders of the relative concentrations of the hydronium ions and hydroxide ions at low, high, and intermediate pH.

Aqueous Solutions and Solubility — Chapter 2

You observed that pure water does not conduct an electric current, which suggests that there are no ions in pure water. That's almost, but not quite, true. At 25 °C, about two molecules in every billion react to form a hydronium ion, $H_3O^+(aq)$, and a **hydroxide ion**, $OH^-(aq)$:

$$H_2O(l) + H_2O(l) \rightleftharpoons H_3O^+(aq) + OH^-(aq) \tag{2.19}$$

$$\ddot{\underset{H}{\text{Ö}}}{:}H\cdots{:}\ddot{\underset{H}{\text{Ö}}}{:}H \rightleftharpoons {:}\overset{-}{\underset{H}{\ddot{\text{Ö}}}}{:}\cdots H{:}\overset{+}{\underset{H}{\ddot{\text{O}}}}{:}H \tag{2.20}$$

A hydroxide ion is formed when a water molecule transfers an H^+ to another molecule. Figure 2.25 shows charge density models representing reaction (2.20). The proton in the hydrogen bond, the dashed line in reaction (2.20), has to move only about 80 pm to form the two ions; no electrons move between the two molecules of water.

> When an –OH group is covalently bonded to another atom, as in alcohols, it is named *hydroxy-*; the OH^- ion is named *hydroxide*.

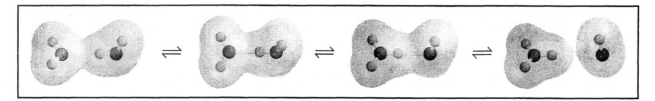

Figure 2.25. Transfer of a proton between two hydrogen-bonded water molecules.
Note the great change in charge distribution as the proton is transferred from one water to another.

The double arrows in Figure 2.25 and reaction equations (2.19) and (2.20) emphasize that the proton transfer is reversible. A hydronium ion can transfer the proton back to the hydroxide ion to form two water molecules. The process of transferring protons among water molecules is constant, and occurs very rapidly, so equilibrium is maintained between the molecules and ions. The pink and blue wedges and the H^+ and OH^- symbols in Figure 2.24 provide a visual indicator of the relative equilibrium concentrations of hydronium and hydroxide ion at different pH values.

> **WEB Chap 2, Sect 2.12.2–3**
> Molecular level visualizations of proton transfers among H_2O molecules.

The water molecules represented in equation (2.20) and modeled in Figure 2.25 are, in turn, hydrogen bonded to many others. This network of hydrogen-bond connections is represented in Figure 2.26. Figure 2.26(a) and Figure 2.26(b) again show the proton transfer between two hydrogen-bonded water molecules. Figure 2.26(c) shows how a "domino effect" of proton transfers among the hydrogen-bonded network can lead to separated $H_3O^+(aq)$ and $OH^-(aq)$ ions. In pure water, there are equal concentrations of each ion. At 25 °C, $[H_3O^+] = [OH^-] = 1 \times 10^{-7}$ M, which is the middle of the pH scale in Figure 2.24.

> Experiments show that the speeds of reactions of $H_3O^+(aq)$ and $OH^-(aq)$ are 10-100 times faster than the ions could move in water. Transfer of protons from one molecule to the next, as shown in Figure 2.25, explains the observed speeds.

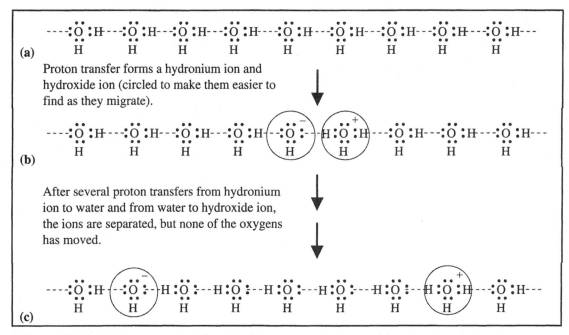

Figure 2.26. Proton transfers in a hydrogen-bonded network of water molecules.
Each line shows the same set of molecules; only the bonding between H and O changes..

Brønsted-Lowry acids and bases. Reactions (2.16) and (2.19) are similar. Hydronium ion, $H_3O^+(aq)$, is formed in each case. The formation of the hydronium ion results from the transfer of a proton (H^+) from dissolved hydrogen chloride gas to water in reaction (2.16) and from one water molecule to another in reaction (2.19). *Any molecule or ion that can transfer or donate a proton to another molecule or ion is called a* **Brønsted-Lowry acid**. *The molecule or ion that accepts the proton is called a* **Brønsted-Lowry base**. An aqueous solution is **acidic**, if a Brønsted-Lowry acid in solution transfers protons to water to give a solution with more $H_3O^+(aq)$ (pH < 7) than is formed by reaction (2.19). An aqueous **basic** solution contains more $OH^-(aq)$ (pH > 7) than is formed by reaction (2.19).

> The definitions for acid and base given here were proposed in 1923 by Johannes Brønsted (Danish chemist, 1879-1947) and independently by Thomas Lowry (English chemist, 1874-1936).

2.71. Consider This *How many moles of hydronium ion are present in water?*

If a compound is dissolved in water and the resulting solution is 10^{-4} M in $H_3O^+(aq)$, is the compound in solution a Brønsted-Lowry acid? Explain why or why not.

The hydrides of Group VI (H_2S, H_2Se, and H_2Te) and Group VII (HF, HCl, HBr. and HI) are Brønsted-Lowry acids that can donate a proton to water, as we have shown in reaction (2.16) for HCl. For example, a saturated solution of H_2S, hydrogen sulfide, in water has a pH of about 4, that is, $[H_3O^+(aq)] \approx 10^{-4}$ M. This hydronium ion concentration is higher than that in pure water,

10^{-7} M, so the solution is acidic. At least some of the hydrogen sulfide molecules have donated one of their protons to water molecules:

$$H_2S(aq) + H_2O(l) \rightleftharpoons H_3O^+(aq) + HS^-(aq) \tag{2.21}$$

You probably know about several other acids besides these hydrides. Indeed, there are many more Brønsted-Lowry acids and in the next section we will introduce a large class of these acids, the oxyacids, that are especially important in biological systems.

2.72. Check This *Reactions of Brønsted-Lowry acids with water*
Write the reaction equations for proton transfer from $H_2Te(aq)$ and $HF(aq)$ to water.

Section 2.13. Acids and Bases in Aqueous Solution

Oxyacids

The great majority of Brønsted-Lowry acids you will meet are **oxyacids**, compounds in which the acidic hydrogen is bonded to oxygen which is, in turn, bonded to a nonmetal atom. Familiar acids like sulfuric (H_2SO_4), nitric (HNO_3), carbonic (H_2CO_3, in carbonated beverages), and acetic ($HC_2H_3O_2$, in vinegar) are oxyacids. The conventional formulas for these acids give you no hint that any of the hydrogens are bonded to oxygen and, indeed, little information about any of the connectivity in the molecules. Alternative structural formulas for oxyacids are shown in these reaction equations for transfer of a proton from sulfuric and carbonic acids to water:

$$(HO)_2SO_2(aq) + H_2O(l) \rightleftharpoons H_3O^+(aq) + HOSO_3^-(aq) \tag{2.22}$$
sulfuric acid   hydrogen sulfate anion

$$(HO)_2CO(aq) + H_2O(l) \rightleftharpoons H_3O^+(aq) + HOCO_2^-(aq) \tag{2.23}$$
carbonic acid   hydrogen carbonate anion

The products of the reactions of oxyacids with water are hydronium ion and an **oxyanion**, the anion formed from an oxyacid by loss of one of its acidic protons. Oxyanions are Brønsted-Lowry bases, that is, proton acceptors. Table 2.7, shows a few common oxyacids and their oxyanions. The table gives both the conventional and structural formulas for each oxyacid as well as the Lewis structure and the names of both the oxyacid and its oxyanion. The atoms of oxyacids and oxyanions are all joined by covalent bonds, including some double bonds.

> Some Lewis structures involving phosphorus, sulfur, and chlorine have more than eight valence electrons around these atoms. Elements of period three and above can accommodate more than eight valence electrons. We will return to discuss these and double bonds in Chapters 5 and 6.

Table 2.7. Some common oxyacids and their oxyanions.

oxyacid name	conventional oxyacid formula	oxyacid Lewis structure	oxyacid structural formula	oxyanion structural formula	oxyanion name
water	H_2O	H-Ö-H	H_2O	HO^-	hydroxide ion
hydronium ion	H_3O^+	[H-Ö(H)-H]$^+$	H_3O^+	H_2O	water
nitric acid	HNO_3	(Lewis structure)	$HONO_2$	NO_3^-	nitrate ion
carbonic acid	H_2CO_3	(Lewis structure)	$(HO)_2CO$	$HOCO_2^-$	hydrogen carbonate ion
hydrogen carbonate ion	HCO_3^-		$HOCO_2^-$	CO_3^{2-}	carbonate ion
acetic acid (ethanoic acid)	$HC_2H_3O_2$	(Lewis structure)	$CH_3C(O)OH$	$CH_3CO_2^-$	acetate ion (ethanoate ion)
perchloric acid	$HClO_4$	(Lewis structure)	$HOClO_3$	ClO_4^-	perchlorate ion
hypochlorous acid	$HClO$	(Lewis structure)	$HOCl$	ClO^-	hypochlorite ion
sulfuric acid	H_2SO_4	(Lewis structure)	$(HO)_2SO_2$	$HOSO_3^-$	hydrogen sulfate ion
hydrogen sulfate ion	HSO_4^-		$HOSO_3^-$	SO_4^{2-}	sulfate ion
phosphoric acid	H_3PO_4	(Lewis structure)	$(HO)_3PO$	$(HO)_2PO_2^-$	dihydrogen phosphate ion
dihydrogen phosphate ion	$H_2PO_4^-$		$(HO)_2PO_2^-$	$HOPO_3^{2-}$	monohydrogen phosphate ion
monohydrogen phosphate ion	HPO_4^{2-}		$HOPO_3^{2-}$	PO_4^{3-}	phosphate ion

2.73. **Check This** *Structures and reactions for Brønsted-Lowry oxyacids and oxyanions*

(a) Each oxyanion in Table 2.7 is formed from its corresponding oxyacid by loss of a proton. Write Lewis structures for the carbonate, sulfate, perchlorate, monohydrogen phosphate and phosphate ions. Use the Lewis structures for the oxyacids and the structural formulas for the oxyacids and oxyanions to aid you.

(b) Write reaction equations for the transfer of a proton to water from nitric acid and from dihydrogen phosphate ion. Write each equation in three forms: with conventional formulas, with structural formulas, and with Lewis structures for all the oxyacids and oxyanions.

Ionic compounds with oxyanions

Oxyanions can form ionic compounds with cations. You investigated several of these in Investigate This 2.30. When one of these compounds is dissolved in water, the acid-base properties of the solution are largely determined by the oxyanion. An important class of these ionic compounds are **hydroxides**, such as NaOH and $Mg(OH)_2$, which contain OH^-, the hydroxide ion. Solutions of these compounds always contain more $OH^-(aq)$ than pure water, so the solutions are basic:

$$Mg(OH)_2(s) \rightleftharpoons Mg^{2+}(aq) + 2OH^-(aq) \qquad (2.24)$$

2.74. Check This   *pH of a saturated magnesium hydroxide solution*

Milk of Magnesia® is a mixture of $Mg(OH)_2(s)$ and water. About 0.01 g of $Mg(OH)_2$ will dissolve in a liter of water. What is $[OH^-(aq)]$ in this solution? What is the approximate pH?

Nucleic acids are oxyanions

The **nucleic acids,** DNA and RNA, are large biomolecules that are ionic in cellular solutions. Each nitrogen base in DNA and RNA is bonded to a sugar molecule, as shown for double-stranded DNA in Figure 2.27. The sugar molecules are, in turn, bonded together by phosphate groups to form the "backbone" of each strand of the nucleic acid, the "uprights" of the ladder we described in Chapter 1, Section 1.9. The phosphate groups in the backbones donate their protons to water, and are ionized to oxyanions. The acidic phosphate groups in the backbone give DNA and RNA their "A" ("acid"). There are two negative phosphate anions for every base pair in the DNA structure; DNA and RNA are highly charged molecules. In the helical structures of the molecules, the sugar phosphate backbones are on the outside of the helix where the anionic sites are in contact with water. These information-carrying molecules can have molar masses in billions of grams, but are still somewhat water soluble because there is a strong attraction between these oxyanions and water molecules. Their water solubility is important because they and the molecules they interact with are all in the aqueous medium of living cells.

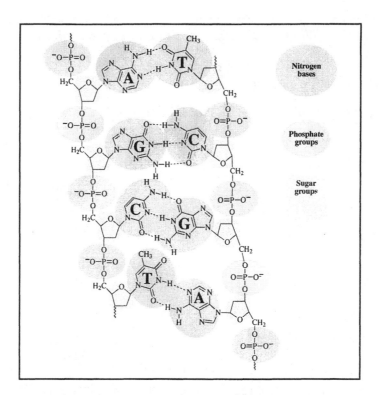

Figure 2.27. Short segment of the ionic DNA molecule.
The phosphate groups are negatively charged and are attracted to water molecules.

Carboxylic acids

Many Brønsted-Lowry oxyacids are **carboxylic acids**, acids that contain the **carboxyl group**, . When a proton leaves a carboxylic acid group, the **carboxylate ion**, , is formed. The reaction for acetic (ethanoic) acid forming acetate (ethanoate) ion is:

$$CH_3C(O)OH(aq) + H_2O(l) \rightleftharpoons CH_3C(O)O^-(aq) + H_3O^+(aq) \quad (2.25)$$

$$\quad (2.26)$$

Most of the acids in biological systems, such as the amino acids (Figure 1.30 in Chapter 1), are carboxylic acids. These acids are sometimes called *organic* acids because the carboxyl group is often bonded to carbon-containing groups that are found in living (organic) systems. In Chapter 6, we will discuss why the hydrogen in a carboxylic acid group (and in other oxyacids as well) is acidic; for now, remember that compounds with a carboxylic acid group are Brønsted-Lowry oxyacids, proton donors.

2.75. Consider This *What are the reactions in an aqueous solution of carbon dioxide?*

(a) When carbon dioxide gas dissolves in water, it can react with water:

$$CO_2(g) + H_2O(l) \rightleftharpoons (HO)_2CO(aq) \tag{2.27}$$

Does reaction (2.27) help explain the acidity of CO_2 solutions and its moderate solubility in water? Explain your responses.

(b) Your results in Check This 2.66 show that about 0.035 mol of $CO_2(g)$ dissolves in a liter of water at room temperature. The dissolved gas could undergo reaction (2.27) followed by reaction (2.23):

$$(HO)_2CO(aq) + H_2O(l) \rightleftharpoons H_3O^+(aq) + HOCO_2^-(aq) \tag{2.23}$$

If reaction (2.23) goes to completion, what would be the concentration of hydronium ion in this solution? What would be the approximate pH of this solution? How does the pH you measured in Investigate This 2.63 compare to this calculated value? What conclusion can you draw from this comparison about the extent of reaction (2.23)?

(c) Does your conclusion in part (b) help explain the observed electrical conductivity of the solution of dissolved carbon dioxide relative to the other solutions in the investigation? Explain why or why not.

Extent of proton-transfer reactions

Reaction equation (2.27) shows that carbon dioxide dissolves in and reacts with water to form an oxyacid, carbonic acid. Several other nonmetal oxides also dissolve in and react with water to yield oxyacids. Sulfur trioxide, $SO_3(g)$, for example reacts to give sulfuric acid:

$$SO_3(g) + H_2O(l) \rightleftharpoons (HO)_2SO_2(aq) \tag{2.28}$$

The stoichiometry of reaction (2.28) shows that, if we dissolve 0.1 mol of $SO_3(g)$ in one liter of water, we get a solution that is 0.1 M in sulfuric acid. If we measure the pH of this solution, we find that it is about 1. Figure 2.24 shows that this corresponds to a hydronium ion concentration, $[H_3O^+(aq)]$, of 10^{-1} M = 0.1 M. Since the hydronium ion concentration is equal to the concentration of sulfuric acid we prepared, we can conclude that the transfer of a proton from sulfuric acid to water, reaction equation (2.22), is complete. That is, in aqueous sulfuric acid solutions, all the sulfuric acid molecules transfer their first proton to water to form hydronium ion and the hydrogen sulfate anion. This is the situation represented at the molecular level in Figure 2.28(a).

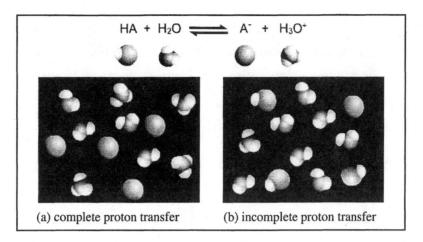

Figure 2.28. Representations of the extent of proton transfer from an acid, HA, to water.

The majority of Brønsted-Lowry acids do not transfer their protons completely to water in aqueous solution. This is the situation represented at the molecular level in Figure 2.28(b). For example, you have found that about 0.035 mol of $CO_2(g)$ dissolves in a liter of water at room temperature. If reactions (2.27) and (2.23) go to completion, the hydronium ion concentration, $[H_3O^+(aq)]$, in this solution would be 0.035 M. That is, $[H_3O^+(aq)]$ would be between 0.1 M (10^{-1} M) and 0.01 M (10^{-2} M), so the pH of the solution would be in the range 1–2. Your pH measurements in Investigate This 2.63 show that a solution of carbon dioxide in water has a pH in the range 4–5. This means that $[H_3O^+(aq)]$ is much lower than it would be if reactions (2.27) and (2.23) went to completion. At equilibrium, only small amounts of $H_3O^+(aq)$ and $HOCO_2^-(aq)$ ions are formed in the solution, which accounts for the low conductivity of the carbon dioxide solution, compared to the solution of $HCl(g)$ in water, hydrochloric acid. Hydrochloric acid, like sulfuric acid, is another acid that transfers its protons completely to water. We will discuss these proton transfers more quantitatively in Chapters 6 and 9.

2.76. **Check This** *Proton transfer from acetic (ethanoic) acid to water*

(a) When the electrical conductivity of 0.1 M aqueous solutions of hydrochloric and acetic acids are tested, the acetic acid solution conducts less well. What can you conclude about the extent of reaction (2.25) compared to reaction (2.16)? Explain your reasoning.

(b) Which solution will have the lower pH, that is, be more acidic? Explain your reasoning.

A Brønsted-Lowry base: ammonia

We have discussed the results of Investigate This 2.63 for all the gases except ammonia, $NH_3(g)$. Ammonia is a small polar molecule that can form hydrogen bonds with four water molecules, so its high solubility is not surprising. The solution conducts electricity, but poorly, so not many ions can be present. The solution has a pH greater than 7, so it is basic. The only source of hydroxide ion in these solutions is water itself, so water molecules must transfer protons to ammonia, which acts as a Brønsted-Lowry base:

$$H_2O(l) + NH_3(aq) \rightleftharpoons OH^-(aq) + NH_4^+(aq) \tag{2.29}$$

(2.30)

Reaction (2.29) produces a Brønsted-Lowry acid, ammonium ion ($NH_4^+(aq)$), and a Brønsted-Lowry base, hydroxide ion, as illustrated in Figure 2.29(a) with space-filling models and in Figure 2.29(b) with charge density models. The formation of ions explains why the solution conducts an electrical current and the formation of hydroxide ion makes the solution basic. The double arrows remind us that the reaction is reversible, hydroxide and ammonium ions can react to form water and ammonia. The rather poor electrical conductivity and only moderately basic solution (pH about 11) are indicators that reaction (2.29) does not go to completion, so produces low concentrations of $OH^-(aq)$ and $NH_4^+(aq)$ ions.

WEB Chap 2, Sect 2.13.4 Study and analyze molecular level animations of NH_3 reacting with H_2O.

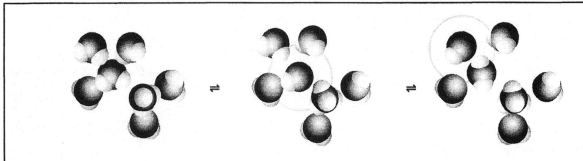

(a) The hydroxide ion is circled to make it easier to find as it is formed and "migrates" by further proton transfer.

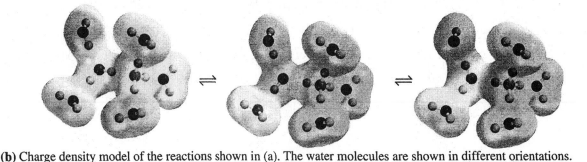

(b) Charge density model of the reactions shown in (a). The water molecules are shown in different orientations.

Figure 2.29. Formation and migration of an ammonium and hydroxide ion by proton transfers.

Chapter 2 — Aqueous Solutions and Solubility

Review of properties of solutions of gases

We have now interpreted all of the results from Investigate This 2.63, based on how molecules can react with water to produce ions in solution:

- Distilled water (with dissolved N_2 and O_2 — and a tiny bit of CO_2) does not conduct an electric current because so few ions are present.

- Aqueous solutions of $HCl(g)$ conduct electricity well and have high acidity (low pH) because reaction (2.16) produces a large number of ions; the reaction goes to completion.

- Aqueous solutions of $CO_2(g)$ conduct electricity poorly and have only moderate acidity (a higher pH than for $HCl(aq)$), because reaction (2.23) produces fewer ions; the reaction does not go to completion.

- Aqueous solutions of $NH_3(g)$ are basic (pH > 7) and conduct electricity poorly, because reaction (2.29) produces ions, $NH_4^+(aq)$ and $OH^-(aq)$ (which makes the solution basic), but does not go to completion, so the concentration of ions is low.

Brønsted-Lowry acid–base pairs

All of the proton-transfer reactions we have described so far involve the reaction of a Brønsted-Lowry acid with a Brønsted-Lowry base to produce a new Brønsted-Lowry acid and Brønsted-Lowry base:

$$HOSO_3^-(aq) + HOCO_2^-(aq) \rightleftharpoons SO_4^{2-}(aq) + (HO)_2CO(aq) \qquad (2.31)$$

$$\text{acid}_1 \qquad \text{base}_2 \qquad \text{base}_1 \qquad \text{acid}_2$$

The hydrogen sulfate ion (identified as acid_1 in equation (2.31)) serves as a Brønsted-Lowry acid, donating a proton to the Brønsted-Lowry base, the hydrogen carbonate ion (identified as base_2). The acid–base reaction produces a new Brønsted-Lowry acid, carbonic acid (called acid_2), and a new Brønsted-Lowry base, sulfate ion (called base_1). In each acid–base pair (pair 1: $HOSO_3^-$ and SO_4^{2-}; and pair 2: $(HO)_2CO$ and $HOCO_2^-$), the acid differs from the base only by having a proton that the base does not have. Acid–base pairs such as this are called **conjugate acids and bases**.

2.77. Check This *Identifying Brønsted–Lowry conjugate acid–base pairs*

Identify the Brønsted–Lowry acid and the Brønsted–Lowry base on each side of the proton transfer reactions (2.16), (2.21), (2.22), (2.23), and (2.25). For each conjugate pair, identify the proton (H^+) that differentiates the acid on one side of the reaction arrows from its conjugate base on the other side.

Another conjugate acid-base example is the proton transfer from one water molecule to another:

$$H_2O(l) + H_2O(l) \rightleftharpoons OH^-(aq) + H_3O^+(aq) \qquad (2.19)$$
$$\text{acid}_1 \quad \text{base}_2 \quad \text{base}_1 \quad \text{acid}_2$$

One water molecule serves as a Brønsted-Lowry acid, donating a proton to the other water molecule, the Brønsted-Lowry base. The acid-base reaction produces a new Brønsted-Lowry acid, H_3O^+, and a new Brønsted-Lowry base, OH^-. Note that water can act as both an acid (proton donor) and a base (proton acceptor).

2.78. Check This *Species that can both donate and accept protons*

(a) In what system besides water itself have we written water as a proton donor? Identify the conjugate acid-base pairs in the reaction.

(b) Find examples of other species in Table 2.7 that can act as both a Brønsted-Lowry acid and a Brønsted-Lowry base.

2.79. Investigate This *What happens when $H_3O^+(aq)$ and $OH^-(aq)$ are mixed?*

Do this as a class investigation and work in small groups to discuss and analyze the results. Add 5.00 mL of 1.00 M hydrochloric acid, HCl(*aq*), and 2-3 drops of phenolphthalein acid-base indicator to 50 mL of magnetically stirred water in a 100-mL beaker. Immerse a pH electrode in the solution. Fill a buret with 1.00 M aqueous sodium hydroxide, NaOH, solution, read and record the liquid volume in the buret, and position the buret over the beaker. Read and record the pH of the solution in the beaker. Open the buret stopcock and allow the NaOH solution to drip slowly into the solution in the beaker. Monitor the pH and when it suddenly changes by 2 or more pH units, stop the flow of base from the buret. Read and record the volume of liquid in the buret and the pH of the solution in the beaker. Continue to drip a further 2 mL of NaOH solution into the beaker and record the pH at the end of the addition. Note any other changes that have occurred during the addition of the base to the acid.

2.80. Consider This *What reactions occur when $H_3O^+(aq)$ and $OH^-(aq)$ are mixed?*

(a) What was the pH range of the solution in Investigate This 2.79 before the sudden pH change? after the pH change? When was the solution acidic? When was it basic?

(b) What reaction do you think is occurring in this solution? Does this reaction explain your observations? How?

Chapter 2 — Aqueous Solutions and Solubility

Stoichiometry of $H_3O^+(aq)$–$OH^-(aq)$ Reactions

The hydronium and hydroxide ions are obviously very important in the acid-base chemistry of water. We have seen that the Brønsted-Lowry acid-base reaction of water molecules with one another, equation (2.19), produces only tiny amounts of hydronium and hydroxide ions in pure water. That is, the equilibrium for this reaction greatly favors the unreacted water molecules. We can reason from this observation that, when hydronium and hydroxide ions are mixed with one another, they will react to use up as many of the ions as possible:

$$H_3O^+(aq) + OH^-(aq) \rightarrow H_2O(l) + H_2O(l) \tag{2.32}$$

We can analyze the data from Investigate This 2.79 by assuming that reaction (2.32) proceeds to completion when solutions containing hydronium ion and hydroxide ion are mixed.

2.81. Worked Example *How many moles of $H_3O^+(aq)$ reacted in Investigate This 2.79?*

How many moles of hydronium ion are present in the HCl solution in Investigate This 2.79?

Necessary information: The data we need, the molarity of acid, 1.00 M, and volume of acid, 5.00 mL, are given in Investigate This 2.79.

Strategy: Since essentially all HCl molecules in aqueous solution donate their protons to water, we assume that the molarity of $H_3O^+(aq)$ in the solution is the same as the molarity of HCl. We use the molarity of the solute to convert volume of solution to moles of solute (hydronium ion).

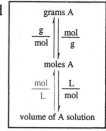

Implementation:

$$0.00500 \text{ L of } 1.00 \text{ M } H_3O^+(aq) = 0.00500 \text{ L}\left(\frac{1.00 \text{ mol}}{1 \text{ L}}\right) = 5.00 \times 10^{-3} \text{ mol } H_3O^+(aq)$$

Does the answer make sense? One liter of the acid solution contains one mole of hydronium ion, so 0.005 L contains 0.005 mol, as we calculated.

2.82. Check This *How many moles of $OH^-(aq)$ reacted in Investigate This 2.79?*

(a) What volume of 1.00 M NaOH solution was required to cause the sudden pH change in Investigate This 2.79? This is the volume of solution that left the buret.

(b) NaOH(s) is an ionic solid that dissolves in water to give $Na^+(aq)$ and $OH^-(aq)$ ions. How many moles of $OH^-(aq)$ ions are present in the volume of solution you determined in part (a)?

(c) How does the number of moles of $OH^-(aq)$ ion added to the acid solution compare to the number of moles of $H_3O^+(aq)$ ion originally in the acid solution (Worked Example 2.81)?

Your calculations in Check This 2.82 show that, when the pH of the solution makes a sudden change, the number of moles of OH⁻(aq) ion added to the acid solution in Investigate This 2.79 is *equal* to the number of moles of H₃O⁺(aq) ion originally present. The **equivalence point** in the addition of one reactant to another occurs when the number of moles of each reactant is stoichiometrically equivalent, that is, when there is just enough of each reactant that both are used up in their stoichiometric reaction. The equivalence point in your reaction between H₃O⁺(aq) and OH⁻(aq) ions was signaled by the sudden change in pH as the base was added to the acid.

2.83. Check This *Limiting reactant in H₃O⁺(aq)–OH⁻(aq) reactions*

(a) Before the equivalence point was reached in Investigate This 2.79, which was the limiting reactant in the reaction between H₃O⁺(aq) and OH⁻(aq)? Does the pH of the solution confirm your choice? Explain why or why not.

(b) Beyond the equivalence point, which was the limiting reactant in the reaction between H₃O⁺(aq) and OH⁻(aq)? Does the pH of the solution confirm your choice? Explain why or why not.

(c) What other change(s), in addition to the sudden change in pH, occurred at the equivalence point? Explain how you might use this(these) change(s) to detect the equivalence point in the reaction between H₃O⁺(aq) and OH⁻(aq) without using a pH electrode and pH meter.

Reflection and projection

The concepts of acids and bases in aqueous solution were introduced to help explain the water solubility of carbon dioxide, ammonia, and hydrogen chloride. Water reacts with itself to give tiny, equal concentrations of hydronium ions, H₃O⁺(aq), and hydroxide ions, OH⁻(aq). And in reverse, mixtures of H₃O⁺(aq) and OH⁻(aq) ions react stoichiometrically to give water. If an aqueous solution has a higher concentration of hydronium ion than pure water, the solution is said to be acidic. Brønsted-Lowry acids donate protons to water to form hydronium ions. Basic solutions have a higher concentration of hydroxide ion than pure water. The extra hydroxide can be added by dissolving ionic compounds like sodium hydroxide, NaOH, or by dissolving a Brønsted-Lowry base that accepts protons from water and produces hydroxide ions. All Brønsted-Lowry acid-base reactions involve two conjugate acid-base pairs. Each pair differs only by the absence of a proton in the base form.

You can use the concepts discussed in this chapter — hydrogen bonding, polar attractions, London dispersion forces, favorable and unfavorable molecular reorientations, ionic solubilities, and acid–base reactions — to understand the water solubilities of many compounds under different solution conditions. We will continue to use these ideas and expand upon them more quantitatively as we proceed through the rest of the book.

Section 2.14. Outcomes Review

In this chapter, we developed a few simple rules that permit you to predict solubilities or relative solubilities of molecules and ionic solids in water. To explain these rules we discussed the factors (energy and molecular reorganization) that control the solubility of solutes in water and found that there is a maximum solubility (saturation) for most solutes when the pure solute and dissolved solute are in equilibrium. We examined two properties of solutions, electrical conductivity and pH, that provide evidence for molecular level representations of aqueous solutions as solutions of hydrated molecules and/or ions. We found that water is a reactant in many solutions, especially in acid-base systems, and that acid-base reactions can affect solubilities. We defined solution concentrations in terms of molarity, $mol \cdot L^{-1}$, and applied mole concepts to solutions by analyzing the stoichiometry of precipitation reactions.

Check your understanding of the ideas in the chapter by reviewing these expected outcomes of your study. You should be able to:

- use an energy diagram to characterize the basic steps in the dissolving process and give a molecular level explanation for the direction of the individual and net energy changes [Sections 2.1, 2.2, and 2.5].

- use molecular models, Lewis structures, and other representations of molecules to show how the three major attractions between like and unlike molecules — hydrogen bonding, polar attractions, and London dispersion forces — affect the solubility of a given molecular solute in water [Section 2.2].

- give a molecular level explanation for the favorable and unfavorable factors that determine the solubility of a given molecular solute or ionic compound [Sections 2.2 and 2.7].

- predict the relative aqueous solubilities of a given set of molecular solutes [Section 2.2].

- show the direction of motion of the molecules and ions in a solution being tested with an electrical conductivity tester [Section 2.3].

- write the chemical formula of any ionic compound, given the charges on the cation and anion and name ionic compounds of common ions, given the chemical formula [Sections 2.4 and 2.13].

Aqueous Solutions and Solubility Chapter 2

- draw an energy diagram for the formation of an ionic crystalline compound from its elemental gas phase atoms and give a molecular level explanation for the direction of the energy changes of the individual steps [Section 2.4].

- make a drawing showing the process of dissolving a polar solute or an ionic compound that shows how water molecules hydrate the dissolved molecules or ions [Sections 2.3 and 2.5].

- draw an energy diagram for the dissolution of an ionic crystalline compound in water and give a molecular level explanation for the direction of the energy changes of the individual steps [Section 2.5].

- use lattice and hydration energies to determine whether a given ionic compound will dissolve exothermically or endothermically in water [Section 2.5].

- predict whether a precipitate will form when two ionic solutions are mixed [Sections 2.6 and 2.7].

- carry out these interconversions: grams ⇔ moles of a compound, moles (grams) of reactant ⇔ moles (grams) of product in a stoichiometric reaction, and volume of a solution of known concentration ⇔ moles (grams) of solute in that volume [Sections 2.8, 2.9, 2.10, and 2.13].

- prepare (give step-by-step instructions for preparing) an aqueous solution of a specified molarity in some solute [Section 2.9].

- determine the limiting reactant in a reaction mixture (solution) and the concentrations of all species formed or remaining in the solution when the reaction is complete [Sections 2.10 and 2.13].

- use conductivity and/or pH data to determine whether a solute undergoes an acid-base reaction with water and, if so, write the equation for the chemical reaction [Sections 2.11, 2.12, and 2.13].

- use the concentration of hydronium ion, hydroxide ion, or the pH to tell whether an aqueous solution is acidic or basic [Sections 2.12 and 2.13].

- write the equation for the reaction between Brønsted-Lowry acids and bases and identify the Brønsted-Lowry conjugate acid-base pairs in any acid-base reaction [Sections 2.12 and 2.13].

- draw a molecular-level diagram and/or explain in words the reactions occurring in a reacting system (dissolution, precipitation, or acid-base) at equilibrium [Sections 2.6, 2.12, and 2.13].

Chapter 2 — Aqueous Solutions and Solubility

Section 2.15. EXTENSION — CO_2 and Le Chatelier's Principle

2.84. Investigate This *What happens when $CO_2(g)$ is bubbled into limewater?*

Do this as a class investigation and work in small groups to discuss and analyze the results. Put 150–200 mL of limewater, a solution of $Ca(OH)_2$ in water, in a 400-mL beaker. Add a few drops bromthymol blue acid–base indicator solution to the beaker. Bubble a stream of $CO_2(g)$ into the solution in the beaker. Observe and record any changes occurring in the beaker. Stop the investigation when no further changes are observed in the solution into which $CO_2(g)$ is being bubbled.

2.85. Consider This *What reactions occur when $CO_2(g)$ is bubbled into limewater?*

(a) Bromthymol blue is a dye that changes color as the pH of the solution changes. The color changes from blue to green to yellow as the solution changes from being basic to being acidic. Did you observe any color changes in Investigate This 2.84? If so, did the solution change from low to high or from high to low pH as $CO_2(g)$ is bubbled in?

(b) Did you observe any precipitate formation in the investigation? If so, what do you think precipitated?

(c) If you observed both color changes and precipitation, what was the correlation, if any, between them? Can you suggest an explanation that fits all your observations?

The carbon cycle

All living things require sources of carbon to build the complex molecules of life. Plants use carbon dioxide from the air or dissolved in water to photosynthesize carbon-containing food molecules. Animals eat the plants to get the carbon in these food molecules. The sources of the carbon dioxide required to build food molecules are the non-living carbon compounds that occur in the crust of the Earth. Figure 2.30 shows that the largest two reservoirs of carbon on the planet are carbon-containing minerals in rocks (such as $CaCO_3$ in limestone and marble) and fossil fuels (the remains of once-living organisms). The next largest carbon reservoir is the dissolved carbon (part of it as dissolved carbon dioxide) in the oceans. Atmospheric carbon dioxide is continuously exchanging with that in the oceans. Figure 2.30 shows how the reservoirs are related to one another and to living systems. The movement of carbon between and among living and non-living sources is called the **carbon cycle**.

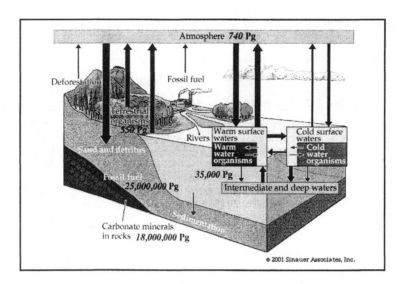

Figure 2.30. Distribution of carbon and its movement among carbon-containing reservoirs. The thickness of the arrows shows the relative amounts of carbon moving between the reservoirs. The numeric values are in petagrams, 1 Pg. = 10^{15} g ≈ 10^9 tons.

Many organisms, such as the mollusks we introduced in the chapter opening, incorporate carbon as carbonates in shells. In turn the shells of creatures that live in the ocean, eventually end up on the ocean bottom. Over time, those carbonates are incorporated into carbonate rock. Large deposits of limestone, which is primarily calcium carbonate, $CaCO_3$, occur almost everywhere on earth. Limestone is quite insoluble in pure water, but it has a larger solubility in water containing dissolved $CO_2(g)$. Let's see why.

Reactions in carbon dioxide-limewater solutions

Analysis of the changes you observed in Investigate This 2.84 will help you understand the conditions for dissolving limestone. The solution of limewater you used contained $Ca^{2+}(aq)$ and $OH^-(aq)$ ions from the dissolved $Ca(OH)_2(s)$. $CO_2(g)$ was bubbled into this solution and some of it dissolved and formed carbonic acid, as we discussed in Section 2.13:

$$CO_2(g) + H_2O(l) \rightleftharpoons (HO)_2CO(aq) \tag{2.27}$$

Carbonic acid reacts with the hydroxide ion from the dissolved calcium hydroxide to form carbonate ions and water in two-steps, both of which are Brønsted-Lowry acid-base reactions:

$$(HO)_2CO(aq) + OH^-(aq) \rightleftharpoons HOCO_2^-(aq) + H_2O(l) \tag{2.33}$$

$$HOCO_2^-(aq) + OH^-(aq) \rightleftharpoons CO_3^{2-}(aq) + H_2O(l) \tag{2.34}$$

Carbonate ions react with calcium ions to precipitate calcium carbonate, as you observed in Investigate This 2.30 and as is predicted by our solubility rules:

$$CO_3^{2-}(aq) + Ca^{2+}(aq) \rightleftharpoons CaCO_3(s) \tag{2.35}$$

Chapter 2 — Aqueous Solutions and Solubility

This series of acid-base and precipitation reactions explains the formation of the white precipitate in Investigate This 2.84.

> **2.86. Check This** *Acid–base changes in carbon dioxide-limewater solutions*
>
> Are your observations on the color changes in Investigate This 2.84 consistent with the composition of the original solution and the chemistry of the series of reactions, (2.27), (2.33), (2.34), and (2.35)? Explain why or why not.

Le Chatelier's principle

As you continued to bubble $CO_2(g)$ through the cloudy solution in Investigate This 2.84, you observed that the solution once again became clear (and a different color from when you began). Carbonic acid continues to be formed by reaction (2.27), but now it has no $OH^-(aq)$ with which to react. To understand what happens to the carbonic acid and the calcium carbonate we can use Le Chatelier's principle.

Henry Louis Le Chatelier (French chemist, 1850-1936) worked on many industrial processes and was interested in ways to assure that reactions would proceed in the directions that produced desired products. Since many such processes involve reversible reactions, he looked for and found patterns in the behavior of equilibrium systems when they were disturbed or changed. **Le Chatelier's principle**, states that *a system at equilibrium responds to a disturbance in a way that minimizes the effect of the disturbance*. The only disturbances that affect equilibria are changes in the *concentrations* of the reactants and/or products and changes in the *temperature* of the system. For example, increasing the concentration of a reactant or product causes the reacting system to respond in such a way as to use up the added molecules.

> Sometimes factors other than concentration and temperature are mentioned as disturbances, but, you will find that they are really concentration or temperature effects.

The system we are considering contains a precipitate of $CaCO_3(s)$ in equilibrium with an aqueous solution containing $Ca^{2+}(aq)$, $CO_3^{2-}(aq)$, $HOCO_2^-(aq)$, and $(HO)_2CO(aq)$. One of the equilibrium reactions in this solution is:

$$(HO)_2CO(aq) + CO_3^{2-}(aq) \rightleftharpoons 2HOCO_2^-(aq) \qquad (2.36)$$

Our continued addition of carbon dioxide to this system forms more carbonic acid, $(HO)_2CO(aq)$, by reaction (2.27), and this is a disturbance to equilibrium reaction (2.36). We have increased the concentration of carbonic acid, $(HO)_2CO(aq)$, one of the reactants so, according to Le Chatelier's principle, the system responds by using up the added reactant until a new equilibrium is established. Carbonic acid is used up by reaction with the carbonate ion, $CO_3^{2-}(aq)$, so some of the carbonate ion is used up as well.

Reducing the concentration of carbonate ion disturbs equilibrium reaction (2.35) by reducing the concentration of one of the reactants. Le Chatelier's principle says that the system will respond by forming more of the missing carbonate ion reactant, which it can do if some of the solid calcium carbonate dissolves. To summarize these changes, we can write the relevant chemical reactions with single arrows instead of equilibrium double arrows, in order to emphasize the directions they take due to these disturbances:

$$CO_2(g) + H_2O(l) \rightarrow (HO)_2CO(aq) \quad (2.27a)$$

$$(HO)_2CO(aq) + CO_3^{2-}(aq) \rightarrow 2HOCO_2^-(aq) \quad (2.36a)$$

$$CO_3^{2-}(aq) + Ca^{2+}(aq) \leftarrow CaCO_3(s) \quad (2.35a)$$

To emphasize that these are the same reactions we wrote before for this system, we have used the same equation numbers with an added "a" to distinguish them from those we wrote as equilibria.

2.87. Check This   *The net reaction for dissolving $CaCO_3(s)$ with added $CO_2(g)$*

Combine reaction equations (2.27a), (2.36a), and (2.35a) appropriately to show that the net equation for dissolving $CaCO_3(s)$ with added $CO_2(g)$ is:

$$CO_2(g) + H_2O(l) + CaCO_3(s) \rightarrow Ca^{2+}(aq) + 2HOCO_2^-(aq) \quad (2.37)$$

Explain the reasoning you used to choose the combination that gives this result.

Water from the surface of the earth contains carbonic acid (dissolved carbon dioxide from the air). As this water finds its way into the rocks beneath the surface, it can dissolve small amounts of limestone by reaction sequences like those we just wrote. Although the process may take millions of years, vast limestone caverns, such as that in Figure 2.31, can be formed. Of more immediate importance, carbonates found in the soil are made soluble. Once these carbonates are converted to hydrogen carbonate ions in solution, the hydrogen carbonate ions can be absorbed by plants and converted to food molecules. As the hydrogen carbonate ion concentration builds up, equilibrium reaction (2.36) is disturbed and this sequence of reactions occurs:

$$(HO)_2CO(aq) + CO_3^{2-}(aq) \leftarrow 2HOCO_2^-(aq) \quad (2.36b)$$

$$CO_2(g) + H_2O(l) \leftarrow (HO)_2CO(aq) \quad (2.27b)$$

The $CO_2(g)$ produced by reaction (2.27b) can also be used by plants and converted to food molecules. And so it goes, round and round the carbon cycle.

Figure 2.31. A limestone cavern with stalactites hanging from the ceiling.

2.88. Check This *Applying Le Chatelier's principle*

Use Le Chatelier's principle to explain how an increasing concentration of hydrogen carbonate in a solution at equilibrium leads to the directions shown for reactions (2.36b) and (2.27b).

2.89. Investigate This *What happens when $Ca^{2+}(aq)$ and $HOCO_2^-(aq)$ react?*

You have two capped tubes, each about one-quarter full of a clear, colorless liquid. One of the tubes contains an aqueous solution of calcium chloride, $CaCl_2$, and the other an aqueous solution of sodium hydrogen carbonate, $NaHCO_3$. Uncap the tubes and pour one of the solutions into the other. Observe and record all evidence that reactions have occurred.

2.90. Check This *Interpreting the reaction of $Ca^{2+}(aq)$ with $HOCO_2^-(aq)$*

(a) What evidence for reactions did you observe in Investigate This 2.89? What kinds of reactions have occurred? Explain how you reach your conclusions.

(b) What reaction products do you think you can identify from your observations? Is the reaction exothermic or endothermic? Explain the reasoning for your choices.

(c) Can you use Le Chatelier's principle and reactions (2.27), (2.35), and (2.36) to explain the formation of the products you identified in part (b)? Show how. Write the reactions in the directions Le Chatelier's principle indicates they will go and combine them to get the net reaction equation for the changes that occurred in Investigate This 2.89.

Recall the question we asked at the opening of the chapter: Why doesn't the calcium carbonate in seashells simply redissolve and return to the sea? Part of the answer is that seawater contains dissolved calcium and carbonate ions, so little *net* dissolution of the calcium carbonate can occur. But another part of the answer is that it *does* redissolve, over millions of years, by processes like reaction (2.37) in limestone caves formed from the shells and in streams flowing to the seas.

Index of Terms

acidic, 2-61, 2-63
alkali metals, 2-21
alkaline earths, 2-21
anion, 2-18
balanced chemical equation, 2-23
basic, 2-61, 2-63
Brønsted–Lowry acid, 2-63
Brønsted–Lowry base, 2-63
carbon cycle, 2-77
carboxyl group, 2-67
carboxylate ion, 2-67
carboxylic acids, 2-67
cation, 2-17
chemical equation, 2-22
concentration, 2-42
conjugate acids and bases, 2-71
Coulomb's law, 2-25
dissolved, 2-6
endothermic, 2-8
entropy, 2-8, 2-12
equilibrium, 2-35
equivalence point, 2-74
exothermic, 2-8
favorable factor, 2-8, 2-39
formula unit, 2-43
halogens, 2-21
heterogeneous, 2-6
homogeneous mixture, 2-6
hydrated ions, 2-19
hydration energy, 2-28
hydration layer, 2-19
hydronium ion, 2-60

hydroxide ion, 2-62
hydroxides, 2-66
ion, 2-17
ion–dipole attractions, 2-19
ionic crystal, 2-18
ionic solution, 2-18
lattice, 2-24
lattice energy, 2-25
Le Chatelier's principle, 2-79
limiting reactant, 2-53
miscible, 2-11
molar mass, 2-43
molarity, 2-43
mole, 2-43
monatomic, 2-20
negative ion, 2-18
net ionic equation, 2-34
nucleic acids, 2-66
oxyacids, 2-64
oxyanion, 2-64
pH, 2-61
polyatomic, 2-20
positive ion, 2-17
precipitate, 2-32
products, 2-22
protonation, 2-60
reactants, 2-22
reorganization, 2-12, 2-39
reversible, 2-36
saturated solution, 2-36
solute, 2-6
solution, 2-6

solvation, 2-8

solvent, 2-6

spectator ions., 2-34

stoichiometry, 2-51

unfavorable factor, 2-8, 2-39

Chapter 2 Problems

Section 2.1. Substances in Solutions

2.1. Tasting samples is unsafe and forbidden in the laboratory. If you had a solution of a sugar, how could you determine if the solute is uniformly distributed throughout the solution without tasting samples? Explain your plan.

2.2. Mixtures involving solids and liquids can be classified into two groups: homogeneous solutions and heterogeneous mixtures in which, after a time, the solid settles out of the mixture. When a solid does not truly dissolve in the liquid but forms a heterogeneous mixture, it is called a suspension. A suspension is usually shaken to redistribute the solid prior to use. Go to your local grocery and/or drug store and identify 5 products that are solutions and 5 products that are suspensions.

2.3. Draw energy diagrams for these processes. Label each energy change as endothermic or exothermic and explain why you draw it this way.
(a) A solvent changing from the liquid state to the gaseous state.
(b) Gaseous solvent molecules solvating a solute to form a liquid solution.

2.4. (a) When a certain liquid molecular substance dissolves in water, the solution feels *cool*. Sketch an energy diagram that shows the relationships among energy theoretically needed to separate the molecules in the liquid state, energy released when the molecules dissolve in water, and the net energy change in this solution process.
(b) When a certain liquid molecular substance dissolves in water, the solution feels *warm*. Sketch an energy diagram that shows the relationships among energy theoretically needed to separate the molecules, the energy released when the molecules dissolve, and the net energy change in this solution process.

2.5. The overall solution process for solution of calcium chloride, $CaCl_2$, in water is exothermic. Draw an energy diagram for this process.

Section 2.2. Solutions of Polar Molecules in Water

2.6. Can you predict whether a substance will be soluble in water by looking at its line formula? Can you predict whether a substance will be soluble in water by looking at its structural formula? Explain your responses.

Aqueous Solutions and Solubility   Chapter 2

2.7. Explain what the expression "like dissolves like" means. Illustrate your explanation with appropriate examples.

2.8. For each of the following compounds, write out the structural formula, using a reference handbook to find the structures you don't know. Circle all the polar bonds found in each structure. Show the direction of the bond polarity for each one. Do not consider the C-H bond to be polar.

   (a) testosterone

   (b) acetylsalicylic acid

   (c) methyl salicylate

2.9. (a) Use Lewis structures to help explain why ethanol, CH_3CH_2OH, is miscible with water.

   (b) How do you predict the solubility of 1-pentanol, $CH_3(CH_2)_4OH$, in water will compare to that of ethanol in water? Explain.

2.10. Explain in terms of intermolecular attractions each of the following "like dissolves like" observations.

   (a) Methanol (CH_3OH) is not miscible with cyclohexane (C_6H_{12}).

   (b) Naphthalene ($C_{10}H_8$) is insoluble in water.

   (c) Naphthalene is soluble in benzene (C_6H_6).

   (d) 1-Propanol ($CH_3CH_2CH_2OH$) is miscible with water.

2.11. Predict the relative solubility of gasoline, C_8H_{18} in water. Explain the reasoning for your prediction.

2.12. (a) Write the Lewis structures for 1-hexanol, $CH_3(CH_2)_5OH$, and 1,6–hexanediol, $HO(CH_2)_6OH$, and, in each molecule, identify the region or regions where hydrogen bonding with water can occur.

   (b) Which compound in (a) do you predict is more soluble in water? Explain.

2.13. Predict the relative water solubilities (from most soluble to least soluble) for each set of structures. Make models to help visualize the structures.

   (a) $CH_3CH_2CH_2CH_2OH$
       $CH_3CH_2CH_2CH_2CH_2OH$
       $HOCH_2CH_2CH_2CH_2OH$
       $HOCH_2CH_2CH_2OH$

   (b) $CH_3CH_2CH_2CH_2NH_2$
       $(CH_3)_2CHCH_2CH_2NH_2$
       $(CH_3)_3CCH_2CH_2NH_2$
       $(CH_3CH_2)_3CCH_2NH_2$

2.14. Sugars are natural products that are generally quite soluble in water. Sucrose, table sugar, is a common source of sugar in food. Fructose and lactose are two other sugars that are important in living systems. Use library resources or the internet to find the structural formulas for these three sugars. On the basis of their structures, explain their high solubility in water. (Your response should include the reference(s) to where you found the structures.)

2.15. The fats in our bodies are composed of relatively nonpolar molecules that are almost insoluble in water. Vitamins are often classified as "fat soluble" or "water soluble." Consider the structures of vitamins A and C shown here. Which of these vitamins do you expect to be more soluble in aqueous systems and which in fatty tissues in the body? Use the structures to explain your reasoning.

2.16. (a) Vitamin B is water soluble and vitamins D, E, and K are fat soluble. (See Problem 2.15.) Based on this information, explain which vitamins could be stored in your body and which should be included in your daily diet.

(b) From your answers for Problem 2.15, would vitamins A and C be stored in your body or should they be included in your daily diet?

(c) With the ready availability of vitamin supplements, cases of hypervitaminosis, an illness caused by an excessive amount of vitamins, are now being diagnosed by medical doctors. Explain for which vitamins hypervitaminosis is likely to occur?

Section 2.3. Characteristics of Solutions of Ionic Compounds in Water

2.17. In Investigate This 2.10, what had to be present in the solution in order for the light bulb to glow?

2.18. WEB Chap 2, Sect 2.3.1 has movies representing the $Fe^{3+}(aq)$ and $NO_3^-(aq)$ ions. How many water molecules are shown in the hydration layer for each ion? What are the similarities and differences in the arrangement and orientation of the water molecules around each of the ions? How do you explain the similarities and differences?

Aqueous Solutions and Solubility — Chapter 2

2.19. Solution A was prepared by mixing 0.5 g of ethanoyl chloride (acetyl chloride), $CH_3C(O)Cl$, with 100 mL of water. Solution B was prepared by mixing 0.5 g of 2-chloroethanol, $ClCH_2CH_2OH$, with 100 mL of water. Solution A conducts an electric current but solution B does not. What can you conclude about the contents of each solution? Explain the reasoning for your answer.

2.20. What ions are present in solution when these solids dissolve? Identify each type of ion as either a cation or an anion.

- (a) $BaCl_2$
- (b) KCl
- (c) Na_3PO_4
- (e) NH_4Cl
- (f) Na_2S
- (g) $MgSO_4$

2.21. (a) Imagine that one of your friends who is not taking this chemistry course says that salt solutions must conduct electrons, just like wires, because you can replace part of an electric circuit (as shown in the pictures in Investigate This 2.10) with a salt solution and the current will still flow. How will you answer your friend and explain how electric charge continues to flow without a flow of electrons through the solution?

(b) WEB Chap 2, Sect 2.3.2. Would this interactive molecular level representation of electrical conductivity in ionic solution help your explanation? Why or why not?

2.22. Solid sodium chloride, NaCl, does not conduct electricity but an aqueous solution of sodium chloride is a good conductor. Solid mercuric chloride, $HgCl_2$, does not conduct electricity and neither will its aqueous solution even though the solid is soluble. Offer a possible explanation for the difference in behavior of the aqueous solutions.

2.23. Imagine that you are a positively charged ion surrounded by large numbers of polar molecules like our simple ellipsoids with positive and negative ends Chapter 1 Figures 1.15 and 1.16. Why would you have a problem feeling the attraction of a negatively charged ion or the repulsion of another positively charged ion? Explain your reasoning. Use drawings, if they help clarify your explanation.

Section 2.4. Formation of Ionic Compounds

2.24. Which of the following do you predict to conduct electricity when dissolved in water? Explain your reasoning in each case.

- (a) $MgBr_2$
- (b) CH_3OH
- (c) $NaOH$
- (d) CH_3OCH_3
- (e) KNO_3
- (h) $CH_3CH_2CH_2CH_3$

2.25. Predict the most likely charge when the following elements form monatomic ions. Explain the rationale for your choice in each case.

    (a) alkali metals   **(c)** alkaline earth metals

    (b) oxygen family   **(d)** halogens

2.26. Write the chemical formula for the ionic compound formed by the combination of the following ions.

    (a) magnesium cation + bromide anion

    (b) calcium cation + nitrate anion

    (c) magnesium cation + sulfate anion

    (d) potassium cation + oxide anion

2.27. Name these ionic compounds.

    (a) Na_2SO_4 **(c)** $(NH_4)_2CO_3$

    (b) $MgCl_2$ **(d)** Al_2S_3

2.28. Write the formulas for these ionic compounds.

    (a) barium nitrate   **(c)** calcium oxide

    (b) ammonium phosphate   **(d)** potassium sulfate

2.29. Name these ionic compounds.

    (a) MgS   **(c)** NH_4NO_3

    (b) Na_3PO_4 **(d)** LiOH

2.30. Write the formulas for these ionic compounds.

    (a) calcium iodide   **(c)** potassium carbonate

    (b) sodium fluoride   **(d)** barium hydroxide

2.31. Complete this grid, giving both the formula and the name of the compound formed between each pair of ions.

	CO_3^{2-}	PO_4^{3-}	F^-
Mg^{2+}			
NH_4^+			
Al^{3+}			
Na^+			

Aqueous Solutions and Solubility Chapter 2

2.32. Find from suitable references (or labels on containers) the chemical formulas and write chemical names for the following ionic substances.
  (a) Milk of Magnesia®
  (b) Epsom salt
  (c) Plaster of Paris
  (d) Caustic soda
  (e) Soda ash

2.33. Equation (2.1) can be broken down into two steps: (1) loss of an electron by a sodium atom, Na → Na$^+$ + e^-, and (2) gain of an electron by a chlorine atom, Cl + e^- → Cl$^-$. Write the appropriate reaction equations for formation of the common cations or anions of these elements.
  (a) potassium
  (b) calcium
  (c) sulfur
  (d) sulfur
  (e) bromine

2.34. In terms of electrical attraction [Coulomb's Law, equation (2.2)], explain why the ionization energy for *all* elemental atoms *always* has a positive value. For example, the energy required for the reaction, Na → Na$^+$ + e^-, is the ionization energy, $\Delta E_{ionization}$ = 496 kJ·mol^{-1}, for sodium atoms.

2.35. Examine the lattice energies in Table 2.3. Are these data consistent with Coulomb's law? Explain the reasoning for your answer.

2.36. (a) Based on Coulomb's Law, in which crystal, KBr or CaBr$_2$, would the greatest forces of attraction and repulsion be observed? Explain your reasoning. Assume that the distance separating the charges is the same for both crystals.
  (b) Do the data in Table 2.3 support your answer in part (a)? If so, explain how. If not, explain why not.

2.37. The lattice energy for calcium chloride, CaBr$_2$, crystals is 2176 kJ·mol^{-1}. The gaseous reaction forming the ions from the atoms,
$$Ca(g) + 2Br(g) \rightarrow Ca^{2+}(g) + 2Br^-(g),$$
requires 966 kJ·mol^{-1}. Draw an energy diagram, analogous to Figure 2.14, and use it to find $\Delta E_{xtal\ form}$ for the formation of ionic crystals of CaBr$_2$ from the gaseous atoms.

2.38. Consider these lattice energies for some ionic solids. All values are in kJ·mol⁻¹.

	F⁻	Cl⁻	Br⁻
Li⁺	1046	861	818
Na⁺	929	787	751
K⁺	826	717	689

(a) Use these data to discuss how the lattice energy changes with the size of the anion, keeping the size of the cation constant. *Hint:* The size of the ions in a group (column) of the periodic table increases as one goes down the group.

(b) Use these data to discuss how the lattice energy changes with the size of the cation, keeping the size of the anion constant.

(c) What generalization can be drawn about the size of ions and the lattice energies of their salts?

(d) Use your generalization from part (c) to predict how the lattice energy of CsI compares with that of NaCl.

2.39. Consider this energy diagram for the formation of one mole of ionic crystals of $MgCl_2$.

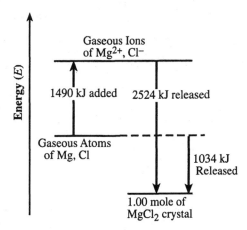

(a) What is the lattice energy for this compound? Explain how you get your answer.

(b) How much energy is required for the reaction: $Mg(g) + 2Cl(g) \rightarrow Mg^{2+}(g) + 2Cl^-(g)$? Explain how you get your answer.

(c) What is the energy associated with the formation of ionic crystals from the gaseous atoms? Explain how you get your answer.

(d) How do the three energies associated with the formation for $MgCl_2$ compare with those for NaCl, given in Figure 2.14?

2.40. (a) The energy required to remove electrons from gaseous silver atoms to form gaseous silver cations is 731 kJ·mol^{-1}. When gaseous iodine atoms gain electrons to form gaseous iodide anions, 296 kJ·mol^{-1} of energy is released. Calculate the energy for this reaction:

$$Ag(g) + I(g) \rightarrow Ag^+(g) + I^-(g)$$

(b) The lattice energy for AgI(s) crystals is 887 kJ·mol^{-1}. Draw an energy diagram analogous to Figure 2.14 and use it to find $\Delta E_{xtal\ form}$ for the formation of ionic crystals of AgI from the gaseous atoms.

2.41. The lattice energy for potassium bromide, KBr, crystals is 689 kJ·mol^{-1}. The formation of separate gaseous atoms of potassium, K(g), and bromine, Br(g), from the ionic crystal would require 594 kJ·mol^{-1}. How much energy is required for the reaction of the gaseous potassium and bromine atoms to form ions in the gas phase?

$$K(g) + Br(g) \rightarrow K^+(g) + Br^-(g)$$

Explain how you get your answer and draw an energy diagram analogous to Figure 2.14 illustrating your answer.

2.42. What is the lattice energy for magnesium fluoride, MgF$_2$? The formation of ionic crystals of MgF$_2$ from the gaseous atoms releases 1424 kJ·mol^{-1}. The reaction of the atoms to form ions in the gas phase, Mg(g) + 2F(g) → Mg$^{2+}(g)$ + 2F$^-(g)$, requires 1533 kJ·mol^{-1}. Explain how you get your answer and draw an energy diagram analogous to Figure 2.14 illustrating your answer.

Section 2.5. Energy Changes When Ionic Compounds Dissolve

2.43. When ammonium acetate, NH$_4$C$_2$H$_3$O$_2$ [= (NH$_4^+$)(C$_2$H$_3$O$_2^-$)], is dissolved in water, the mixture becomes quite cold. (Ammonium acetate is the salt used in some cold packs.)
(a) Is the dissolving of ammonium acetate endothermic or exothermic? Explain.
(b) What are the relative magnitudes of the crystal lattice energy and hydration energy for ammonium acetate? Use an energy diagram to explain the reasoning for your answer.
(c) Write formulas for the ions in solution using standard chemical notation.
(d) Sketch the hydrated ions in the way we have tried to show molecular level interactions in Figure 2.9.

2.44. (a) When LiCl dissolves in water, is the process exothermic or endothermic? Use an energy diagram to explain the reasoning for your answer.
(b) When KBr dissolves in water, is the process exothermic or endothermic? Use an energy diagram to explain the reasoning for your answer.

2.45. Lithium sulfate, Li_2SO_4, is quite soluble in water (261 g·L^{-1}) while calcium sulfate, $CaSO_4$, is essentially insoluble (4.9 mg·L^{-1}). What ion-ion and/or ion-dipole interactions are responsible for this difference? Be as specific as you can.

2.46. There are many ways to describe or represent what happens when sodium sulfate, Na_2SO_4, dissolves in water. First give an explanation in words. Next, write an ionic equation to represent the solution process. Then use a molecular level representation to illustrate what happens when sodium sulfate dissolves in water. (See Figures 2.9, 2.17, and 2.22.)

Section 2.6. Precipitation Reactions of Ions in Solutions

NOTE: Whenever you write a chemical reaction equation, remember to include the appropriate state notation, *(s)*, *(l)*, *(g)*, or *(aq)*, for each species in your equation.

2.47. When aqueous solutions of potassium phosphate, K_3PO_4, and calcium bromide, $CaBr_2$, are mixed, a white precipitate is formed. When tested for electrical conductivity, both starting solutions test positive. Following the mixing and precipitation, the product solution also tests positive for electrical conductivity.

(a) Prepare a table modeled after Table 2.4 to summarize what has happened.

(b) What new combinations of cations and anions are possible following mixing? One of these new combinations is the precipitate and the other is soluble in water. Which is which? Explain your reasoning.

(c) Draw a molecular level representation similar to Figure 2.17 to illustrate what happens when the two solutions are mixed.

(d) Write a complete ionic equation describing the reaction that occurs when these solutions are mixed.

(e) Write a net ionic equation describing the reaction that occurs when these solutions are mixed.

2.48. (a) When aqueous solutions of potassium chloride, KCl, and sodium bromide, NaBr, are mixed, no precipitate is formed. What can you conclude about the water solubility of NaCl*(s)* and KBr*(s)*? Explain your reasoning.

(b) Draw a molecular level representation similar to Figure 2.17 to illustrate what happens when the two solutions are mixed.

(b) Write a complete ionic equation describing what occurs when these solutions are mixed. What would you write for the net ionic equation? Explain.

Aqueous Solutions and Solubility — Chapter 2

2.49. $Ba^{2+}(aq)$ is *extremely* toxic to humans. However, when physicians need to x-ray the gastrointestinal (GI) tract – stomach and intestines, they fill the patient's GI tract with barium sulfate and water. How can it be that the patient is not harmed by this procedure?

2.50. Differences in solubility can be used to help separate cations from solutions where they are mixed together. The process is called selective precipitation. Consider this table of solubilities, and then suggest a sequence of precipitation reactions to separate Ag^+, Ba^{2+}, and Fe^{3+} from solution. Explain your approach and write a net ionic equation for each reaction that takes place.

	Cation		
Test Solution	$Ag^+(aq)$	$Ba^{2+}(aq)$	$Fe^{3+}(aq)$
NaCl	ppt	no ppt	no ppt
NaOH	ppt	no ppt	ppt
Na_2SO_4	no ppt	ppt	no ppt

2.51. Aluminum nitrate, $Al(NO_3)_3$, is soluble in water. So is sodium oxalate, $Na_2C_2O_4$. When an aluminum nitrate solution is mixed with a sodium oxalate solution, a precipitate forms.
(a) What is the precipitate? State the reasoning for your prediction.
(b) Write a net ionic equation for the reaction that occurs.

2.52. A solution of lithium nitrate is mixed with a solution of sodium phosphate. A white precipitate is observed to form.
(a) What is the white precipitate? State the reasoning for your prediction.
(b) Write a net ionic equation for the reaction that occurs.

2.53. What does it mean if there is a forward arrow over a backward arrow, "⇌," in an equation?

Section 2.7. Solubility Rules for Ionic Compounds

2.54. A solution of cadmium chloride, $CdCl_2$, is mixed with a solution of ammonium sulfide, $(NH_4)_2S$. A yellow-orange precipitate is observed to form.
(a) What is the orange-yellow precipitate? State the reasoning for your prediction.
(b) Write a net ionic equation for the reaction that occurs.

2.55. Predict the products of each of the following reactions between aqueous solutions. If no visible change will occur, write NO APPARENT REACTION to the right of the arrow. Give the reasoning for your prediction in each case. Write the balanced complete ionic reaction equation and the net ionic reaction equation for each case where reaction occurs.

(a) barium chloride*(aq)* + sodium sulfate*(aq)* →

(b) silver nitrate*(aq)* + magnesium chloride*(aq)* →

(c) strontium nitrate*(aq)* + potassium nitrate*(aq)* →

(d) ammonium phosphate*(aq)* + calcium bromide*(aq)* →

2.56. Write balanced net ionic equations for reactions that would be suitable for laboratory preparation of the following solid ionic compounds. Suggest compounds whose aqueous solutions you could use to carry out these preparations.

(a) $BaSO_4$ (c) $Ca_3(PO_4)_2$

(b) $AgCl$ (d) CaC_2O_4

Section 2.8. Concentrations and Moles

2.57. You have prepared 1 L of a 0.1 M solution of NaOH. Next, you accidentally spilled about 200 mL of this solution. What has happened to the concentration of the remaining solution?

2.58. You have been asked to assist with a chemical inventory of a General Chemistry stockroom and have found a 0.5-L bottle about half full of a solution labeled, "0.5 M $CaCl_2$."

(a) What does this label tell you about the solution?

(b) Can you tell about how many moles of $CaCl_2$ are in the bottle? If so, show how. If not, tell what further information you need to answer the question.

(c) Can you tell about how many grams of $CaCl_2$ are in the bottle? If so, show how. If not, tell what further information you need to answer the question.

2.59. Refer to the molecular structure of vitamin C, shown in Problem 2.15.

(a) What is the molecular formula of vitamin C?

(b) What is a molar mass of vitamin C? Explain your work.

(c) How many moles of vitamin C are present in a 500-mg tablet of the vitamin? Explain your reasoning.

(d) How many molecules of vitamin C are present in each 500-mg tablet? Explain your reasoning.

Aqueous Solutions and Solubility   Chapter 2

2.60. Calculate the mass (in grams) of the following. Show your reasoning clearly.

(a) 2.5 mole of the artificial sweetener aspartame, $C_{14}H_{18}N_2O_5$

(b) 0.040 mole of aspirin, $C_9H_8O_4$

(c) 2.5 x10^{23} molecules of cholesterol, $C_{27}H_{46}O$

(d) 1.2 x10^{22} molecules of caffeine, $C_8H_{10}N_4O_2$

2.61. How many atoms of carbon are there in 5 mg of niacin? Show your reasoning clearly.

niacin

2.62. Bacteria generally contain a single molecule of DNA that encodes all their genetic information. What is the concentration, mol·L^{-1}, of DNA in a spherical bacterium that has a diameter of 10^{-6} m = 1 μm? Clearly explain how you arrive at your answer.

2.63. Blood serum is typically about 0.14 M in NaCl. Calculate the number of sodium ions in 50 mL of blood serum. Show your reasoning clearly.

Section 2.9. Mass-Mole-Volume Calculations

2.64. Calculate the number of grams of solute present in each of the following solutions. Show your reasoning clearly.

(a) 350 mL of 0.105 M $K_2Cr_2O_7$

(b) 50 mL of 1.0 M $FeCl_3·6H_2O$

(c) 0.3 L of 1.70 M KCl

2.65. Calculate the molar concentration of solute present in each of the following solutions. Show your reasoning clearly.

(a) 120 mL containing 4.5 grams of NaCl

(b) 0.25 L containing 1.3 g of NH_4Cl

(c) 1.3 L containing 1.85 g of $AgNO_3$

Chapter 2   Aqueous Solutions and Solubility

2.66. (a) 5.405 g glucose, $C_6H_{12}O_6$, is dissolved in enough water to make 1.000 L of solution at 20 °C. What is the molarity of glucose in this solution? Show your reasoning clearly.
(b) How many mL of the solution prepared in part (a) will you need in order to obtain 0.950 millimoles of glucose at 20 °C? Show your reasoning and work clearly and completely.

2.67. Two students were asked to prepare a 1.00 M solution of $CuSO_4$. One student found a bottle, labeled "$CuSO_4 \cdot 5H_2O$". He weighed 159.60 g of this hydrated copper sulfate, transferred it to a 1-L volumetric flask and dissolved it in a small quantity of water. Then, he added more water until the solution just reached the calibration mark etched on the neck of the flask and thoroughly mixed the contents of the flask. The second student followed exactly the same procedure, but she used the anhydrous salt of copper sulfate, $CuSO_4$. Which student prepared the solution with the correct concentration? Calculate the molar concentration of $CuSO_4$ in each solution.

2.68. You need about 170 mL of 0.10 M NaOH for an experiment. The concentration of this solution has to be fairly exact. Describe how to prepare the solution.

2.69. Normal saline, a solution given by intravenous injection, is a 0.90% (mass to volume %) sodium chloride solution. How many grams of sodium chloride are required to make 250. mL of normal saline solution? What is the molarity of this solution? The density of this solution is the same as water, $1.00 \text{ g} \cdot \text{mL}^{-1}$. Show your reasoning clearly.

2.70. When urine is analyzed, the normal range for urea, $(NH_2)_2CO$, one of the solutes in urine, is 13-40 $\text{g} \cdot (24 \text{ hr})^{-1}$. (For urinalysis, a patient's urine is collected over a 24-hour period to be sure that the sample is representative.) A patient's laboratory tests show a urea content of 25 $\text{g} \cdot (24 \text{ hr})^{-1}$. Suppose the normal output of urine for a patient of this age is 2.5 $\text{L} \cdot (24 \text{ hr})^{-1}$. What is the molarity of the urea in the patient's urine? Explain your reasoning.

2.71. One of the ionic compounds in sports drinks is potassium dihydrogen phosphate, KH_2PO_4. The label on one of these drinks tells us that 240 mL of the solution contains 30 mg of potassium. KH_2PO_4 is the only ingredient in the solution that can provide this potassium. How many grams of KH_2PO_4 are dissolved in 240 mL of the solution? What is the molarity of the KH_2PO_4, in this solution? Clearly show and explain, all the work you do to solve this problem.

Aqueous Solutions and Solubility   Chapter 2

Section 2.10. Reaction Stoichiometry in Solutions

2.72. How many moles each of carbon, hydrogen, and oxygen atoms are present in two moles of ammonium acetate, $NH_4C_2H_3O_2$? What is the total number of moles of atoms in two moles of the compound? What is the total number of moles of *ions* in two moles of the compound?

2.73. A student is trying to prepare artificial kidney stones in the laboratory. How many grams of calcium phosphate can he make by mixing 125. mL of 0.100 M calcium chloride with 125. mL of 0.100 M sodium phosphate? Explain your reasoning and any assumptions you make in solving this problem.

2.74. What volume of 0.100 M SO_3^{2-}(aq) is needed to react exactly and completely with 24.0 mL of 0.200 M Fe^{3+}(aq)? The equation that represents the reaction that occurs is:
$$2Fe^{3+}(aq) + SO_3^{2-}(aq) + 3H_2O(l) \rightarrow 2Fe^{2+}(aq) + SO_4^{2-}(aq) + 2H_3O^+(aq)$$

2.75. Assume that you mix 50.0 mL of a solution that is 0.45 M Na_2SO_4 with 50.0 mL of a solution that is 0.36 M $BaCl_2$.
 (a) How many moles of each of the four ions, Na^+, SO_4^{2-}, Ba^{2+}, and Cl^-, are present in the mixture? Explain your reasoning clearly.
 (b) If the SO_4^{2-}(aq) in the mixture reacts with Ba^{2+}(aq) to give $BaSO_4$(s), how many moles of Ba^{2+}(aq) are required to react with all the SO_4^{2-}(aq) in the mixture? Explain your reasoning clearly.
 (c) If the Ba^{2+}(aq) in the mixture reacts with SO_4^{2-}(aq) to give $BaSO_4$(s), how many moles of SO_4^{2-}(aq) are required to react with all the Ba^{2+}(aq) in the mixture? Explain your reasoning clearly.
 (d) Is Ba^{2+}(aq) or SO_4^{2-}(aq) the limiting reactant in this mixture? Explain how you make this choice.

2.76. Predict what precipitate will form when each of the following aqueous solution mixings is carried out. Determine the limiting reagent for each reaction and the mass of the precipitate (assuming that all precipitation reactions go to completion). If there is no precipitate, then write NO APPARENT REACTION and explain your reasoning.
 (a) Mix 125 mL of 0.15 M $BaBr_2$ with 125 mL of 0.15 M Na_3PO_4.
 (b) Mix 85 mL of 0.40 M NH_4Cl with 65 mL of 0.50 M KNO_3.
 (c) Mix 85 mL of 0.40 M $(NH_4)_2S$ with 65 mL of 0.50 M $ZnCl_2$.
 (d) Mix 15.0 mL of 0.20 M $AgNO_3$ with 15.0 mL of 0.40 M NaBr.

Chapter 2   Aqueous Solutions and Solubility

2.77. When 50. mL of an aqueous 0.1 M $SrCl_2$ solution are mixed with 50. mL of an aqueous 0.1 M Na_3PO_4 solution, a white precipitate is formed.

(a) How many moles of chloride anion remain in solution when the precipitation is complete? How many grams of chloride is this? Explain the reasoning for your answers.

(b) How many moles of each of the other ions remains in solution when the precipitation is complete? Explain the reasoning for your answers.

(c) Write a complete ionic reaction equation for the reaction in the mixture. Use this equation and your results from parts (a) and (b) to show that the solution is electrically neutral after the precipitation is complete. Show your reasoning clearly.

Section 2.11. Solutions of Gases in Water

2.78. Are gases very soluble in water? Explain your reasoning.

2.79. Predict whether the noble gases (He, Ne, Ar, Kr, and Xe) have a low solubility in water (less than 1 $g \cdot L^{-1}$) or a high solubility in water (greater than 10 $g \cdot L^{-1}$). Explain clearly.

2.80. Would you expect hydrogen chloride gas, $HCl(g)$, to be more or less soluble in hexane than in water. Explain your reasoning.

2.81. (a) Use the data in Table 2.6 for this problem. How many moles of nitrogen gas, $N_2(g)$, dissolve in 10.0 L of water when the temperature is 25 °C and the pressure of the gas is 101 kPa (one atmosphere)?

(b) How many moles of oxygen gas, $O_2(g)$, dissolve in 0.100 L of water when the temperature is 25 °C and the pressure of the gas is 101 kPa (one atmosphere)?

2.82. The solubility of $H_2(g)$ in water at 25 °C is 7.68×10^{-4} mol L^{-1}. When the temperature is decreased to 0 °C, the solubility of hydrogen is 9.61×10^{-4} mol L^{-1}. How do you account for the greater solubility at the lower temperature?

2.83. Hydrogen bromide gas, $HBr(g)$, dissolves in water to form an acidic solution. What is the name of this aqueous solution? *Hint:* What is the analogous solution of $HCl(g)$ called?

2.84. If $HBr(g)$ is bubbled into water until the solution is saturated, the resulting solution is approximately 8.9 M in $HBr(aq)$. The density of the solution is about 1.5 $kg \cdot L^{-1}$. What is the solubility expressed in $g \cdot kg^{-1}$ (as in Table 2.6 for other gases)? Clearly explain the reasoning for your answer.

2.85. $HCl(g)$ is very soluble in diethyl ether ($CH_3CH_2OCH_2CH_3$). Write equations similar to equations (2.16) and (2.17) to describe why the solubility should be so high.

Aqueous Solutions and Solubility Chapter 2

2.86. Assume that the amount of a gas that dissolves in water is directly proportional to its pressure over the solution; the lower the pressure, the less gas dissolved.

(a) Use the data in Table 2.6 to figure out the masses of nitrogen and oxygen that dissolve in 1.0 L of water at 25 °C when air (80% nitrogen and 20% oxygen — mole percents) at a total pressure of 101 kPa dissolves in the water. State all your assumptions explicitly and explain clearly the method you use to arrive at your answer.

(b) What percent of the dissolved mass of gas is oxygen? Show how you get your answer.

(c) Is the mass percent of oxygen in the air greater than, less than, or the same as its mass percent in the gases dissolved in water? Explain your reasoning.

Section 2.12. The Acid-Base Reaction of Water with Itself

2.87. What is an acid?

2.88. What is a base?

2.89. If a solution of acid A has a pH of 1 and a solution of acid B has a pH of 3, what can you tell about the two acid solutions?

2.90. Identify aqueous solutions with these properties as acidic or basic or neither. Explain your reasoning in each case.

(a) $pH < 7$
(b) $[H_3O^+(aq)] = 1.0 \times 10^{-7}$ M
(c) $[OH^-(aq)] > 1.0 \times 10^{-7}$ M
(d) $pH > 7$
(e) $[H_3O^+(aq)] > 1.0 \times 10^{-7}$ M
(f) $[OH^-(aq)] < 1.0 \times 10^{-7}$ M
(g) $[H_3O^+(aq)] < 1.0 \times 10^{-7}$ M
(h) $[OH^-(aq)] = 1.0 \times 10^{-7}$ M

2.91. Calculate the pH of each of the following solutions.

(a) $[H_3O^+(aq)] = 1.0 \times 10^{-2}$ M
(b) $[H_3O^+(aq)] = 1.0 \times 10^{-10}$ M
(c) $[H_3O^+(aq)] = 5.0 \times 10^{-4}$ M
(d) $[H_3O^+(aq)] = 5.0 \times 10^{-8}$ M

2.92. Assuming that the reaction of HCl(g) and water goes to completion to form $H_3O^+(aq)$ and $Cl^-(aq)$, what is the molar concentration of HCl(aq) that will result in solutions having

(a) pH = 4?
(b) pH = 2?

2.93. HCl(g) is named hydrogen chloride, but HCl(aq) is named hydrochloric acid. By analogy, what are the names of HI(g) and HI(aq)? of H_2S(g) and H_2S(aq)?

2.94. Figure 2.24 shows the concentration of hydroxide ion, [OH⁻(aq)], as well as the concentration of hydronium ion, [H₃O⁺(aq)], correlated with the pH of solutions.

(a) When the pH is 3, what are the concentrations of the hydroxide and hydronium ions? What is the numeric value of the mathematical product [H₃O⁺(aq)]·[OH⁻(aq)] at this pH?

(b) For any pH you choose, what is the mathematical product [H₃O⁺(aq)]·[OH⁻(aq)]? Can you think of a reason for this result?

2.95. WEB Chap 2, Sect 2.12.3. Write a brief essay describing the relationship of the two movies to the figure at the bottom of the page (which is similar to Figure 2.26).

Section 2.13. Acids and Bases in Aqueous Solutions

2.96. Phosphorus pentoxide, $P_2O_5(s)$, is a nonmetal oxide which reacts with water to form a solution of phosphoric acid, $(HO)_3PO(aq)$ (or $H_3PO_4(aq)$).

(a) Write the balanced chemical reaction equation for the reaction of phosphorus pentoxide with water.

(b) If 1.42 g of phosphorus pentoxide is mixed with 250. mL of water, what is the molarity of the resulting phosphoric acid solution? Show your reasoning clearly.

2.97. Give a name for each of the following ionic compounds with oxyanions (shown with their conventional formulas). See Table 2.7 for the names of oxyanions. *Hint:* Arsenic, As, is in the same family as P and forms many analogous compounds.

(a) $Ca(HSO_4)_2$ (d) Ce_2SO_4

(b) Na_2CO_3 (e) $KHCO_3$

(c) $Al_2(HPO_4)_3$ (f) Na_3AsO_4

2.98. Draw Lewis structures (showing all nonbonding electron pairs as a pair of dots and all covalent bonds as lines) for the nitrate, ethanoate (acetate), and hydrogen sulfate oxyanions.

2.99. Identify each Brønsted-Lowry acid and base in the following reactions. If necessary, write out the complete balanced ionic equation before identifying the acids and bases. Place an A below each acid and a B below each base.

(a) $H_2S(g) + H_2O(l) \rightleftharpoons HS^-(aq) + H_3O^+(aq)$

(b) $NaOH(aq) + HCl(aq) \rightleftharpoons NaCl(aq) + H_2O(l)$

(c) $NH_3(g) + HCl(g) \rightleftharpoons NH_4^+Cl^-(s)$

Aqueous Solutions and Solubility

2.100. The Lewis structures of $HOCO_2^-$, $(HO)_2PO_2^-$, and $HOPO_3^{2-}$ are omitted from Table 2.7. Draw their Lewis structures (showing all nonbonding electron pairs as a pair of dots and all covalent bonds as lines).

2.101. Identify each Brønsted-Lowry acid and base in the following reactions. If necessary, write out the complete balanced ionic equation before identifying the acids and bases.

(a) $NO_2^-(aq) + H_3O^+(aq) \rightleftharpoons HNO_2(aq) + H_2O(l)$

(b) $2H_3O^+(aq) + 2ClO_4^-(aq) + Mg^{2+}(OH^-)_2(s) \rightleftharpoons Mg^{2+}(aq) + 2ClO_4^-(aq) + 2H_2O(l)$

(c) $HNO_3(aq) + Al^{3+}(OH^-)_3(s) \rightleftharpoons$

(d) $HCN(aq) + NaOH(aq) \rightleftharpoons$

2.102. (a) The sulfur dioxide, $SO_2(g)$, molecule has a permanent dipole moment. What is the shape of the molecule? Explain the reasoning for your answer.

(b) When sulfur dioxide, a nonmetal oxide, dissolves in water, the resulting solution conducts electricity. How can this be explained? Be sure to include an appropriate chemical reaction equation to justify your answer.

2.103. (a) Table 2.7 lists the name of the $HOCO_2^-$ ion (or HCO_3^-) as hydrogen carbonate ion. This ion also has the common name "bicarbonate," used in substances such as bicarbonate of soda, $NaHCO_3$. Explain how this name can be rationalized.

(b) TSP is the common name for a cleaning product containing sodium and phosphate ions. What is the chemical formula for the major ingredient in TSP and what do the letters TSP represent?

2.104. When ammonia dissolves in water, it does so as the result of an acid-base reaction. Two possible acid-base reactions of ammonia and water are:

$$H_2O(l) + NH_3(g) \rightleftharpoons H_3O^+(aq) + NH_2^-(aq)$$
$$H_2O(l) + NH_3(g) \rightleftharpoons OH^-(aq) + NH_4^+(aq)$$

(a) Identify the Brønsted-Lowry acids and bases in each reaction by placing an A below each acid and a B below each base.

(b) Use reasoning based on the relative electronegativities of nitrogen and oxygen to predict which equation represents the actual acid-base reaction when ammonia gas dissolves in water. (You can check your prediction by recalling that, in Investigate This 2.63, you discovered that an aqueous ammonia solution has a pH > 7.)

2.105. When methylamine, $CH_3NH_2(g)$, dissolves in water, a weak electrical conductivity is observed. Explain this observation using a balanced equation in your answer. Omit any ions/molecules that do not directly participate in the reaction.

2.106. Ethylene glycol, $HOCH_2CH_2OH$ (used in automobile antifreeze products), is miscible with water in all proportions. Will the resulting solution be basic, acidic, or neutral? Explain. Will the resulting solution display electrical conductivity? Explain.

2.107. What volume of 0.075 M sulfuric acid, $(HO)_2SO_2(aq)$, solution will be required to reach the equivalence point of the reaction with each of the following basic solutions? *Hint:* Each sulfuric acid molecule can provide two hydronium ions to the solution.
(a) 1.00 g of $KOH(s)$ dissolved in 75 mL of water
(b) 1.00 g of $KOH(s)$ dissolved in 150 mL of water

2.108. Assume that you have a one pound (454 g) container of drain cleaner, mostly solid sodium hydroxide, that you wish to get rid of by reaction with vinegar, about 0.9 M ethanoic (acetic) acid.
(a) Write a balanced chemical reaction equation for the reaction between the drain cleaner and vinegar.
(b) What is the minimum volume of vinegar required to react completely with the drain cleaner? Explain your reasoning clearly.

2.109. Rain drops dissolve gaseous oxides of nitrogen and sulfur (formed by both natural processes and by burning fossil fuels) and form acidic solutions (acid rain), such as nitric and sulfuric acids [see equation (2.28)]. Acid rain has caused a small pond to become so acidic that most of its aquatic life has died. A community group has made a proposal to restore the pond by adding enough lime, $CaO(s)$ (quicklime), to react with the hydronium ion by this reaction stoichiometry:
$$2H_3O^+(aq) + CaO(s) \rightarrow 3H_2O(l) + Ca^{2+}(aq)$$
They have asked for your help to figure out how much lime to use.
(a) The volume of the pond is about 4.5×10^4 m^3 (1 m^3 = 1000 L) and the $H_3O^+(aq)$ concentration is 5.0×10^{-5} M (pH = 4.30). How many moles of hydronium ion does the pond contain? Explain clearly the procedure you use.
(b) How many moles of lime have to be added to react with 90% of the hydronium ion present? How many kilograms of lime is this? If lime is purchased in 50 pound bags, how many bags will be needed? Explain clearly, so the community group can understand.
(c) Estimate what the pH of the pond will be after the lime is added.

Aqueous Solutions and Solubility   Chapter 2

Section 2.15. EXTENSION — CO$_2$ and Le Chatelier's Principle

2.110. Represent each of these statements as a complete balanced chemical equation.

(a) Carbonic acid is formed when carbon dioxide reacts with water.

(b) Calcium carbonate (limestone) reacts with carbonic acid to form an aqueous solution of calcium hydrogen carbonate.

(c) Calcium hydrogen carbonate reacts with calcium hydroxide to form calcium carbonate precipitate.

(d) Calcium hydrogen carbonate reacts with sodium hydroxide to form calcium carbonate precipitate and the water-soluble salt sodium carbonate.

(e) The mixing of aqueous solutions of sodium hydrogen carbonate and sodium hydroxide is exothermic. A reaction has occurred but there is no precipitate.

2.111. A gas is evolved when calcium carbonate is placed in an aqueous solution of HCl(g).

(a) What is the gas? Explain how you reached this conclusion.

(b) Write the net ionic equation for the reaction that produces the gas.

2.112. The names stalactite and stalagmite for the structures that grow, respectively, down from the ceiling and up from the floor of a limestone cave (Figure 2.31) are derived from the Greek word meaning "to drip." Indeed, if you examine the tip of a stalactite, you will often find a drop of liquid. The liquid is an aqueous solution containing calcium cations and hydrogen carbonate anions. See Check This 2.87.

(a) As the water evaporates from this drop, what happens to the concentration of calcium cations? of hydrogen carbonate anions? Explain your reasoning.

(b) What reaction does Le Chatelier's principle predict will occur in the evaporating drop of solution in part (a)? Clearly explain your choice.

(c) Does your answer in part (b) help explain the growth of stalactites? How about stalagmites (which grow directly under stalactites)? Give our reasoning clearly.

2.113. Use explanations based on Le Chatelier's principle to explain or make predictions in each of the following cases.

(a) Consider This 2.75 examined the solubility of carbon dioxide in water. Why is the solubility of carbon dioxide greater in an aqueous sodium hydroxide solution than in water itself?

(b) Calcium sulfate is slightly soluble in water. If sodium sulfate (solid) is added to a saturated aqueous solution of calcium sulfate, what will happen to the concentration of calcium cation, [Ca^{2+}(aq)]?

Chapter 2   Aqueous Solutions and Solubility

General Problems

2.114. WEB Chap 2, Sect 2.2.3 and 5. Sugar (sucrose) crystals are hard and they crunch and break when you put pressure on them. Grease is soft and easily smeared on a surface using little pressure. How do these molecular level representations of sucrose and grease explain the observed macroscopic behavior of these substances? Clearly relate your explanation to the structures shown.

2.115. (a) About 2 g of calcium sulfate, $CaSO_4(s)$, dissolve in a liter of water. What are the molarities of $Ca^{2+}(aq)$ and $SO_4^{2-}(aq)$ in a saturated solution of calcium sulfate? Is seawater saturated with calcium sulfate? (See Consider This 2.51 for the composition of seawater.) Explain your response.

(b) An ionic compound is less soluble in a solution that already contains either its cation or anion. Is this effect consistent with Le Chatelier's principle? Explain why or why not.

(c) Does the effect described in part (b) influence your response to part (a)? How?

(d) Does the effect described in part (b) help explain why the calcium carbonate in seashells (see the chapter opening) does not redissolve in the sea? Explain your response.

2.116. A sample of salt water with a density of 1.02 g·mL^{-1} contains 17.8 ppm (by mass) of nitrate, $NO_3^-(aq)$. Calculate the molarity of nitrate ion in the sample of salt water. (ppm = parts per million).

2.117. (a) Assume that seawater may be represented by a 3.50% by weight aqueous solution of NaCl which has a density of 1.025 g·mL^{-1}. What is the molarity of sodium chloride in this "seawater"?

(b) Is your result in part (a) consistent with your answer to Consider This 2.51(b)? Explain why or why not.

2.118. A student prepared a solution for her biochemistry laboratory by weighing 5.15 grams of a compound and dissolving it in 10.0 grams of water. The concentration of this solution was 2.7 M, and its density was 1.34 g·mL^{-1}. Which of the following compounds did the student use to prepare the solution? Explain your reasoning clearly.

(a) $(NH_4)_2SO_4$ (c) CsCl

(b) KI   (d) $Na_2S_2O_3$

2.119. What percentage (approximate) of the water molecules is protonated in aqueous solutions with these pH's? Explain your reasoning. *Hint:* The molarity of water in pure water and dilute aqueous solutions is about 55.5 M.

(a) 7   (b) 6   (c) 4

2.120. To cool themselves, many animals sweat when the weather is warm (see Chapter 1, Section 1.10). Chickens, however, do not have sweat glands, so they pant to help cool themselves. In hot weather, chickens pant a lot and lose more carbon dioxide than normal (they hyperventilate). The level of dissolved carbon dioxide in their blood decreases and the hens lay eggs with thinner and more fragile shells.

(a) Why are the shells less sturdy than usual? What reaction(s) is(are) being affected?

(b) Egg farmers use a simple and inexpensive method to keep their hens' dissolved-carbon-dioxide levels normal in hot weather. What do you think they do?

2.121. Esterification (which is discussed in Chapter 6) is one of the most important reactions of carboxylic acids in biological systems. A simple example is the reaction of acetic (ethanoic) acid with ethanol to form ethyl acetate (a common solvent found in fingernail polish remover) and water in this equilibrium reaction:

$$CH_3C(O)OH + HOCH_2CH_3 \rightleftharpoons CH_3C(O)OCH_2CH_3 + H_2O$$
    acetic acid   ethanol   ethyl acetate

Use Le Chatelier's principle to predict and clearly explain the outcome of these reaction conditions:

(a) Starting with 0.1 mole of acetic acid and 0.1 mole of ethanol, would more, less, or the same amount of ethyl acetate be formed, if water is added to the reaction mixture?

(b) Would a mixture of 0.2 mole of acetic acid and 0.1 mole of ethanol form more, less, or the same amount of ethyl acetate as a mixture of 0.1 mole of acetic acid and 0.1 mole of ethanol?

2.122. WEB Chap 2, Sect 2.6.3. The chemical equations we write to represent precipitation reactions usually look like this one:

$$Ag^+(aq) + Cl^-(aq) \rightleftharpoons AgCl(s)$$

This representation gives us little clue about what might be happening at the molecular level during the precipitation. The animated movie showing the interaction of chloride and silver ions provides one way to visualize the precipitation process. Describe the steps in the process and illustrate your description with chemical reaction equations. That is, try to translate the molecular level representation to a symbolic representation.

Chapter 3. Origin of Atoms

Section 3.1. Spectroscopy and the Composition of Stars and the Cosmos 3-6
    Spectroscopy 3-6
    Continuous and line spectra 3-8
    Line spectra of elements and stars 3-9
    Elemental abundance in the universe 3-11
    Trends in elemental abundance 3-13

Reflection and projection   *3-14*

Section 3.2. The Nuclear Atom 3-14
    The nuclear atom 3-15
    Protons and atomic number 3-15
    Neutrons and mass number 3-16
    Isotopes 3-17
    Ions 3-19

Section 3.3. Evolution of the Universe: Stars 3-20
    The Big Bang Theory 3-20
    The first nuclear fusions 3-21
    The life and death of a star 3-22
    The birth of a new star 3-24

Reflection and projection   *3-25*

Section 3.4. Nuclear Reactions 3-25
    Emissions from nuclear reactions 3-25
    Balancing nuclear reactions 3-26
    Positron-electron annihilation 3-29
    Radioactive decay 3-31
    Half life 3-32
    Radioisotopes and human life 3-35

Section 3.5. Nuclear Reaction Energies 3-37
    Mass is not conserved in nuclear reactions 3-37
    Comparison of nuclear and chemical reactions 3-38
    Nuclear binding energies 3-41
    Fusion, fission and stable nuclei 3-43
    Nuclear chain reactions 3-45

Reflection and projection — *3-47*

Section 3.6. Cosmic Elemental Abundance and Nuclear Stability 3-48
    Trends in elemental abundance 3-49
    Correlations between abundance and nuclear binding energy 3-50

Section 3.7. Formation of Planets: The Earth 3-51
    The age of the Earth 3-53
    The elements of life 3-54

Reflection and projection — *3-56*

Section 3.8. Outcomes Review 3-56

Section 3.9. EXTENSION — Isotopes: Age of the Universe and a Taste of Honey 3-58
    Age of stars and the universe 3-58
    Stable isotopes and pure honey 3-61

Origin of Atoms Chapter 3

The Crab Nebula, a relatively near neighbor in the constellation Taurus, is the remnant of a supernova explosion about 6500 light years from Earth. The stellar explosion occurred about 7500 years ago. Light from the supernova reached the Earth and was observed and recorded by Chinese astronomers in the year 1054. Elements were scattered into interstellar space by the explosion, which was so energetic that the new "star" could be seen during the day for more than three weeks. The graph shows the present-day abundance of elements in the universe.

Chapter 3. Origin of Atoms

"Twinkle, twinkle, little star,
How I wonder what you are!
Up above the world so high,
Like a diamond in the sky!"

Jane Taylor (1783-1824), The Star

3.1. Consider This *Where do atoms come from?*

(a) Where do the atoms in your body come from?

(b) In part (a), you probably thought of several different sources for the atoms in your body. Where did the atoms in these sources come from?

(c) If you continue working back, as in part (b), toward the origin of atoms, where do you end up?

As you have seen in the previous chapters, a unique feature that distinguishes the Earth from all the other planets and moons in our solar system is the presence of abundant surface water at a temperature that keeps most of it in the liquid state. You also know that about 70% of your own body is water. Where did all this water come from? Consider This 3.1 pushes the question back further: *Where did the elemental atoms in water, hydrogen and oxygen, come from in the first place?* And the question applies to all the elements you have met so far: Where did the atoms of carbon, nitrogen, sodium, phosphorus, and other elements come from? For all the elements except hydrogen, and most helium and lithium, the answer is: *the stars*. Indeed, as the song says, "We are stardust." In this chapter, you'll see how literally true this statement is and also how a star is "like a diamond in the sky."

> "We are stardust," is a phrase from Joni Mitchell, *Woodstock* (popularized by Crosby, Stills, Nash, and Young).

A great deal of what we know about our solar system, stars, galaxies, and the universe is based on spectroscopic evidence gathered by astronomers over the past several centuries. We will introduce spectroscopy in this chapter and find in Chapter 4 that it is a thread that ties together cosmic models and the atomic models we will discuss there. The results of more than a century of spectroscopic study and modeling of the universe are summarized in the cosmic elemental abundance graph shown on the facing page. After an examination of these data, we will briefly discuss our present model for the birth and evolution of the universe that has led to

Origin of Atoms Chapter 3

the present abundance of elements in the universe and on Earth. Elemental atoms are formed in nuclear reactions, so a good deal of the chapter is devoted to these reactions and their energies.

Section 3.1. Spectroscopy and the Composition of Stars and the Cosmos

3.2. Investigate This *What is the effect of a diffraction grating on white light?*
Completely cover the stage of an overhead projector with a sheet of thin cardboard that has a "slit" about 6×100 mm cut into its center. Project the image of the slit vertically on the projection screen. Hold a sheet of transmission diffraction grating in front of the projector lens with the grating lines parallel to the length of the slit. Write a description of what you observe.

3.3. Consider This *How does a diffraction grating affect different colors of light?*
(a) Are different colors of light affected differently as they pass through the diffraction grating in Investigate This 3.2? If so, what color of light is most affected? Explain why you answer as you do.
(b) What do you observe about the symmetry of the image on the screen? How might you explain your observation?

Spectroscopy

Isaac Newton (English natural philosopher and mathematician, 1642-1727) was the first scientist to observe that white light could be dispersed by a prism into the **spectrum** of colors (wavelengths of light) shown in Figure 3.1. (We'll discuss the concepts of waves, light, and the origin of diffraction in Chapter 4. For our purpose here, your observation that light can be spread into a spectrum by a diffraction grating is all we need.) Newton used light from the sun for his experiments, so **spectroscopy** (spectrum + *skopeein* = to watch), the observation of spectra, began with light from a star. Modern spectroscopic experiments are usually carried out using diffraction gratings to disperse the light, as you did in Investigate This 3.2 and as is illustrated in Figure 3.2.

> **Spectrum** (singular; plural = **spectra** -- from Latin for vision) was originally used to refer to the rainbow of colors created by dispersion of white light; see Investigate This 3.9 and Figures 3.4 and 3.5. Spectrum now refers to the entire range of electromagnetic wavelengths we will meet in Chapter 4.

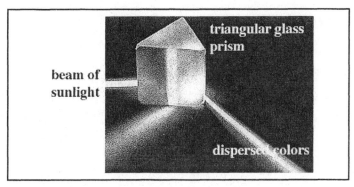

Figure 3.1. A prism disperses white light into the visible spectrum.
Violet light is bent most from the original light beam.

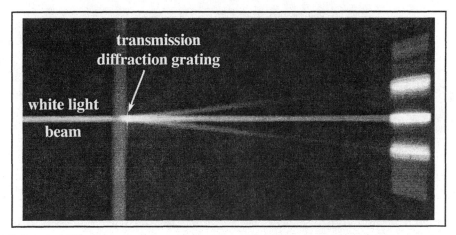

Figure 3.2. A diffraction grating disperses white light into the visible spectrum.
Violet light is bent least from the original light beam.

3.4. Investigate This *What do spectra of different emission sources look like?*

Use a small diffraction grating or hand-held spectroscope to view the light emitted from an incandescent light bulb, a fluorescent light bulb, and two or more atomic discharge lamps, such as a neon lamp, set up by your instructor. Record your observations and work in small groups to develop descriptions of what you observe from each light source.

3.5. Consider This *How would you characterize the spectra from different sources?*

(a) What is the same in the various spectra you observed in Investigate This 3.4? What is different? How might you explain any differences?

(b) Is there a correlation between the spectra you observe and the appearance of the emission from each of the sources? Explain your response with specific examples.

Continuous and line spectra

Gases energized by an electric current passing through them often emit visible light. The glow from neon lights and the discharge tubes in Investigate This 3.4 are examples of this phenomenon; scientists have investigated the spectra of many such emissions. The instruments used for these experiments are called **spectroscopes** or **spectrographs**, Figure 3.3, if they record the spectrum photographically. Usually black-and-white film is used in spectrographs, so that all emissions show up as black (exposed film) against a white background (unexposed film).

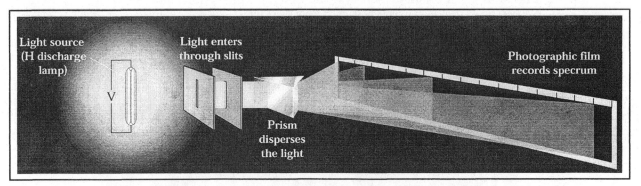

Figure 3.3. Diagram of the parts of a simple spectrograph.

The results of these experiments were just what you found in Investigate This 3.4. The visible emissions from some glowing systems resemble the **continuous spectrum** shown in Figures 3.1, 3.2, and 3.4(a); all the wavelengths of the spectrum are present. Other glowing systems, as you observed, emit only a few wavelengths of light; these are called **line spectra**. A few examples of visible emission line spectra are shown in Figure 3.4(b)-(d). Light enters a spectrograph through long, narrow rectangular slits, as you see in Figure 3.3. It is the image of the slits that is observed. If light of a single color (wavelength) enters the slits, the image you see is a long, narrow rectangle, *a line*. A line spectrum shows that the emission source is emitting only certain wavelengths of light.

Chapter 3 — Origin of Atoms

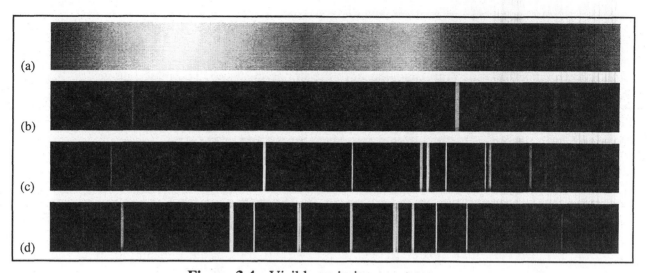

Figure 3.4. Visible emission spectra.
(a) is a continuous spectrum to compare with line spectra of (b) hydrogen, (c) helium, and (d) carbon atoms.

Line spectra of elements and stars

The emission wavelengths in a line spectrum are characteristic of the element(s) in the sample that is(are) emitting. *Each element has a unique line spectrum*; its presence in a sample can be identified by its line spectrum. The continuous visible spectrum of the sun that Newton saw proved to be much more interesting when it was investigated with better spectrographs. There are brighter lines at several wavelengths in the sun's spectrum. When scientists compared these brighter lines with the line spectra of known elements, they found that they could match emission lines from the sun with emissions lines from elements on Earth. Thus, spectroscopy could

> One prominent emission line from the sun could not be assigned to any element known on Earth, so it was assumed to be a new element and named "helium" (*helios* = sun). Soon helium was also discovered on Earth; its emission line spectrum matched the emission line from the sun.

be used to analyze the elemental composition of the sun. Further improvements in telescopes and spectrographs now permit us to do the same kind of analysis of far more distant stars and other cosmic structures, such as nebulas like the one pictured in the chapter opening.

3.6. Worked Example *Spectroscopic analysis of stellar composition*

Figure 3.5 shows the spectral emission lines from the surfaces of two stars (without the continuous background emission) and line spectra for several elements. (The line spectra are negative photographic images; they are white where they are exposed and dark where not exposed.) Do hydrogen atoms occur on star X?

Necessary Information: We need the emission lines from the star and from hydrogen atoms, both of which are given in Figure 3.5.

Implementation: A simple way to compare the spectra is to line them up with one another (or use a straightedge aligned at the same wavelengths on the two scales in Figure 3.5) to see if the lines in the atomic spectrum also appear in the stellar spectrum. The match up is:

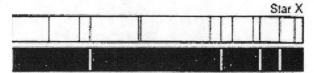

The four lines in the hydrogen atom spectrum match four lines in the stellar spectrum, so we can conclude that hydrogen atoms are present on star X. The other emission lines in the spectrum of the star show that other elements are present as well (see Check This 3.7).

Does the Answer Make Sense? In Section 3.3, we will look at our present model for the beginning and evolution of our universe to see how scientists make sense of data like these.

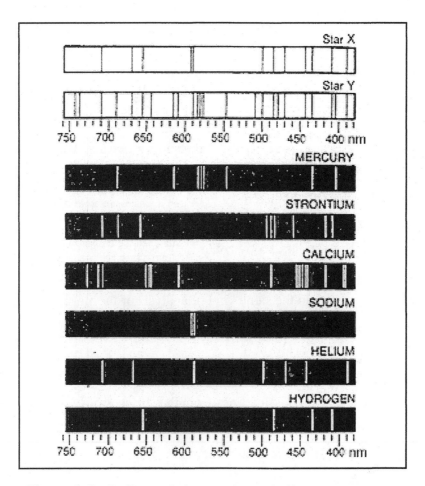

Figure 3.5. Stellar and elemental atomic line spectra.

3.7. **Check This**   *Spectroscopic analyses of stellar composition*

Use the data in Figure 3.5 to answer these questions. Give the reasoning for your answers.

(a) Does hydrogen occur on star Y?

(b) What is the composition of star X?

(c) What is the composition of star Y?

(d) What elements are common to both stars?

Elemental abundance in the universe

Determining the abundance of the elements in the universe (or even our little corner of it) is not an easy task. Spectroscopic studies provide most of the evidence for this determination, but they are limited in the information they can provide. It is easier and more accurate to find the *ratio* of the abundance of one element to another, rather than try to determine the absolute amount of either one. Figure 3.6 illustrates how the ratio of the amounts of two elements, A and B, might be determined from the relative brightness of their emission lines. In a laboratory sample, Figure 3.6(a), the emission lines are equally bright when there are equal amounts of A and B. By comparing these emissions to the relative brightness of the emissions of these elements from a star, Figure 3.6(b), we can determine the amount of B relative to the amount of A in the star.

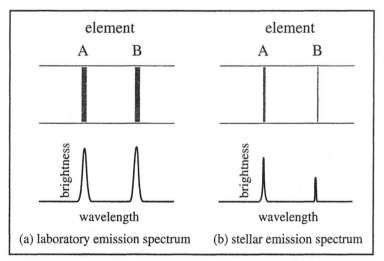

Figure 3.6. Laboratory and stellar emission lines from elements A and B.
(a) represents the emissions from equal amounts of elements A and B; (b) are the same emissions from a star. The peaks on the graph beneath each spectrum show the brightness of the corresponding line. The brighter the emission, the more abundant the element.

3.8. Check This *Relative abundance of elements in Figure 3.6*

If you arbitrarily assign the value 100 to the stellar abundance of element A in Figure 3.6(b), what value would you give the abundance of element B?

From the results of many observations similar to those in Figure 3.6 plus other kinds of data and models of the universe, scientists make estimates of the *relative* abundance of all the elements in the universe. Figure 3.7 shows these relative abundances for the first two-thirds of the periodic table and brings together a great deal of information about the universe. In order to present the information compactly, the figure presents the data on a logarithmic scale and gives all the abundances relative to silicon. The logarithmic scale is required because the lightest elements are 100-billion (10^{11}) times more abundant than heavier elements. There is no way to express such enormous variations on a linear scale. Silicon is chosen as the reference because it is quite abundant on Earth, moderately abundant in the universe, and easily detected.

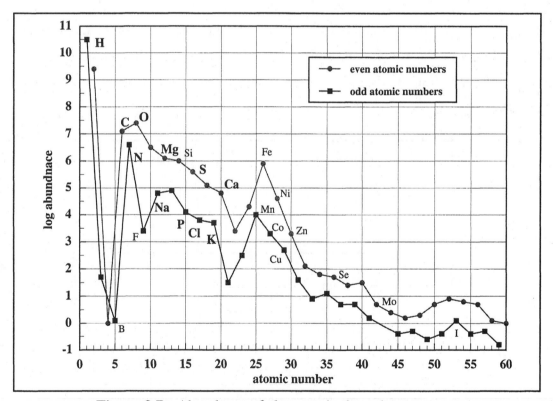

Figure 3.7. Abundance of elements in the universe.
All abundances are relative to 10^6 atoms of silicon, atomic number 14. Elements shown in larger bold type are required by all organisms on Earth in substantial amounts. Elements in smaller regular type are required in smaller amounts or by only a few organisms. The points are connected only to highlight the trends in the data.

Chapter 3 — Origin of Atoms

3.9. Worked Example — *Abundance of carbon relative to iron in the universe*

Use the data in Figure 3.7 to estimate the abundance of carbon relative to iron in the universe.

Necessary Information: The data in the figure are given on a logarithmic scale. One unit on a log scale represents a 10-fold change (an order of magnitude) in the value.

Implementation: The values for C and Fe on the plot are close to 7 and 6, respectively. This means that the number of carbon atoms in the universe is about 10 times as large as the number of Fe atoms (about 10^7 compared to 10^6).

Does the Answer Make Sense? The abundances increase as we go up on the abundance axis. The position of carbon is higher on the plot than the position of iron, so carbon has the higher abundance, which is what we found from the numerical values.

3.10. Check This — *Relative abundances of elements in the universe*

Use the data in Figure 3.7 to do these problems.

(a) Estimate the relative abundance of Ni to Co.

(b) What is the most abundant element in the universe?

(c) Which elements are *at least* 10 times more abundant than Si?

(d) Which elements are between 10 and 100 times *less* abundant than Si?

3.11. Consider This — *What are the trends in elemental abundance?*

(a) In Figure 3.7, what is the general trend in abundance as the atomic number increases?

(b) Do you see any major exceptions to the trend you described in part (a)?

Trends in elemental abundance

The general trend in Figure 3.7 is decreasing abundance of the elements with increasing atomic number. The trend continues for the rest of the heavier elements that are not shown in the figure. There are two striking exceptions to the relatively smooth downward trend of the abundances as a function of atomic number. The abundances of Li, Be, and B (atomic numbers 3, 4, and 5) are less than we would expect, if they behaved like the elements just before and just after them, and there is a peak in abundance around atomic number 26 (iron) where the abundances are larger than we would expect for a smoothly decreasing curve. We also note that the abundances of the elements with even atomic numbers are higher than nearby elements with odd atomic numbers.

Several questions are raised by the data in Figure 3.7. Why do the elemental abundances follow (or deviate from) the pattern we see? Have the abundances always been as they are today? If they have changed over time, are they still changing? What is(are) the origin(s) of the elements? Many kinds of data, including some of those in Figure 3.7, are the basis for our present model of the formation and evolution of the universe and the elements that comprise it. In Section 3.3, we will give a brief description of this model and its consequences for elemental atom formation and then go on in succeeding sections to look at some properties of atomic nuclei that help explain more of the details of Figure 3.7.

Reflection and projection

Light passing through a prism or diffraction grating is diffracted or broken up into its constituent wavelengths (colors). The fundamental discovery that each element has a unique emission line spectrum gave scientists a tool for analyzing the atomic composition of any object, including stars and other luminous objects in the heavens, that emit light. One result of stellar analysis is that hydrogen and helium are found in essentially all stars and throughout the universe. Since these are the lightest elements and the simplest atoms, it makes sense that the atomic history of the universe might begin with these atoms (or their nuclei).

It's a remarkable achievement to be able to analyze a sample that is millions or even billions of light years from Earth (or to discover a new element). The results are part of the experimental evidence that led to the development of the model for elemental synthesis and evolution we will discuss in the next sections of the chapter. Before we see how atoms of the elements are formed in stars, we will review some familiar atomic concepts in a little more detail than in Chapter 1. You may find that you do not need this review and can skip directly to Section 3.3.

Section 3.2.   The Nuclear Atom

3.12. **Consider This**   *What do you already know about atoms?*
List at least six things you know about the composition and structure of atoms. Work in a small group to combine your individual lists into a single list that includes all the things the group knows about atomic structure. Discuss your group list with the entire class and your instructor.

The nuclear atom

In our nuclear atomic model, all atoms are composed of the same three types of subatomic building blocks: **electrons**, **protons**, and **neutrons**, the masses and charges of which are given in Table 3.1. Protons and neutrons make up the tiny, dense **nucleus** (plural, **nuclei**) that occupies about 1 part in 10^{15} of the volume of an atom. Recall from Chapter 1 (see Figure 1.5) that, if an atom were the size of a baseball stadium, the nucleus would be about the size of this "o". The remainder of an atom's volume is "empty" space containing enough light, fast moving electrons to balance the positive nuclear charge.

Table 3.1. Properties of subatomic particles.

Particle	Symbol	Unit charge, e*	Mass, g	Mass, u**
electron	e^-	1–	9.10939×10^{-28}	5.48680×10^{-4}
proton	p	1+	1.672623×10^{-24}	1.00728
neutron	n	0	1.674929×10^{-24}	1.00866

* The charges here are given *relative* to the magnitude of the charge on the electron, which is called the **elementary charge** = 1.60218×10^{-19} C (coulombs).

** An **atomic mass unit**, u, is defined as exactly 1/12 the mass of a carbon-12 atom:
u = 1.66054×10^{-24} g. Relative atomic masses, inside front cover, may be interpreted as molar masses in grams or as atomic masses in atomic mass units.

Protons and atomic number

The nuclei of *all* the atoms of a particular element have the same number of protons; for example, all oxygen atoms have eight protons. Atoms of different elements have different numbers of protons in their nuclei; whereas all oxygen atoms have eight protons, all carbon atoms have six protons. The number of protons is therefore the determining characteristic of an element. The number of protons in a nucleus is called the **atomic number**, Z. Each proton is positively charged, so the atomic number is equal to the positive charge on the nucleus.

The periodic table on the end papers of the book shows the atomic number above each elemental symbol. Since each element has a unique atomic number, the elemental symbol itself specifies the atomic number of the element, but for emphasis and clarity, we sometimes write the atomic number (equal to the nuclear charge) as a *subscript to the left of the elemental symbol*. For example, hydrogen may be written as $_1$H, helium as $_2$He, and carbon as $_6$C, as illustrated in Figure 3.8.

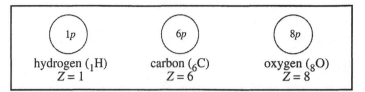

Figure 3.8. Nuclear structure and nomenclature: atomic number.
The atomic number (Z) is the number of protons in the element and is unique to each element.

Neutrons and mass number

Neutrons and protons each have a mass of about one u, as you can see in Table 3.1. The **mass number**, A, of an atom is equal to the sum of the number of protons, Z, *plus* the number of neutrons, N: $A = Z + N$. Thus, the mass number represents the total number of nuclear particles, collectively called **nucleons**, of an element. When the mass number of a nucleus is to be specified, it is written as a *superscript to the left of the atomic symbol*. For example, the helium nucleus is represented as ^4_2He to specify that the nucleus consists of two protons ($Z = 2$) and a total of four nucleons, two protons and two neutrons ($A = Z + N = 4$). When the name of an element is written out, we specify its mass number, if necessary, by writing the mass number after the name, as in helium–4. Figure 3.9 illustrates our three examples.

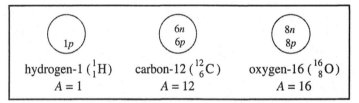

Figure 3.9. Nuclear structure and nomenclature: mass number.
The atomic mass (A) is the number of protons plus the number of neutrons (N): $A = Z + N$.

3.13. Worked Example   *Writing atomic symbols*

How many protons, electrons, and neutrons are found in an atom of potassium-41? Write its atomic symbol, including its mass number and atomic number. Note that we have specified the mass number of the potassium nucleus by naming it "potassium-41."

Necessary Information: We will use the periodic table as well as the definitions of mass number, A, and atomic number, Z.

Implementation:

Determine the number of protons: The atomic number, Z, is equal to the number of protons in the nucleus of an atom. In the periodic table, we find that potassium's atomic number is 19. Therefore, the potassium-41 nucleus contains 19 protons ($Z = 19$).

Determine the number of electrons: Since an atom of potassium-41 is neutral, the number of electrons is equal to the number of protons. A potassium-41 atom contains 19 electrons.

Determine the number of neutrons: The mass number, A, is the sum of the atomic number, Z, plus the number of neutrons, N: $A = Z + N$. We rearrange this equation to solve for the number of neutrons: $N = A - Z$. The atom name, potassium-41, gives the mass number, $A = 41$. Therefore:

number of neutrons = $N = A - Z = 41 - 19 = 22$

Write the atomic symbol: Use the periodic table to find the atomic symbol, K, for potassium. Write the mass number as a superscript in front of K. Write the atomic number as a subscript in front of K. The atomic symbol, including the mass number and atomic number, is $^{41}_{19}K$.

Does the answer make sense? Double-check to be sure that $A = Z + N$, so you can be confident of your arithmetic.

3.14. Check This   *Writing atomic symbols*

How many protons, neutrons, and electrons are found in the following atoms? Write their atomic symbols, including their mass numbers and atomic numbers.

(a) lead-206   (b) cobalt-59

Isotopes

The number of neutrons can differ among nuclei having the same number of protons and this difference affects the mass of the atom. Atoms with the same atomic number, Z, but different mass numbers, A, are known as **isotopes.** For example, there are three isotopes of hydrogen: ^1_1H has no neutrons in the nucleus, ^2_1H has one neutron, and ^3_1H has two neutrons. All three are hydrogen atoms because they have only one proton in their nuclei. A sample of hydrogen atoms naturally occurring on Earth consists of 99.985 % ^1_1H and 0.015 % ^2_1H; ^3_1H does not occur naturally; it is synthesized in nuclear reactors. We specify which isotope we are referring to by writing the mass number after the name: helium-4, hydrogen-3, or carbon-12 (also abbreviated as He-4, H-3, and C-12). Isotopic nuclei are illustrated in Figure 3.10.

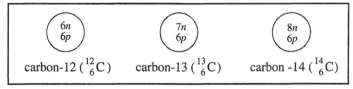

Figure 3.10. Nuclear structure and nomenclature: isotopes.
Isotopes of an element all have the same number of protons but different numbers of neutrons.

There is no significant difference in the size of an atom when the number of neutrons is different. This is because the size of an atom is determined not by the tiny nucleus but by the much larger volume in which its electrons move, as we saw in Chapter 1 and will elaborate upon in Chapter 4. The chemical reactions of isotopes of any element are nearly indistinguishable, because *chemical properties are largely determined by the arrangement of the electrons, not the nuclei.* Substitution of one isotope for another in a reaction almost never changes the reaction, so using isotopic tracers to follow the fate of particular atoms in a chemical change can provide information about the pathway for the reaction. Also, subtle variations in the way the isotopes of an element take part in a reaction

> Geochemists make extensive use of isotopic analyses to determine the age of geological samples and to reconstruct the conditions, such as temperature, under which samples were formed.

under different conditions can be used to tell what the conditions were when the reaction occurred. One example of this use of isotopes is given in Section 3.9.

3.15. Worked Example *Identifying isotopes*

(a) Determine the pair of isotopes from the atoms in this list (X represents an unknown element.):

$$^{54}_{24}X \quad ^{54}_{26}X \quad ^{56}_{25}X \quad ^{56}_{26}X$$

(b) Identify the element that has these isotopes.

Necessary Information: We will use the periodic table as well as the definitions of an isotope, mass number, A, and atomic number, Z.

Implementation: (a) We need to identify the pair of atoms that have the same atomic number, Z, but different mass number, A. In the above list, the pair of isotopes is $^{54}_{26}X$ and $^{56}_{26}X$.

(b) To determine the element, we note that iron, Fe, has an atomic number of 26. Two isotopes of iron are $^{54}_{26}Fe$ and $^{56}_{26}Fe$.

Does the answer make sense? These two isotopes of iron differ only in the number of neutrons that each contains. $^{54}_{26}Fe$ contains 28 neutrons ($N = A - Z = 54 - 26 = 28$), while $^{56}_{26}Fe$ contains 30 neutrons ($56 - 26 = 30$).

3.16. Check This *Identifying isotopes*

Identify the pair of isotopes and explain why you reject the others:

(i) $^{242}_{94}Pu$ and $^{242}_{96}Cm$ (ii) $^{134}_{56}Ba$ and $^{138}_{56}Ba$ (iii) $^{27}_{13}Al$ and $^{27}_{14}Si$

Chapter 3 — Origin of Atoms

Ions

In an electrically neutral atom of an element, the number of electrons is equal to the number of protons; the net charge is zero. As you saw in Chapter 2, an **ion** is formed if the number of electrons is either fewer than or greater than the number of protons and the net charge on the ion is indicated as a *superscript to the right of the atomic symbol*. For example, the hydrogen cation is a positively charged ion because the atom has lost an electron; it can be written as $_1H^+$. The hydrogen anion is a negatively charged ion because the atom has gained an electron; it can be written as $_1H^-$. You have seen that it's possible for an atom to lose or gain more than one electron; for example, $_{12}Mg^{2+}$ has two fewer electrons than it has protons. Charge is nearly always shown if it is not zero, but the A and Z are usually omitted in chemical formulas and equations. For nuclear reaction equations, the A and Z (or at least A) are shown, but physicists often omit the charge.

> By convention, 1+ and 1– charges on ions are shown as + and –; the numeral one is not used.

3.17. Worked Example *Counting subatomic particles in ions*

How many protons, neutrons, and electrons are present in the ion, $^{34}_{16}S^{2-}$?

Necessary Information: We will need the definitions of ion; mass number, A; and atomic number, Z.

Implementation:

Determine the number of protons and neutrons: Using the explanation given in Worked Example 3.3, the $^{34}_{16}S^{2-}$ nucleus contains 16 protons and 18 neutrons.

Determine the number of electrons: The superscript 2– shows that $^{34}_{16}S^{2-}$ is a negatively charged ion, containing two more electrons than protons. We already know that $^{34}_{16}S^{2-}$ contains 16 protons. Therefore, the number of electrons in $^{34}_{16}S^{2-}$ is 18 (16 + 2 = 18).

Does the answer make sense? Given $^{34}_{16}S^{2-}$, we can add the number of protons ($Z = 16$) and neutrons ($N = 18$) to check the mass number, A: 16 + 18 = 34. Since this ion has a negative two charge, it contains two more electrons than protons, for a total of 18 electrons.

3.18. Check This

Explain the difference(s) between $^{206}_{82}Pb^{2+}$ and $^{208}_{82}Pb^{4+}$.

Origin of Atoms   Chapter 3

Section 3.3.   Evolution of the Universe: Stars

> *3.19.* **Consider This**   *How much of the universe is hydrogen and helium?*
> Over 99% of the atoms in the universe are hydrogen or helium. Helium atoms make up about 7% of this total. Use the data in Figure 3.7 to prove or disprove these statements.

Most stars contain hydrogen and helium, as you found for the two stars in Check This 3.7. These are the simplest elements and they are ubiquitous in the present-day universe. Whatever model we develop to explain the origin and evolution of the universe must account for all this hydrogen and helium.

The Big Bang Theory

The most widely accepted scientific model for the beginning of the universe is called the **Big Bang Theory**, according to which the universe began in a rapid expansion from an initial state of nearly infinite mass at nearly infinite density and temperature. Though evidence supports this theory, the "laws" of energy and matter that we are familiar with on present day Earth fail to explain how the universe could have been so dense and so hot. Nor do they explain what happened within the first one-trillionth of a second (10^{-12} s = 1 picosecond, ps). Nevertheless, the Big Bang is the only scientific theory yet devised that explains the observable features of the universe as we know them today.

In trying to understand how these features evolved, the expansion is most significant. During the expansion, which continues to this day, the universe has cooled from an initial temperature of $>10^{27}$ K to its modern-day average value of about 3 K. As matter cooled and continued to expand, **condensations** occurred, some of the particles condensed, that is, stuck together and became more dense or concentrated. In order, for particles to stick together, forces attracting them to one another must be greater than the kinetic forces (motions) that keep them apart. Lower temperatures reduced the kinetic energy (slowed the motion), so bits of matter stuck together as the temperature went down. Figure 3.11 summarizes the history of the universe as a series of condensations from the Big Bang to the formation of the Earth. In this section we will focus on the evolution of stars and, in Section 3.7, return to the formation and history of the Earth.

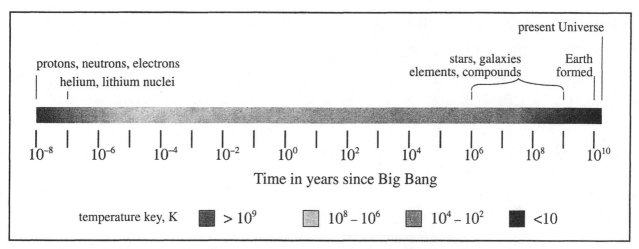

Figure 3.11. A timeline for the condensations in the evolution of the universe.
The temperature indicated by the intensity of color is the *average* temperature of the universe. There are a vast number of local variations, especially within and near stars and planets.

The first nuclear fusions

A few seconds after the Big Bang, when the temperature had dropped to approximately 10^9 K, protons and neutrons could condense to form small amounts of $^2H^+$, $^3He^{2+}$, $^4He^{2+}$ and $^7Li^{3+}$ nuclei by reactions like these:

$$^1p^+ + {}^1n^0 \rightarrow {}^2H^+ + \gamma \text{ (gamma radiation; discussed in Section 3.4)} \tag{3.1}$$

$$^2H^+ + {}^2H^+ \rightarrow {}^3He^{2+} + {}^1n^0 \tag{3.2}$$

$$^3He^{2+} + {}^2H^+ \rightarrow {}^4He^{2+} + {}^1p^+ \tag{3.3}$$

These reactions are **nuclear fusions**, the building of heavier nuclei from two (or more) lighter particles or nuclei. Nuclear fusion often creates new elements, as in reaction (3.2) where an isotopic helium, He-3, nucleus is produced from two isotopic hydrogen, H-2, nuclei.

For nuclear fusion to occur, the positively charged nuclei have to be able to get close enough together to fuse into a single nucleus. They have to collide with enough kinetic energy to overcome their mutual coulombic repulsion [equation (2.2), Section 2.4]. A minute or two after the Big Bang, the temperature had dropped further ($<10^7$ K), the nuclei didn't have enough kinetic energy to overcome this repulsion and these nuclear fusions stopped. Most nuclei were still $^1H^+$ ($^1p^+$, protons). As the expansion continued for about 10^5 years, the temperature fell below a few thousand kelvin, electrons became associated with atomic nuclei to form ions and atoms:

$$^4He^{2+} + e^- \rightarrow {}^4He^+ \tag{3.4}$$

$$^1H^+ + e^- \rightarrow {}^1H \tag{3.5}$$

The universe today is filled with vast clouds of hydrogen left over from the first seconds of its birth and the later condensation of protons with electrons.

3.20. Check This *Correlating reactions and temperatures in atom formation*

In atom formation, what would be the approximate range of temperatures in which each of these reactions might occur? Explain your reasoning.

(a) $^6Li^{2+} + e^- \rightarrow {}^6Li^+$

(b) $^2H^+ + {}^4He^{2+} \rightarrow {}^6Li^{3+}$

The life and death of a star

Carbon exists throughout the present-day universe, but there was no carbon present when fusion stopped in the early universe. Synthesis of carbon and other heavier elements had to await the formation of stars. For billions of years, clouds of hydrogen atoms (mixed with smaller amounts of helium and lithium) have spanned the expanding universe. Within these clouds, gravitational attraction pulled atoms together, making some regions of the clouds more dense and hotter than the rest. Collisions in these denser regions became more frequent, and occurred with ever-increasing kinetic energy, as gravity pulled the atoms closer and closer together, until nuclear fusion could again occur. This is how the first stars were born.

Within these local concentrations of hydrogen that became stars, the temperature rose to 1.5×10^7 K. Atoms lost their electrons and the hydrogen nuclei (protons) now had enough kinetic energy to overcome the coulombic repulsion of like charges. The combination of high-energy collisions and high density of nuclei produced nuclear fusion, beginning with this net reaction:

$$4{}^1H^+ \rightarrow {}^4He^{2+} + 2e^+ \quad \text{(energy released)} \quad (3.6)$$

Reaction (3.6), which we will discuss further in the next sections, is an example of "nuclear burning," or more specifically in this case, "hydrogen burning" because the reaction, once started, releases enough energy to keep the temperature high enough to cause more hydrogen nuclei to fuse.

The conversion of four hydrogen nuclei into one helium nucleus resulted in an overall reduction of the number of particles in the star. Just as reducing the amount of air in a balloon causes it to get smaller, fewer particles in a star produce less outward push to oppose the collapse of the star. Consequently, the core of the star became smaller due to the inward pull of gravity. Total collapse was prevented by the energy released from the fusion reactions at the core of the star, which gave the helium nuclei additional kinetic energy and more outward push (like heating the air in a balloon) to counterbalance the gravitational force.

Ultimately the temperature of the star's core rose to 10^8 K, at which temperature the energies were high enough to start helium burning to produce carbon-12 by a rapid sequence of fusion reactions involving the almost-simultaneous collision of three helium–4 nuclei:

$$^4\text{He}^{2+} + {}^4\text{He}^{2+} + {}^4\text{He}^{2+} \rightarrow {}^{12}\text{C}^{6+} \qquad \text{(energy released)} \qquad (3.7)$$

If the star was massive enough, the net conversion of helium nuclei to carbon nuclei led to a second collapse and further heating of the interior of the star to about 10^9 K. At this temperature, the kinetic energies of the nuclei were large enough for carbon burning:

$$^{12}\text{C}^{6+} + {}^{12}\text{C}^{6+} \rightarrow {}^{20}\text{Ne}^{10+} + {}^4\text{He}^{2+} \qquad \text{(energy released)} \qquad (3.8)$$

Another series of collapses, heatings, and nuclear burnings occurred in the star and progressed from neon to oxygen to silicon burning in less than a year. Silicon burning lasted about a day and involved stepwise reactions of silicon nuclei with helium nuclei, protons, and neutrons to form nuclei of the elements near iron in the periodic table. An unbalanced representation of this process is:

> The scenario outlined here is based on our present cosmological model of star evolution. The time scale of events is related to the calculated values for the speeds of the processes taking place.

$$^{28}\text{Si}^{14+} + ({}^4\text{He}^{2+} + {}^1p^+ + {}^1n^0) \rightarrow {}^{56}\text{Fe}^{26+} \text{ (+ nearby nuclei)} \qquad \text{(energy released)} \qquad (3.9)$$

This rapid series of burnings (fusions) caused a catastrophic collapse of the core of the star, as its temperature rose well above 10^9 K, and it exploded. The exploding star is called a **supernova**. The remains of these massive explosions, one of which is shown in the chapter opening, scatter atoms for great distances and continue to emit energy in the form of light and other radiation for many thousands of years. In the cataclysm of the supernova very high concentrations of neutrons were produced and these reacted with iron and its neighboring nuclei to form all the higher atomic number nuclei by addition of neutrons and subsequent loss of electrons from unstable nuclei. The first stages in this process for some iron nuclei were:

$$^{56}\text{Fe}^{26+} + 3{}^1n^0 \rightarrow {}^{59}\text{Fe}^{26+} \qquad (3.10)$$

$$^{59}\text{Fe}^{26+} \rightarrow {}^{59}\text{Co}^{26+} + e^- \qquad (3.11)$$

The final result of these processes repeated billions of times and still continuing, is the presence of all the elements (up to $Z = 92$) in the universe today. Nuclear furnaces, stars and supernovae, produced and continue to produce all the elements heavier than hydrogen. So, in addition to the twinkling light we see, stars are producing heavier atoms, including carbon.

> Diamonds are pure carbon. All the atoms in all diamonds, natural or synthetic, were made in stars. The nursery rhyme at the beginning of the chapter is right: twinkling stars are "like a diamond in the sky."
>
>

Origin of Atoms   Chapter 3

The birth of a new star

All the elements that formed in the star's interior during its billions of years as well as the heavy elements formed in the supernova explosion scattered throughout the galaxy in which it occurred. The remains of exploded stars drift through galaxies, including our own, as enormous clouds of gas and dust. The surprising fact is that galaxies contain far more of this debris than all the matter in their stars. These interstellar gas clouds are not all alike. Those of greatest interest for our story are the cooler dust clouds that furnish the raw material for the formation of new stars and planetary systems. These clouds are approximately 10^{14} times less dense than the Earth's atmosphere.

> Our knowledge of the composition of these dust clouds, like our knowledge of the stars, is based mostly on spectroscopic evidence.

3.21. Check This   *Mass of hydrogen atoms in our Milky Way galaxy*

Assume that our galaxy, the Milky Way, is a circular disk about 20,000 light years thick with a radius of about 40,000 light years.

(a) What is the volume of our galaxy in m^3? Show how you arrive at your answer. *Hint:* A light year is the distance light travels in one year. Light travels at a speed of 3×10^8 m·s^{-1}.

(b) A cubic meter of air at the surface of the Earth contains about 2.4×10^{25} molecules which have a mass of about 1.2 kg. What is the mass of a cubic meter of hydrogen atoms that has a density 10^{14} times lower than the Earth's atmosphere?

(c) If hydrogen atoms, at the density you found in part (b), fill our galaxy, what is their total mass? Explain how you get your answer.

(d) Masses in the universe are sometimes given in units of "solar mass," that is, the mass of our star, the Sun, which is about 2.0×10^{30} kg. According to your result in part (c), how many solar masses of hydrogen atoms are present in our galaxy? Show how you get your answer.

(e) One estimate of the number of stars in our galaxy is 2×10^{11}. Does your result in part (d) provide any justification for the statement we made about the relative amounts of matter in stars compared to that in the gas clouds that fill the galaxy? Explain why or why not.

In the course of time, the atomic nuclei in the dust clouds cool enough to pick up electrons and form atoms and molecules. The dust clouds are rich in water and small organic molecules—the essentials for life. As new stars form, many are surrounded by enormous clouds of dust. When a large enough mass of this dust cools sufficiently, gravitational collapse leads to collections of dust grains that are drawn to one another to form larger and larger lumps of matter. Over time, planets form from these growing lumps as they hurtle through space sweeping up more dust. All the while, this matter is being held in orbit around the new star. One of these new stars is our Sun and one of the planets is Earth, whose story we will continue in Section 3.7.

Reflection and projection

The history of the universe is a fascinating story that can lead to useful insights about the universe today. We have included a brief synopsis of the story to provide an explanation for the origins of elemental atoms and to show how spectroscopic analysis provided much of the information required to develop that explanation. The Big Bang theory provides a model for the generation of the very lightest elements within the first few moments of the history.

Over the succeeding several billion years, these initially formed nuclei have interacted with one another by gravitational attraction and by collisions to form stars. In stars, the density and temperature are high enough for nuclei to fuse to form heavier nuclei, whose atoms are observed spectroscopically at the surface of the stars. We have written several of these nuclear reactions, but have not indicated what the rules are for writing and balancing them. We have also pointed out the large amounts of energy released in nuclear reactions, but have said nothing about the source of this energy. In order to understand more about the evolution of the elements and their abundance in the universe and on Earth, we need to examine nuclear reactions and energetics in more detail.

Section 3.4. Nuclear Reactions

Emissions from nuclear reactions

Before we try to balance nuclear fusion reactions, we need to consider some of their unique products. Both chemical and nuclear reactions can produce heat, sound, and light. Nuclear reactions also produce various particles and forms of energy. Two examples are **positrons**, e^+, which have the same mass but opposite charge of electrons, e^-, and **neutrinos**, v_e ("little neutrons"), which are neutral particles with high energies but almost zero mass. Nuclear particles you are more likely to have heard of are **alpha particles**, α, which are helium atom nuclei; **beta particles**, β, which are electrons; and neutrons, $_0^1 n^0$. Energy in the form of high-energy **gamma radiation**, γ, is also emitted by many nuclear reactions. Table 3.2 summarizes the emissions from nuclear processes.

> WEB Chap 3, Sect 3.4.1-2
> Study animations of the changes that occur in nuclear emissions.

> Sometimes you hear about "alpha rays" or "beta rays." Alpha and beta particles travel in straight lines from their source, like a ray of light, and they have high energies, so they were first thought to be forms of radiation (see Chapter 4). They were soon shown to be affected by electric and magnetic fields, indicating that they are charged particles, but the names stuck.

Table 3.2. Emissions from nuclear reactions.

Emission	Symbol	Contribution to atomic number, Z	Contribution to mass number, A
alpha	$\alpha = {}^{4}_{2}He^{2+}$	+2	+4
beta	$\beta^{-} = {}^{0}_{-1}e^{-}$	−1	0
gamma*	γ	0	0
positron	$\beta^{+} = {}^{0}_{+1}e^{+}$	+1	0
neutrino	ν_e	0	0
neutron	${}^{1}_{0}n^{0}$	0	1

* Gamma emissions (radiation) have different amounts of energy that depend upon the process that produces them; these differences are not relevant for our purposes.

3.22. Check This *Nuclear emissions*

(a) WEB Chap 3, Sect 3.4.2. Isotopic nuclei that emit beta particles do not emit positrons, and *vice versa*. Do the emissions shown in these animations exemplify this observation? Explain why or why not.

(b) Do these animations suggest that an isotopic nucleus might emit either an alpha or a beta particle? Explain why or why not.

Balancing nuclear reactions

The sum of the overall sequence of fusion reactions that produce helium nuclei from hydrogen nuclei in newly-forming stars, equation (3.6) is written in more complete notation as:

$$4\,{}^{1}_{1}H^{+} \rightarrow {}^{4}_{2}He^{2+} + 2\,{}^{0}_{+1}e^{+} \qquad \text{(energy release)} \qquad (3.12)$$

To balance a nuclear reaction, you need to know that *charge, mass number (A), and atomic number (Z) are all conserved in nuclear reactions*. You might find it useful to denote an electron as ${}^{0}_{-1}e^{-}$ and a positron as ${}^{0}_{+1}e^{+}$, as we have done in equation (3.12), to remind yourself of their respective contributions to the atomic numbers and mass numbers (zero) in a reaction, as shown in columns 3 and 4 of Table 3.2.

WEB Chap 3, Sect 3.4.3-4 Practice balancing nuclear reactions with these interactive animations.

In reaction (3.12), the sum of the charges on each side of the arrow is 4+. The sum of the mass numbers is four on each side; the mass of the positron is tiny and contributes zero to the mass number count. Each positron contributes +1 to the sum of the atomic numbers in the products. When this +2 contribution (from two positrons) is combined with the atomic number of helium, the sum of the atomic numbers in the products is 4, which is the same as in the reactants. A helium nucleus contains two neutrons, so reaction (3.12) must have converted protons, $_1^1p^+$, to neutrons, $_0^1n^0$:

$$_1^1p^+ \rightarrow \,_0^1n^0 + \,_{+1}^0e^+ \tag{3.13}$$

3.23. Worked Example *Balancing nuclear reactions: particles emitted*

Balance this nuclear reaction by finding the particle emitted.

$$^{14}C \rightarrow \,^{14}N + ?$$

Necessary Information: We will use the periodic table to find atomic numbers (and charge on the nucleus).

Strategy: Write the reactant and product nuclei with their mass numbers, atomic numbers, and charges shown explicitly. Add these quantities on each side of the reaction arrow and determine what values the unknown particle must have, in order to make the sums the same on each side. Find the particle with these characteristics in Table 3.2 and complete the problem by writing the reaction including this particle.

Implementation: $^{14}_6C^{6+} \rightarrow \,^{14}_7N^{7+} + ?$

	known reactants	known products	unknown
mass number sum	14	14	0
atomic number sum	6	7	–1

We require a particle that contributes zero mass and an atomic number of –1 (a negatively charged particle). The particle in Table 3.2 that fits this description is the electron (beta particle), $_{-1}^0e^-$. The balanced nuclear reaction is:

$$^{14}_6C^{6+} \rightarrow \,^{14}_7N^{7+} + \,_{-1}^0e^-$$

A carbon-14 nucleus forms a nitrogen-14 nucleus by emitting a beta particle.

Does the answer make sense? Making "sense" of the balanced equation, means checking *again* to see that the sums of the charges, mass numbers, and atomic numbers are the same on both sides.

3.24. Check This *Balancing nuclear reactions: particles emitted*

Balance these nuclear reactions by finding the particle emitted in each case.

(a) $^{150}Gd \rightarrow \,^{146}Sm + ?$

(b) $^{65}Zn \rightarrow \,^{65}Cu + ?$

Origin of Atoms   Chapter 3

3.25. Worked Example   *Balancing nuclear reactions: nuclear product*

Balance this nuclear reaction by finding the elemental nucleus formed.
$$^{232}\text{Th} \rightarrow ? + \alpha$$

Strategy: This kind of problem is solved in exactly the same way as in Worked Example 3.23. For the alpha particle, use the notation $^{4}_{2}\text{He}^{2+}$.

Implementation: $^{232}_{90}\text{Th}^{90+} \rightarrow ? + ^{4}_{2}\text{He}^{2+}$

	known reactants	known products	unknown
mass number sum	232	4	228
atomic number sum	90	2	88

We require a product with a mass number of 228 and an atomic number of 88 (an 88+ charge). The element with atomic number 88 (from the periodic table) is radium, so the unknown product is radium-228 and the balanced nuclear reaction is:

$$^{232}_{90}\text{Th}^{90+} \rightarrow {^{228}_{88}}\text{Ra}^{88+} + {^{4}_{2}}\text{He}^{2+}$$

Does the answer make sense? Check the sums of the charges, mass numbers, and atomic numbers on each side of the reaction *again*.

3.26. Check This   *Balancing nuclear reactions: nuclear product*

Balance these nuclear reactions by finding the elemental nucleus formed in each case.

(a) $^{18}\text{F} \rightarrow ? + \beta^+$

(b) $^{214}\text{Bi} \rightarrow ? + \beta^-$

3.27. Worked Example   *Balancing nuclear reactions*

The first human transmutation of one element into another used this reaction:
$$^{14}\text{N} + ? \rightarrow {^{17}}\text{O} + {^{1}}\text{H}$$

Balance the reaction to discover the particle used to produce the transmutation.

Strategy: Again, the basic approach is just as in the previous examples. The variation is that we don't know whether the missing reactant is another elemental nucleus or one of the particles from Table 3.2. Our analysis will give us the answer.

Implementation: $^{14}_{7}\text{N}^{7+} + ? \rightarrow {^{17}_{8}}\text{O}^{8+} + {^{1}_{1}}\text{H}^{+}$

	known reactants	known products	unknown
mass number sum	14	18	4
atomic number sum	7	9	2

We require a reactant with a mass number of 4 and an atomic number of 2 (a 2+ charge). The alpha particle, $^{4}_{2}\text{He}^{2+}$, fits this description, so the balanced nuclear reaction is:

$$^{14}_{7}\text{N}^{7+} + ^{4}_{2}\text{He}^{2+} \rightarrow ^{17}_{8}\text{O}^{8+} + ^{1}_{1}\text{H}^{+}$$

Does the answer make sense? Check the sums of the charges, mass numbers, and atomic numbers on each side of the reaction *again*.

3.28. Check This *Balancing nuclear reactions*

Balance these nuclear reactions by finding the missing elemental nucleus or particle in each case.

(a) $^{13}\text{C} + n \rightarrow ?$

(b) $^{1}\text{H} + ^{1}\text{H} \rightarrow ^{2}\text{H} + ?$

(c) $^{249}\text{Cf} + ^{18}\text{O} \rightarrow ? + 4n$ (This is the way we use particle accelerators to synthesize most elements beyond uranium.)

(d) $^{20}\text{Ne} + ? \rightarrow {}^{24}\text{Mg}$

3.29. Check This *Element-forming nuclear reactions*

(a) Write a balanced nuclear reaction forming nitrogen-14 from carbon-12.

(b) Write a series of balanced nuclear reactions adding alpha particles, one at a time, to form magnesium-24 from carbon-12.

(c) Write a balanced nuclear reaction forming magnesium-24 from two carbon-12 nuclei.

Positron-electron annihilation

Equation (3.12) does not finish this reaction sequence in stars. We haven't accounted for the fate of the positrons. Remember that there are electrons in stars. Positrons and electrons destroy (annihilate) one another and produce energy (gamma radiation):

$$^{0}_{+1}e^{+} + ^{0}_{-1}e^{-} \rightarrow 2\gamma \qquad (3.14)$$

The energy released in reaction (3.14) is 1.0×10^{8} kJ·mol^{-1}. Accounting for positron-electron annihilation in reaction (3.12) gives the net reaction:

$$4\,{}^{1}_{1}\text{H}^{+} + 2\,{}^{0}_{-1}e^{-} \rightarrow {}^{4}_{2}\text{He}^{2+} \qquad \text{(energy released)} \qquad (3.15)$$

The source of the energy released in nuclear reactions is the topic of Section 3.5.

Here on Earth, we take advantage of reaction (3.14) in *positron emission tomography* or PET scans. In a PET scan, positron emitting nuclei, usually ^{18}F [see Check This 3.26(a)], are used to identify active areas of the brain in the following way. Active brain cells require glucose for energy. A PET scan subject is injected with a compound that acts like glucose but contains ^{18}F. Once the compound has been taken up by brain cells, positrons emitted by the ^{18}F react with nearby electrons to produce γ radiation, which is detected by an array of detectors around the person's head. A computer converts this information into a picture, as in Figure 3.12, showing the location of the active brain cells, that is, those that are using the glucose-like substance. PET scans can be used to locate regions of a patient's brain that are malfunctioning, due to injury, disease, or genetics, in order to diagnose problems and help determine effective treatments.

> Tomography (*tomos* = slice) means pictures of brain "slices." The images of the locations of the positron-emitting nuclei are constructed as though slices of the brain were being examined.

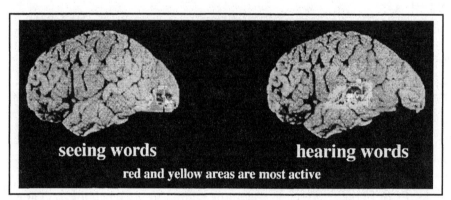

Figure 3.12. PET scans showing active areas of the brain during seeing and hearing activities.

3.30. Investigate This *How do the emissions from a radioactive sample differ?*

Do this as a class investigation and work in small groups to analyze the results. The most common kind of household smoke detectors contain a tiny amount of radioactive americium-241 dioxide. Use this source from a dismantled smoke detector as a sample and a radiation survey meter, Figure 3.11, that is sensitive to alpha, beta, and gamma emissions.

(a) (b)

Figure 3.13. Radiation survey meter and dismantled smoke detector with its label.

Measure and record the radioactivity of the sample in whatever units the meter provides. Insert a piece of thin cardboard between the sample and the probe and measure and record the amount of radioactivity detected. Repeat the measurement and recording after adding a thin sheet of aluminum to the cardboard between the sample and the probe and then after adding a sheet of lead.

3.31. **Consider This** *What is(are) the emission(s) from decay of americium-241?*

(a) Were there substantial changes in the radioactivity when any of the materials was inserted between the source and the detector in Investigate This 3.30? For which ones, if any?

(b) The emissions from radioactive sources have different abilities to penetrate matter. Alpha particles are stopped by a sheet of cardboard, but it requires a sheet of aluminum to stop beta particles. Gamma radiation is only stopped by a dense metal like lead. Based on your results from Investigate This 3.30, what is(are) the emission(s) from americium-241? Give your reasoning.

Radioactive decay

The only nuclear reaction we have discussed so far is fusion, but the nuclear reaction that is probably most familiar is **radioactive decay**. Radioactive decay transforms one nucleus, a **radioisotope**, into a different nucleus by emission of one or more of the particles in Table 3.2. The "radio" in these terms arises because all these emissions were once thought to be *radiation*. The nuclear emissions you have heard about before were probably from radioisotopes; an example is the americium-241 in the smoke detector in Investigate This 3.30. All the nuclear reactions in Worked Examples 3.23 and 3.25 and in Check This 3.24 and 3.26 are radioactive decays. In a radioactive decay process the reactant nucleus is often called the **parent** and the product nucleus is called the **daughter**. For example, the nuclear equation for the radioactive decay of the uranium-235 nucleus is:

$$^{235}_{92}\text{U}^{92+} \rightarrow \,^{231}_{90}\text{Th}^{90+} + \,^{4}_{2}\text{He}^{2+} \tag{3.16}$$

The uranium-235 parent decays to form the thorium-231 daughter by alpha emission.

3.32. Check This *Daughter nuclei in radioactive decays*

Determine the daughter nucleus and write the balanced nuclear reaction for these decays.

(a) Thorium-231 decays by beta emission.

(b) $^{241}\text{Am} \rightarrow ?$ + (decay product(s) from Investigate This 3.30 and Consider This 3.31)

3.33. Investigate This *How can you count a moderately large sample?*

For this investigation use a sample of 150-200 (exact number unknown) coins or candies, such as M&M's®. Place your sample in a covered container, shake and invert the container to thoroughly mix the sample, and then spill it out on a large flat surface. Remove all the coins that show heads (or all the M&M's® whose "m" is showing). Place the remaining sample in the container and repeat the shaking, spilling, and selection. Repeat this procedure, keeping track of how many times you do it, until 12 ± 3 coins or candies remain. Work in your group to discuss how to use your data to determine the number of coins or candies in the original sample.

3.34. Consider This *Can you count backward to determine the original sample size?*

(a) Assume that each object you used in Investigate This 3.33 has an equal probability of ending up in either of its two states (heads or tails; "m" or no "m") when it is spilled from the container. What fraction of the sample you spill out will be removed at each round of the procedure? Explain the reasoning for your answer.

(b) Is it possible to use your answer in part (a) to work backward from the final small sample to determine the original sample size? If so, explain how and determine the original sample size. To check your procedure, count the original objects.

Half life

Radioisotopes decay by well-defined reactions, like equation (3.16), and each radioisotope has a definite half life for its radioactive decay. The **half life** is the time required for one half of a radioactive sample to decay to its products; the half life of uranium-235 decaying by reaction

WEB Chap 3, Sect 3.4.5-9
Use interactive animations to determine the half life of a radioisotope.

(3.16) is 7.1×10^8 years. The half lives of different radioactive nuclei are unique and can often be used as way of dating different materials. Radioisotope half lives vary from much less than a second to more than 10 billion years. Figure 3.14 shows graphically the fraction of a radioactive sample remaining after a given number of half lives.

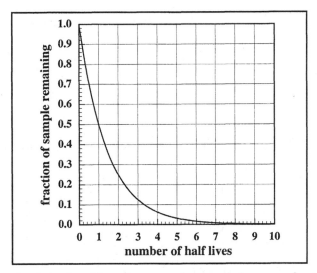

Figure 3.14. Fraction of original radioactive sample remaining as a function of half life.
After 10 half lives, a little less than one-thousandth of the original sample remains.

3.35. **Consider This** *What fraction of a sample remains after n half lives?*

If, at time zero, a sample contains N_0 radioactive nuclei, then, after one half life, the number of nuclei remaining unreacted, N_1, will be $(1/2)N_0$. After another half life has elapsed, half of this remaining sample will have reacted. The number of nuclei remaining after two half lives, N_2, will be $(1/2)(1/2)N_0 = (1/2)^2 N_0$.

(a) Show that:
$$\text{fraction of sample remaining after } n \text{ half lives} = f_n = \frac{N_n}{N_0} = (1/2)^n \tag{3.17}$$

(b) Does equation (3.17) describe the curve in Figure 3.14? Explain why or why not.

(c) How is equation (3.17) related to your answers and results in Consider This 3.34? Can you use equation (3.17) to determine the number of objects in your original sample in Investigate This 3.33? Explain.

3.36. **Worked Example** *Sample age related to carbon-14 half life*

All living plants contain about the same percentage of carbon-14. When the plant dies, no more carbon-14 is added; the isotope decays by beta particle emission (see Worked Example

3.23) with a half life of 5730 years. A sample of wood from an archaeological excavation had about 20% as much carbon-14 as living wood. About how old is the sample?

Necessary Information: We will use equation (3.17).

Strategy: If we know the number of half lives since the sample of wood was no longer part of a growing plant or tree, we can multiply by the length of a half life to get the age of the sample. We will determine the number of half lives required for the sample to have decayed to 20% of its original C-14 by applying equation (3.17).

Implementation: One way to use equation (3.17) is to take the logarithm of both sides of the equation:

$$\ln(f_n) = \ln(1/2)^n = n \cdot \ln(1/2) = -0.693n \quad (3.18)$$

We know the decimal fraction of C-14 remaining, 0.20, so we can calculate the number of half lives, n:

$$n = \frac{\ln(f_n)}{-0.693} = \frac{-1.61}{-0.693} = 2.3 \text{ half lives}$$

The time since the wood was no longer part of a living plant is:

$$\text{sample age} = \left(\frac{5730 \text{ yr}}{1 \text{ half life}}\right)(2.3 \text{ half lives}) = 13 \times 10^3 \text{ years}$$

Does the answer make sense? One-quarter (25%) of a radioactive sample remains after two half lives have gone by. It must take a little longer for another 5% to decay, leaving only 20% (0.20) of the original sample, and this is what we found; the number of half lives is a bit more than two. You can also check the result using Figure 3.14, which is reproduced here with a horizontal and vertical line showing how to read the graph to go from fraction remaining to number of half lives elapsed.

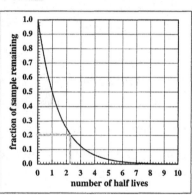

3.37. Check This *Radioisotope decay and half life*

Phosphorus-32, which is extensively used in biochemical experiments, decays by beta emission with a half life of 14.3 days. Phosphorus-32 waste can be disposed of as trash, if it has been stored in radioactive waste containers until less than 1% of the original sample is left. About how long must a sample of phosphorus-32 be stored before it can be placed in the trash? Explain your reasoning.

> ***3.38. Check This*** *Half-life determination*
>
> **(a)** WEB Chap 3, Sect 3.4.6-9. Draw a decay curve for an element undergoing radioactive decay. Clearly indicate on the curve how you would measure the half life of the element. *Hint:* Equation (3.17) might be helpful.
>
> **(b)** Clearly indicate on the curve in part (a) how you would show that the half life is a constant, no matter what the size of the sample.

Radioisotopes and human life

Radioisotopes can be harmful to living things. For example, strontium-90, one of the many radioactive products from nuclear weapons explosions decays by emission of a beta particle:

$$^{90}_{38}\text{Sr} \rightarrow {}^{90}_{39}\text{Y} + {}^{0}_{-1}e^- \tag{3.19}$$

During the 1950s, when nations were testing nuclear weapons above ground, strontium-90 was present in the particles that fell on plants, was eaten by cows, and got into their milk. Milk is a good source of calcium which is essential for bone growth, especially in children. Since strontium and calcium are chemically quite similar (in the same family of the periodic table), some of the strontium-90 ended up in children's bones. The beta particles emitted could destroy some of the cells in the bone marrow where red blood cells are produced and could cause anemia or leukemia.

Radioisotopes are also helpful. For example, they are used extensively in medical diagnosis and treatment, as we saw above for PET scans. The radioisotope most commonly used in medical diagnosis is technetium-99m, $^{99m}_{43}\text{Tc}$. The "m" stands for metastable and means the nucleus is not

> The name technetium comes from the Greek word for "artificial." All Tc isotopes are unstable. The longest lived has a half life of 2.6×10^6 years, so any of the natural element that was present when the Earth formed has long ago decayed. Tc was the first synthetic element made.

in its most stable form; it decays by gamma emission with a half life of 6 hours:

$$^{99m}_{43}\text{Tc} \rightarrow {}^{99}_{43}\text{Tc} + \gamma \tag{3.20}$$

$^{99m}_{43}\text{Tc}$ is the decay product of molybdenum-99 (which has a half life of about 67 hours):

$$^{99}_{42}\text{Mo} \rightarrow {}^{99m}_{43}\text{Tc} + {}^{0}_{-1}e^- \tag{3.21}$$

Molybdenum-99 is produced in nuclear reactors by neutron bombardment of Mo-98:

$$^{98}_{42}\text{Mo} + {}^{1}_{0}n \rightarrow {}^{99}_{42}\text{Mo} \tag{3.22}$$

The Mo-99 is packaged in small Tc-99m generators, Figure 3.15; many hospitals have these generators and routinely use $^{99m}_{43}\text{Tc}$. A few radioisotopes commonly used in medical applications are listed in Table 3.3. Radioisotopes that are medically useful can also be biologically damaging, so, whenever they are used, radioactive elements and their compounds have to be handled and disposed of with special precautions (see Check This 3.37).

Figure 3.15. A generator containing $^{99}_{42}Mo$ which produces $^{99m}_{43}Tc$ for medical purposes.

Table 3.3. Some radioisotopes used in biomedical applications.

Radioisotope	Emission	Half life	Uses
$^{99m}_{43}Tc$, technetium-99m	γ	6 hrs	imaging brain, lung, liver, spleen, bone
$^{18}_{9}F$, fluorine-18	β^+	110 mins	PET scans
$^{131}_{53}I$, iodine-131	β^-	8 days	treatment of the thyroid
$^{67}_{31}Ga$, gallium-67	γ	78 hrs*	treatment of lymphomas
$^{51}_{24}Cr$, chromium-51	γ	28 days*	blood tests
$^{201}_{81}Tl$, thallium-201	γ	73 hrs*	diagnosing heart defects
$^{60}_{27}Co$, cobalt-60	β^-, γ	5.26 yrs	radiation therapy for cancer

* Ga-67, Cr-51, and Tl-201 decay by capturing one of their core electrons to transform a proton to a neutron, *e.g.*, $^{67}Ga + e^- \rightarrow {}^{67}Zn + \gamma$.

3.39. Check This *Electron capture reactions*

(a) Write the electron capture reaction for Ga-57 (see note to Table 3.3) with full symbols for all nuclei and particles to show that mass number, atomic number, and charge are conserved.

(b) Write the electron capture reaction for Tl-201.

Chapter 3 — Origin of Atoms

Section 3.5. Nuclear Reaction Energies

Mass is not conserved in nuclear reactions

The unimaginably large energies produced by the stars, as well as nuclear explosions and nuclear power reactors on Earth are evidence of the energy that can be released in nuclear reactions. The source of this energy arises from a fundamental property of nuclear reactions: *mass is not conserved in nuclear reactions*. Nuclear reactions convert some mass to energy or *vice versa*. From his relativity theory, Albert Einstein (German-born physicist, 1879-1955) derived the **mass-energy equivalence** relationship:

$$\Delta E = \Delta m \cdot c^2 \qquad (3.23)$$

In equation (3.23), c is the **speed of light**, 3.00×10^8 m·s^{-1} and, if the change in mass, Δm (final minus initial mass), is in kg, the equivalent amount of energy, ΔE, is in joules, J.

3.40. Check This *Energy equivalence of mass*

(a) If 0.1 mg (= 1×10^{-7} kg; about the mass of a tiny grain of salt) of mass is converted to energy, what is ΔE? *Hint:* The final mass is zero and the initial mass is 0.1 mg, so Δm is –0.1 mg.

(b) Is the process in part (a) exothermic or endothermic? Explain how you know.

3.41. Worked Example *Energy release in nuclear fusion*

What is the energy change, ΔE, in the net reaction of four protons and two electrons to form a helium nucleus?

$$4\,{}^{1}_{1}H^+ + 2\,{}^{0}_{-1}e^- \rightarrow {}^{4}_{2}He^{2+} \qquad \text{(energy released)} \qquad (3.15)$$

What is the energy change when 1.00 g of the reactants undergo reaction (3.15)?

Necessary Information: We have the masses of the reactants in Table 3.1 and use a reference handbook to find that the mass of a ^{4}He^{2+} nucleus is 6.64466×10^{-24} g.

Strategy: Use the particle mass data, in kilograms, to find the total mass of reactants and products and then take the difference to find, Δm, the change of mass in the reaction. Then use the mass-energy equivalence relationship, equation (3.23), to calculate ΔE, the energy change. Convert this energy for four protons and two electrons to the energy change that one gram (0.00100 kg) of reactants would produce.

Implementation:

reactant mass = 4·(1.67262 × 10^{-27} kg) + 2·(9.1094 × 10^{-31} kg) = 6.69230 × 10^{-27} kg

product mass = 6.64466 × 10^{-27} kg

Δm = 6.64466 × 10^{-27} kg – 6.69230 × 10^{-27} kg = –0.04764 × 10^{-27} kg

Origin of Atoms   Chapter 3

energy equivalent = $\Delta E = (-0.04764 \times 10^{-27} \text{ kg}) \cdot (3.00 \times 10^8 \text{ m·s}^{-1})^2 = -4.29 \times 10^{-12}$ J

We can convert to one gram (0.00100 kg) of reactants, using the equivalence:

6.69×10^{-27} kg of reactants = -4.29×10^{-12} J

energy from 1.00 g of reactant = $(0.00100 \text{ kg}) \cdot \left(\dfrac{-4.29 \times 10^{-12} \text{ J}}{6.69 \times 10^{-27} \text{ kg}} \right) = -6.41 \times 10^{11}$ J

Does the Answer Make Sense: The reaction is highly exothermic. We have said several times that the energies released in nuclear reactions are large. The –640 billion joules we calculated for one gram of reactants is a large amount of energy (enough to vaporize 250 metric tons of water). Note that the mass data we used have six significant figures, but the difference between them is small and has only four significant figures.

3.42. Check This   *Energy release in nuclear fusion*

(a) WEB   Chap 3, Sect 3.5.1. Calculate the energy change in the nuclear fusion reaction represented in this animation. The masses of C-12 and O-16 nuclei are, respectively, 19.92101×10^{-24} g and 26.55291×10^{-24} g.

(b) How is the energy change represented in the animation?

3.43. Check This   *Nuclear reaction energies*

(a) Show that reactions (3.1), (3.2), and (3.3) can be combined to give the net reaction for the nucleosynthesis of helium-4 nuclei in the first few seconds after the Big Bang:

$$2\,{}^1p^+ + 2\,{}^1n^0 \rightarrow {}^4\text{He}^{2+} \tag{3.24}$$

Hint: Consider how many times reaction (3.1) must occur to provide the reactants for the succeeding reactions.

(b) What is the energy change in the nuclear reaction represented by equation (3.24)?

(c) How does your result in part (b) compare to the energy change calculated in Worked Example 3.41 for nucleosynthesis of helium-4 by a different pathway? What could be responsible for a difference in energy between the two pathways? *Hint:* Consider the conversion of a proton, ${}^1p^+ = {}^1_1\text{H}$, to a neutron, ${}^1n^0$.

Comparison of nuclear and chemical reactions

Nuclear reactions are very different from chemical reactions. Chemical reactions involve only electrons. Nuclear reactions (except electron capture) involve only nuclei and are essentially independent of what any electrons in the vicinity are doing. The biggest observable difference between chemical and nuclear reactions is that nuclear reactions involve far greater amounts of energy. Table 3.4 compares the energy released in chemical reactions and two kinds of nuclear reactions: the nuclear fusion reactions we have considered so far, which build heavier nuclei from lighter nuclei and **nuclear fission** reactions, which split heavier nuclei into two lighter nuclei plus neutrons.

> Fission is from the Latin *findere* = to split.

Table 3.4. Comparison of chemical and nuclear reaction energies.
The energy is for the reaction of one gram of reactant in each case.

Reaction	ΔE, kJ
Combustion of methane (one gram of CH_4) $CH_4(g) + 2O_2(g) \rightarrow CO_2(g) + 2H_2O(l)$	-56
Spontaneous nuclear fission of uranium–235 $^{235}U^{92+} \rightarrow {}^{90}Sr^{38+} + {}^{143}Xe^{54+} + 2{}^{1}n^0$ (one example)	-7.0×10^7
Nuclear fusion of hydrogen–1 to form helium–4 $4{}^{1}H^+ + 2e^- \rightarrow {}^{4}He^{2+}$	-64×10^7

3.44. Worked Example *Mass loss in nuclear and chemical reactions*

Assuming that the energies in Table 3.4 represent mass lost in the reactions, compare the mass loss when one gram of hydrogen is converted to helium by nuclear fusion to the mass loss when one gram of methane is burned.

Strategy. We apply the Einstein mass-energy equivalence relationship to the two relevant energies in Table 3.4.

Implementation:

$$\text{mass loss in fusion} = \Delta m = \frac{\Delta E}{c^2} = \frac{-6.4 \times 10^{11} \text{ J}}{\left(3.00 \times 10^8 \text{ ms}^{-1}\right)^2} \approx -7 \times 10^{-6} \text{ kg} = -7 \times 10^{-3} \text{ g}$$

$$\text{mass loss in combustion} = \Delta m = \frac{-5.6 \times 10^4 \text{ J}}{\left(3.00 \times 10^8 \text{ ms}^{-1}\right)^2} \approx -6 \times 10^{-13} \text{ kg} = -6 \times 10^{-10} \text{ g}$$

Does the Answer Make Sense: The energies are different by about seven orders of magnitude, so the mass losses should be different by seven orders of magnitude, as they are.

3.45. Check This *Mass loss in nuclear fission*

Assuming that the energy of the nuclear fission reaction in Table 3.4 represents mass lost in the reaction, calculate the mass loss when one gram of U-235 reacts as shown. How does this mass loss compare to the that for the fusion of hydrogen nuclei to yield helium?

The results in Worked Example 3.44 show that the mass loss for hydrogen-to-helium nuclear fusion, reaction (3.15), is about 7 mg (7×10^{-3} g) for each gram of hydrogen reacted. This is a mass that is easily measurable with a good laboratory balance. The mass loss in the chemical reaction, combustion of methane, is about 0.6 ng (6×10^{-10} g), a mass too small to be measured on any balance. All of our usual mole and mass calculations for chemical reactions assume that *mass is conserved in chemical reactions*. The assumption is valid because the reaction energies are so small that their mass equivalence is not observable.

3.46. Worked Example *Energy released in a hypothetical nuclear reaction*

How much energy would be released in this hypothetical nuclear reaction?

$$13p + 14n \rightarrow {}^{27}Al^{13+}$$

The mass of an Al-27 nucleus is 44.7895×10^{-24} g.

Necessary Information: We have the Einstein mass-energy equivalence relationship and the masses of the reactant protons and neutrons are given in Table 3.1.

Strategy: We find the difference in mass between the products and reactants in the reaction and convert the mass change to an energy change.

Implementation:

Mass change in forming one Al-27 nucleus:

$\Delta m = 44.7895 \times 10^{-27}$ kg $- [13(1.67262 \times 10^{-27}$ kg$) + 14(1.67493 \times 10^{-27}$ kg$)]$

$= -0.4038 \times 10^{-27}$ kg

Energy equivalent of mass change in forming one Al-27 nucleus:

$\Delta E = (-0.4038 \times 10^{-27}$ kg$) \cdot (3.00 \times 10^8$ m·s$^{-1})^2 = -3.63 \times 10^{-11}$ J

Does the Answer Make Sense? Comparing this result with Check This 3.43(b), we see that there is an increased mass change (loss) as the size (mass) of the synthesized nucleus increases. Do Check This 3.47 to see if this pattern continues.

3.47. Check This   Energy released in a hypothetical nuclear reaction

How much energy would be released in this hypothetical nuclear reaction?

$$82p + 126n \rightarrow {}^{208}\text{Pb}^{82+}$$

The mass of a Pb-208 nucleus is 345.2788×10^{-24} g. Does your result fit the pattern of increasing mass loss for increasing mass of the synthesized nucleus? Explain.

Nuclear binding energies

The masses of most nuclei are known to at least six decimal places. You can do the kind of calculations in Worked Example 3.46 and Check This 3.47 to determine the mass loss and energy released in the hypothetical nuclear reaction forming any nucleus from its component protons and neutrons. Since mass is lost and energy is released in these nuclear syntheses, the energy of the nucleus is lower (more negative) than the separated protons and neutrons, as illustrated in Figure 3.16. The energy zero is the separated protons and neutrons. The nuclei are more stable than their separated components. The energy released in the formation of a nucleus from its protons and neutrons is known as the **nuclear binding energy**.

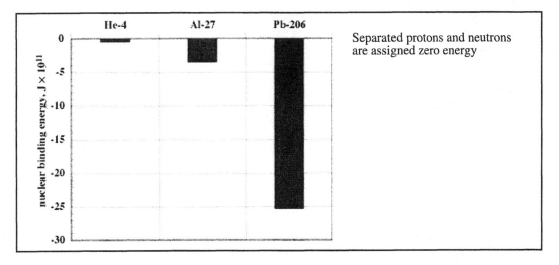

Figure 3.16. Energy released in the synthesis of He, Al, and Pb nuclei.

The greater the mass of the nucleus, the more mass is converted to energy in its formation from protons and neutrons. In heavier nuclei more protons and neutrons are bound together, so it makes sense that more energy will be released when they bind, as shown in Figure 3.16. But, is the same amount of energy released every time a new proton or neutron is added? To answer this question, we divide the total binding energy for each nucleus by the number of nucleons (protons plus neutrons = mass number, A) in the nucleus. For Pb-208, we divide -26.2×10^{-11} J by 208 nucleons and get -0.126×10^{-11} J·nucleon^{-1}. This procedure gives us the **binding energy per nucleon** shown in Figure 3.17 for our example nuclei. The binding energy per nucleon is comparable for all three nuclei and is about -0.1×10^{-11} J·nucleon^{-1} for the formation of a single nucleus. As you see, the binding energy per nucleon does vary a bit. The intermediate mass nucleus, Al-27, is somewhat more stable than the lower and higher mass nuclei, He-4 and Pb-208.

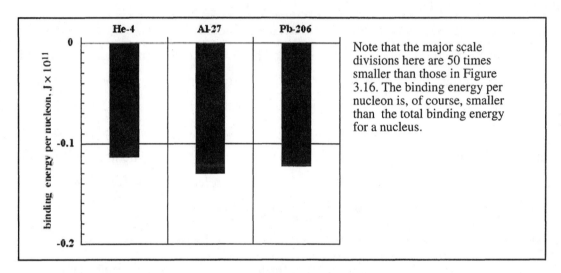

Figure 3.17. Energy released per nucleon in the synthesis of He, Al, and Pb nuclei.

You will usually find nuclear binding energies given per mole rather than per individual nucleus. A mole of nuclei is Avogadro's number of nuclei, so we must multiply the binding energies per nucleon (for a single nucleus) by Avogadro's number, 6.022×10^{23} mol^{-1}, to get the molar binding energy. For Pb-208 we have:

$$\text{binding energy per nucleon} = (-0.126 \times 10^{-11} \text{ J}) \cdot (6.022 \times 10^{23} \text{ mol}^{-1})$$

$$= -76.0 \times 10^{10} \text{ J·mol}^{-1} = -76.0 \times 10^{7} \text{ kJ·mol}^{-1}$$

Figure 3.18 shows the binding energies per nucleon, in kJ·mol^{-1}, for all nuclei. The figure provides a basis for comparison among nuclei; *the more stable nuclei are lower on the diagram*.

3.48. Check This   *Binding energy per nucleon for carbon*

The mass of a carbon-12 nucleus is 19.92101×10^{-24} g. What is the binding energy per nucleon, in kJ·mol^{-1}, for carbon-12? Compare your answer with the value in Figure 3.18.

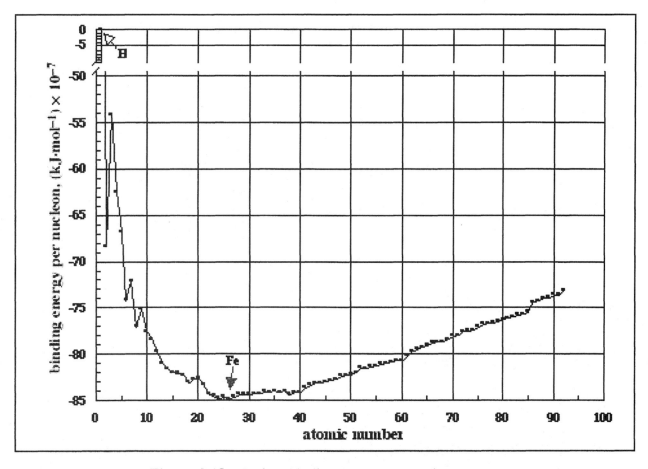

Figure 3.18. Nuclear binding energy per nucleon.
Zero energy on this plot is for hydrogen, Z = 1. The lowest energy, most stable, nucleus is iron, Z = 26.

3.49. Consider This *Trends in nuclear stability*

(a) In Figure 3.18, what is the trend in nuclear stability up to iron-56?

(b) What is the trend in nuclear stability after iron-56?

Fusion, fission and stable nuclei

Figure 3.18 shows that iron-56 is the most stable nucleus; it lies in an energy valley that includes the metals in the middle of the fourth period of the periodic table. On the low mass (lower atomic number) side of the valley, the products of nuclear fusion reactions, such as ^{12}C (from three ^4He) or ^{16}O (from ^{12}C and ^4He), are more stable than the reactants. Figure 3.19 illustrates the energy changes in these two reactions. Highly exothermic reactions, giving products that are more stable than the reactants, are usually favored to proceed. Such reactions

> **WEB Chap 3, Sect 3.5.4**
> Study animations based on the nuclear binding energy curve.

account for the stellar synthesis of the nuclei up to iron and neighboring metals, as outlined in Section 3.3.

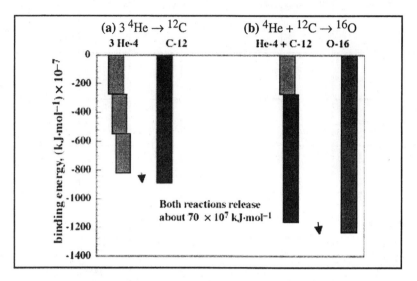

Figure 3.19. Net energy release in two nuclear fusion reactions.
Arrows indicate the energy released by the reaction.

3.50. Check This *Energy release in nuclear fusion reactions*

Show how the energies represented by the bars in Figure 3.19 are obtained and check the statement that both reactions release about the same energy. *Hint:* The data and results from Check This 3.42 and 3.43 could be useful.

On the high mass (higher atomic number) side of the stability valley in Figure 3.18, nuclear fusion is not a favored process. The heavier the nucleus, the less stable it is, so fusion reactions are endothermic. This is the reason why, in the evolution of the elements, Section 3.3, the heavier nuclei are formed mainly by neutron capture and subsequent beta particle emission, rather than nuclear fusion. For heavier nuclei, fission reactions that produce lighter, more stable nuclei from heavier ones are exothermic. Two of the naturally-occurring isotopes of uranium, ^{235}U and ^{238}U, undergo **spontaneous nuclear fission** to produce two lighter nuclei and two or more neutrons as products. A representative reaction for ^{235}U is:

$$^{235}U^{92+} \rightarrow {}^{90}Sr^{38+} + {}^{143}Xe^{54+} + 2\,{}^{1}n^{0} \tag{3.25}$$

The energy released in reaction (3.25) is shown in Table 3.4 (for one gram of ^{235}U reacting) and illustrated in Figure 3.20 (for one mole of ^{235}U reacting). Figure 3.20 shows that the strontium-90 and xenon-143 nuclei formed in this fission reaction are more stable than the reactant uranium-235 nucleus. Several isotopes of plutonium, an element synthesized in nuclear reactors, also undergo spontaneous nuclear fission.

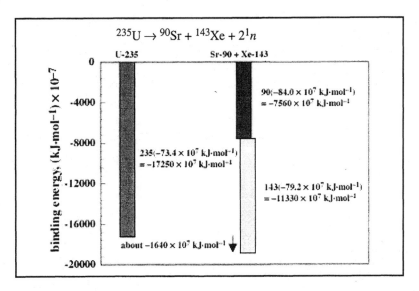

Figure 3.20. Net energy release in a spontaneous nuclear fission reaction.
The binding energies per nucleon used here are calculated for the isotopic nuclei in the reaction. They differ slightly from those in Figure 3.18, which are calculated for the most abundant isotope of each element.

3.51. Check This   *Fusion or fission?*

Indicate whether each of these elemental nuclei would be more likely to undergo fusion or fission reactions and why:

(a) Pu-240   (b) Li-6   (c) Fe-56

Nuclear chain reactions

In addition to spontaneous fission, these heavy nuclei also undergo **collision-induced fission** when a neutron collides with the nucleus and gives it extra energy. An example of a collision-induced fission is reaction (3.25) initiated by collision with a neutron:

$$^{235}_{92}U^{92+} + ^{1}n^0 \rightarrow ^{90}_{38}Sr^{38+} + ^{143}_{54}Xe^{54+} + 3\,^{1}n^0 \qquad (3.26)$$

The same amount of energy is released by this collision-induced fission reaction as is released by the spontaneous fission. Collision-induced fission is important because it speeds up the fission process. Note that more neutrons are produced in reaction (3.26) than are used up. These product neutrons could go on to cause more fission and even more neutrons—a **chain reaction** that is represented in Figure 3.21.

> **WEB Chap 3, Sect 3.5.2-3**
> Collision-induced fission reactions are represented by these animations.

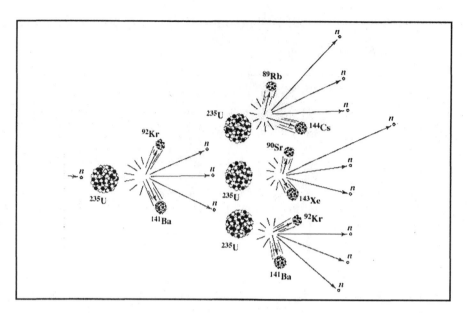

Figure 3.21. Representation of a ^{235}U chain reaction, initiated by collision with a neutron.

3.52. Check This *Collision-induced nuclear fission reactions*

Reaction (3.26) is one of the collision-induced fission reactions represented in Figure 3.21. Write balanced nuclear reaction equations for the other collision-induced fission reactions shown in the figure.

If every neutron produced by the fissions in Figure 3.21 goes on to cause another fission, the amount of energy released triples at each stage. If the mass of the ^{235}U is large enough, most of the neutrons produced collide with another ^{235}U nucleus before they can escape. In this **critical mass**, the chain continues and causes the fission of a large amount of the ^{235}U in less than one second. The energy released in this process is enormous and causes a massive explosion. Such uncontrolled chain reactions are the heart of nuclear bombs. To control this chain reaction, the ^{235}U can be contained in such a way that many of the neutrons produced are absorbed by other materials before they can collide with another ^{235}U nucleus; Figure 3.22 shows how this is done in a nuclear reactor. Controlled chain reactions are at the core of power and medical nuclear reactors. Uranium that has been enriched in the ^{235}U isotope is used in nuclear power reactors to produce electricity.

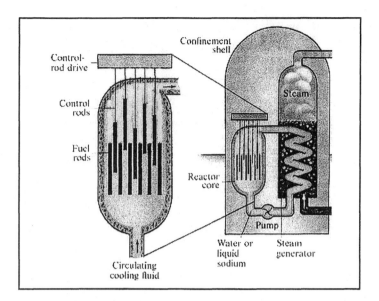

Figure 3.22. A nuclear power reactor.
The rods of neutron-absorbing material, such as Cd, are moved in and out to control the U-235 chain reaction.

3.53. **Consider This**   *Chain reactions*

(a) WEB Chap 3, Sect 3.5.3. Assume that you have a critical mass of U-235 that reacts as shown in this animation. If every neutron produced reacts with another U-235 nucleus within one microsecond, 10^{-6} s, about how long will it take for a mole or more of nuclei, 6.022×10^{23} nuclei, to be reacting at each step? Explain clearly how you get your answer. *Hint:* The sort of reasoning we used in Consider This 3.35 and Worked Example 3.36 can be applied here as well.

(b) Your result in part (a) represents an uncontrolled nuclear chain reaction. Consider now a controlled chain reaction, as in a nuclear reactor, Figure 3.22. Assume that for every 10,000 nuclear disintegrations, 10,001 neutrons go on to cause further reaction. All the other neutrons are lost or absorbed by the control rods. About how many reaction steps will it take for a mole or more of nuclei, 6.022×10^{23} nuclei, to be reacting at each step? Explain clearly how you get your answer. *Hint:* In this case the amount of reaction increases by 1.0001 at each stage.

Reflection and projection

We interrupted our history of the universe to look in more detail at the nuclear reactions that are the energy source for stars and the source of all the elements more massive than hydrogen. Nuclear reactions, including radioactive decay of unstable nuclei, produce enormous amounts of energy and emit particles and radiation that carry off this energy. All of the emitted particles, as well as the nuclear reactants and products have to be accounted for in balancing nuclear reactions which conserve charge, mass number (A = number of nucleons), and atomic number (Z). Nuclear reactions do not, however, conserve mass; the products of many nuclear reactions have less mass than the reactants. The lost mass is converted to energy, which we can calculate using the Einstein mass-energy equivalence relationship. This is the source of the energy from nuclear reactions.

Nuclei have less mass than the sum of the masses of their individual protons and neutrons. The energy equivalent of this mass loss is the nuclear binding energy, which we usually express as binding energy per nucleon, so we can compare one nucleus with another on the same basis. The binding energy per nucleon varies with atomic number and has a minimum at iron-56. Lighter elements generally undergo fusion reactions that produce heavier nuclei that bring them closer to the binding energy minimum. Heavier nuclei are more likely to undergo fission reactions that produce products closer to the binding energy minimum.

Now we will resume our story of the evolution of the universe and its atoms. Our first goal is to try to correlate the observed abundance of the elements in the universe from Section 3.1 with the relative stabilities of their nuclei from Section 3.5. Then we will go on to discuss the formation of planets, in particular, the Earth and its elements.

Section 3.6. Cosmic Elemental Abundance and Nuclear Stability

3.54. Consider This *How are elemental abundance and nuclear stability related?*

This graph is a combination of Figures 3.7 and 3.18 for the low atomic number elements. The left-hand axis is the binding energy per nucleon (plotted in black) and the right-hand axis is the logarithm of cosmic elemental abundance (red and blue plots for even and odd atomic numbers, respectively).

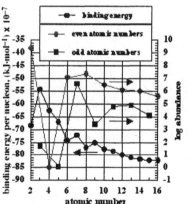

(a) Compare the binding energies of lithium, beryllium, and boron with all the others. Does your comparison help explain the abundances of these three elements? Explain why or why not.

(b) Do binding energies help explain the abundances of nitrogen and fluorine relative to carbon, oxygen, and neon? Explain why or why not.

(c) Do reactions such as (3.7) and (3.8) help explain the difference in abundance of even and odd atomic number elements? Explain why or why not.

Trends in elemental abundance

The evolution of the elements we discussed in Section 3.3 proceeded from lighter to heavier nuclei in sequence. The syntheses of heavier elements depends on the existence of lighter elements that can fuse (or capture neutrons) to make them. The syntheses also depend on a high enough temperature to make them possible. Many stars are not massive enough to go through the complete process outlined in Section 3.3 and these produce few heavier elements. The combination of all these factors creates the general trend of decreasing abundance with increasing atomic number that was shown in Figure 3.7 and repeated in Figure 3.23. The trend continues for the rest of the heavier elements that are not shown in the figures.

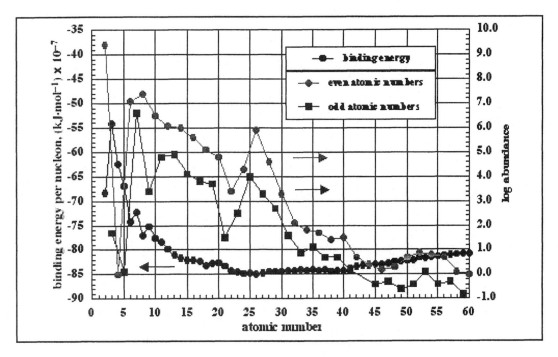

Figure 3.23. Comparison of cosmic elemental abundance with binding energy per nucleon.
Hydrogen is omitted from these plots because its only source in the universe was the Big Bang.

As we noted in Section 3.1, there are two striking exceptions to the relatively smooth downward trend of the abundances as a function of atomic number. The abundances of Li, Be, and B (atomic numbers 3, 4, and 5) are seven to eight orders of magnitude less than we would expect, if they behaved like the elements just before and just after them. Conversely, there is a peak in abundance around atomic number 26 (iron) where the abundances are about three orders of magnitude larger than we would expect for a smoothly decreasing curve. Recall that we started our discussion of nuclear reactions and their energetics by suggesting that they would help us understand more about nuclear evolution. Let's see how the binding energy data help us understand more about the cosmic abundance of the elements.

Origin of Atoms — Chapter 3

Correlations between abundance and nuclear binding energy

As you see in Figures 3.18 and 3.23, the nuclear stabilities (binding energies per nucleon) for the first several elements do not increase uniformly, but form a sort of sawtooth pattern of alternating stabilities. In particular, you see that Li, Be, and B nuclei are less stable than either helium or carbon. Their synthesis from helium is an unfavorable process, because they are less stable than helium. Their fusion reactions to form higher atomic number nuclei are favored, since these reactions all form more stable nuclei. The combination of unfavorable conditions for synthesis, but favored reactions to form heavier, more stable nuclei, explains why the cosmic abundance of Li, Be, and B is so low.

In Section 3.3, we saw that the heaviest nuclei produced in stellar nuclear fusion processes are those in the neighborhood of iron. Now we can see why fusion stops at this point. The most stable nuclei (largest binding energies per nucleon) are those with atomic numbers near 26 (iron). Fusion to form higher atomic number nuclei is not favored and a wholly different process has to occur in exploding stars to form the higher atomic number nuclei. If all the nucleons in the universe were fused to form the most stable nuclei, the products would be the nuclei near atomic number 26. This is why these products have built up in the cosmos and why there is a peak in cosmic elemental abundance at iron.

3.55. **Check This**   *Cosmic elemental abundance and nuclear reactions*

Figures 3.7 and 3.23 show that there is a small peak in abundance between atomic numbers 50 to 60 and the hint of another between atomic numbers 34 to 40. Do the nuclear reactions represented in Figure 3.21 suggest an origin for these peaks in abundance? Clearly explain the reasoning for your response.

3.56. **Check This**   *The higher abundance of elements with even atomic numbers*

(a) The nuclear synthesis reactions we have written, equations (3.7), (3.8), and (3.9), for example, involve combinations of alpha particles and the products of their fusion reactions to produce the lighter nuclei (up to those in the iron region). How, if at all, do these reaction sequences help explain the higher abundance of nuclei with even atomic numbers?

(b) Propose fusion reactions of hydrogen and carbon nuclei that would produce stable nitrogen nuclei and thus provide the beginning of reaction sequences that would form nuclei with odd atomic numbers. *Hint:* N-13 is unstable and decays by electron capture.

The nuclear fusion (and neutron capture) processes that produce the elements never stop. They are going on right now in billions of stars in our own galaxy, including our Sun, and in billions more stars in billions of other galaxies. The relative abundances of elements in the universe are continuously changing (although not rapidly), as the lightest nuclei are being used up to form more of the heavier ones. A few billion years ago a very tiny amount of this matter collected and formed our solar system. We will end our brief history of elemental evolution by looking at the elements on Earth today.

Section 3.7. Formation of Planets: The Earth

3.57. **Consider This** *Where did(does) terrestrial matter come from?*

(a) All the matter on Earth was made in stars (or the Big Bang), but how did it get here? Would you expect the abundances of the elements on Earth to be about the same as the cosmic abundance of the elements? Why or why not?

(b) Is matter still being added to the Earth? What evidence is there to justify your answer?

The Earth. Recall, from the end of Section 3.3, that our solar system was formed from the remains of previous stars. The elements that were born in those stars became part of a growing planet, our Earth, as it circled the Sun and grew by sweeping up the dust and larger particles from which the solar system evolved. Initially, these collision processes created a hot body that was mainly molten rock and metals. When the dust was mostly used up, collisions became less frequent and the Earth began to cool and differentiate by density into a core, a mantle, and a crust, as illustrated in Figure 3.24.

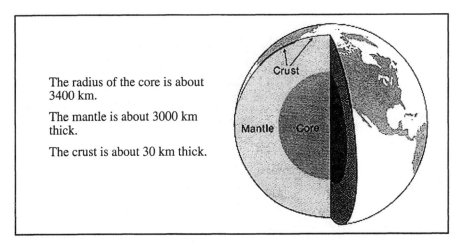

Figure 3.24. Simplified structure of the Earth.
The thickness of the crust is exaggerated; the crust is only about $1/100^{th}$ the thickness of the mantle.

The core is iron and nickel, probably solid at the center but mostly liquid. The mantle is a viscous, molten mixture of various minerals, mostly silicates such as olivine, $FeMgSiO_4$. The thin, solid crust floats on top of the mantle and is moved about (very slowly) by convection currents in the mantle. We live on the outer surface of the crust and the oceans fill the valleys where the crust is thinnest.

We can't be sure of the composition of the original dust cloud from which our solar system formed, but we can guess that it was probably similar to the composition of the universe shown in Figure 3.7. To find out more about the possible composition of the cloud, we can analyze the crust of the Earth to determine its elemental composition. Comparing the cosmic and crustal compositions shown in Figure 3.25 provides clues about the early history of the planet.

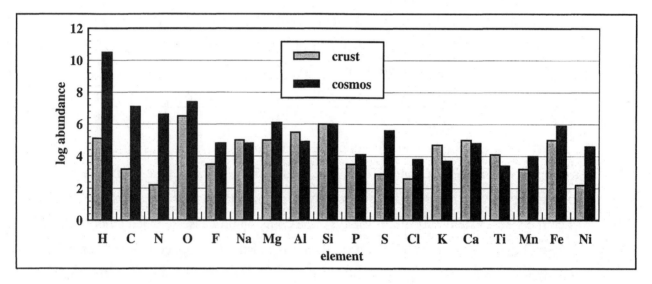

Figure 3.25. The abundance of selected elements in the Earth's crust and in the cosmos.
The cosmic abundances are from Figure 3.7. Abundances are all relative to 10^6 Si atoms.

3.58. **Consider This** *How do crustal and cosmic elemental abundances compare?*
For which of the elements in Figure 3.25 are the crustal abundances two or more orders of magnitude less than the cosmic abundances? For which elements are the crustal abundance's two or more orders of magnitude greater than the cosmic abundances? What explanation can you think of for these experimental observations?

For most of the elements in Figure 3.25, the crustal and cosmic abundances are rather similar. Three notable exceptions are hydrogen, carbon, and nitrogen – all of which are much less abundant in the Earth's crust than in the universe. If we assume that these elements were present at their cosmic abundance when the Earth formed, their loss is one piece of evidence that the

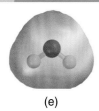

(e)

(a) Charge separation makes a polar molecule a permanent electric dipole.

(b) Coincidence of charges leaves a nonpolar molecule with no permanent electric dipole.

Figures 1.8(e) and 1.17(a)

Charge-density model of a water molecule. In the charge-density models you will see in this text, red denotes parts of molecules that are more negative, and blue represents parts that are more positive. Regions where positive and negative charges balance one another are in green. [pp. 1-18; 1-31]

Figure 1.15

Schematic illustrations of (a) polar and (b) nonpolar molecules. [p. 1-29]

	I	II	III	IV	V	VI	VII	VIII
1	Hydrogen H 2.2							Helium He —
2	Lithium Li 1.0	Beryllium Be 1.6	Boron B 2.0	Carbon C 2.6	Nitrogen H 3.0	Oxygen O 3.4	Florins F 4.0	Neon Ne —
3	Sodium Na 0.93	Magnesium Mg 1.3	Aluminum Al 1.6	Silicon Si 1.9	Phosphorus P 2.2	Sulfur S 2.6	Chlorine Cl 3.2	Argon Ar —
4	Potassium K 0.82	Calcium Ca 1.3						

Figure 1.20

Periodic variation of electronegativity for the first 20 elements. Relative magnitudes for electronegativity values are indicated using the red-green-blue color-coding scheme that was used earlier to indicate charge density. [p. 1-33]

Figures 1.22

Charge-density models of methane, ethane, and hexane. For comparison, Figure 1.17(a) is a charge-density model of the water molecule computed in the same way. [p. 1-38]

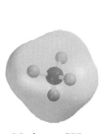

Methane, CH_4

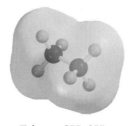

Ethane, CH_3CH_3

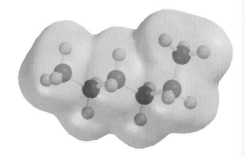

Hexane, $CH_3(CH_2)_4CH_3$

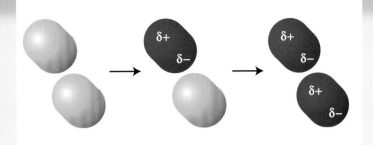

Figure 1.23

Induced-dipole attraction between nonpolar molecules. The plus sign represents the center of positive charge in the molecule. Displacement of the electron distribution in one molecule creates a transient dipole that induces a dipole in the second molecule. [p. 1-39]

$$\underset{\substack{| \\ H-C-H \\ | \\ H}}{\overset{H}{\underset{|}{H-N}}-\overset{H}{\underset{|}{C}}-\overset{O}{\overset{\|}{C}}-O} \;\; H \;\;+\;\; \underset{\substack{| \\ H-C-H \\ | \\ H}}{\overset{H}{\underset{|}{H-N}}-\overset{H}{\underset{|}{C}}-\overset{O}{\overset{\|}{C}}-O} \;\; H \;\;+\;\; \underset{\substack{| \\ H-C-H \\ | \\ H}}{\overset{H}{\underset{|}{H-N}}-\overset{H}{\underset{|}{C}}-\overset{O}{\overset{\|}{C}}-O} \;\; H$$

$$\downarrow$$

$$\underset{\substack{| \\ H-C-H \\ | \\ H}}{\overset{H}{\underset{|}{H-N}}-\overset{H}{\underset{|}{C}}-\overset{O}{\overset{\|}{C}}-\overset{H}{\underset{|}{N}}-\overset{H}{\underset{|}{C}}-\overset{O}{\overset{\|}{C}}-\overset{H}{\underset{|}{N}}-\overset{H}{\underset{|}{C}}-\overset{O}{\overset{\|}{C}}-O} \;\; H \;\;+\;\; 2H_2O$$

Figure 1.30

Linking amino acids to form a chain. The atoms shown in red are part of the structure of all 20 amino acids found in proteins. A double line between atomic symbols represents two pairs of shared electrons (See Chapter 5). [p. 1-49]

Investigate This 1.33

[p. 1-52]

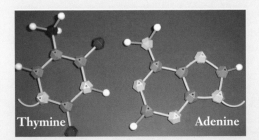

Thymine   Adenine

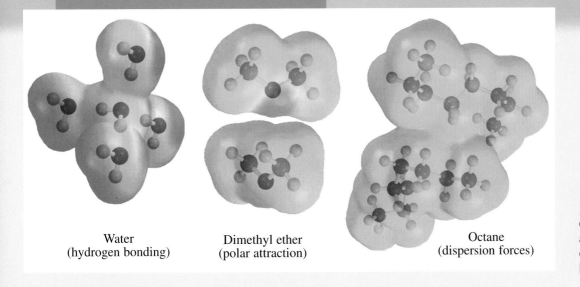

Water
(hydrogen bonding)

Dimethyl ether
(polar attraction)

Octane
(dispersion forces)

From Table 2.1

Comparisons among intermolecular attractions. [p. 2-9]

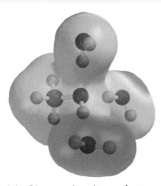

(c) Charge density surface model, showing polarities

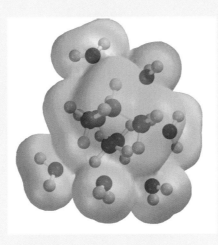

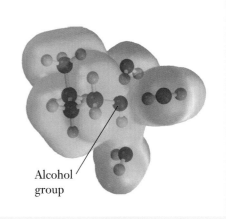

Alcohol group

Figure 2.3(c)

Maximum hydrogen bonding between a methanol molecule and water molecules. [p. 2-11]

Figure 2.4

A hexane molecule in aqueous solution. [p. 2-11]

Figure 2.5

A 1-butanol molecule in aqueous solution. The polar alcohol O—H group in butanol hydrogen bonds with water. The nonpolar end of the butanol, like nonpolar hexane (Figure 2.4), interacts only weakly with water molecules. [p. 2-13]

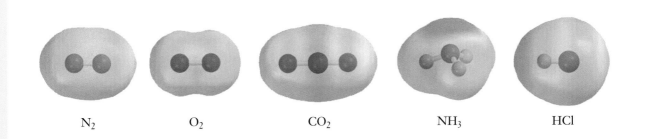

N_2 O_2 CO_2 NH_3 HCl

Figure 2.23

Models showing the polarity of five gaseous molecules. [p. 2-59]

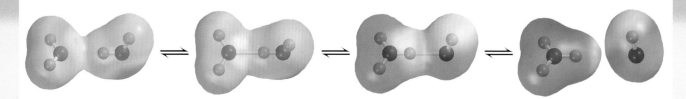

Figure 2.25

Transfer of a proton between two hydrogen-bonded water molecules. Note the great change in charge distribution as the proton is transferred from one water molecule to another. [p. 2-62]

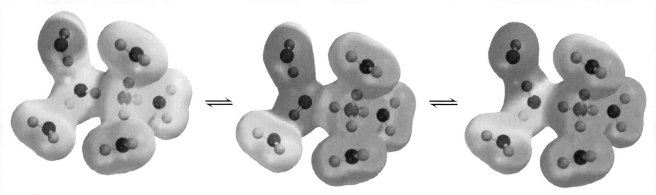

(b) Charge density model of the reactions shown in (a). The water molecules are shown in different orientations.

Figure 2.29b

Formation and migration of an ammonium and a hydroxide ion by proton transfers. [p. 2-70]

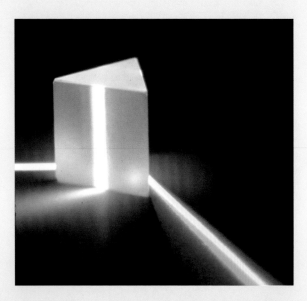

Figure 3.1

A prism disperses white light into the visible spectrum. Violet light is bent most from the original light beam. [p. 3-7]

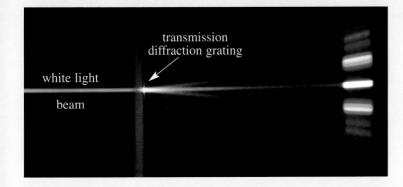

Figure 3.2

A diffraction grating disperses white light into the visible spectrum. Violet light is bent least from the original light beam. [p. 3-7]

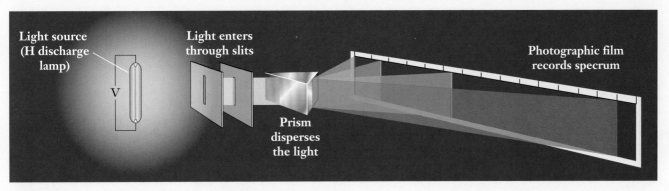

Figure 3.3
Diagram showing the parts of a simple spectrograph. [p. 3-8]

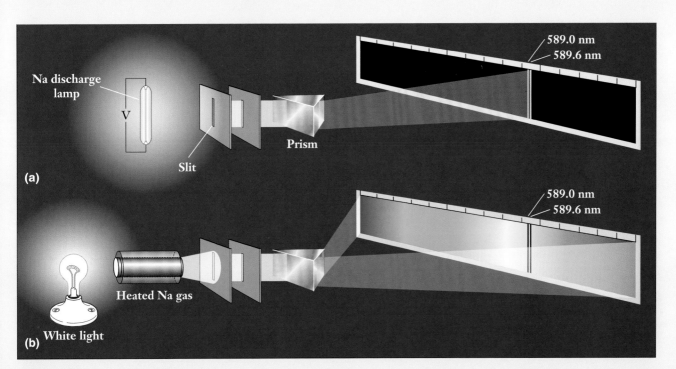

Figure 4.6
Two spectroscopic experiments with sodium atoms. [p. 4-12]

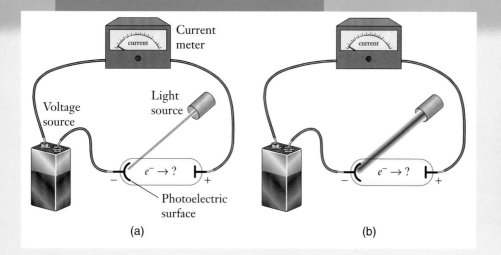

Figure 4.15

Photoelectric experiments with low-frequency light of different intensities. Red represents low-frequency (long wavelength—Figure 4.13) light. The thickness of the light beam shows its intensity (brightness): (a) is low intensity and (b) is high intensity. [p. 4-29]

Figure 4.16

Photoelectric experiments with (a) low- and (b) high-frequency light. Blue, (b), represents higher-frequency (shorter wavelength—Figure 4.13) light than red, (a). [p. 4-29]

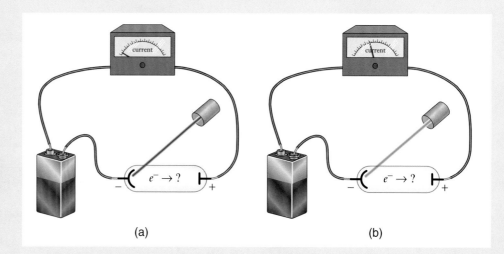

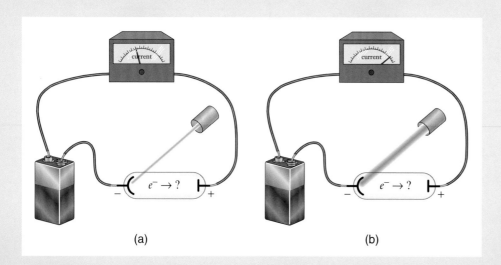

Figure 4.19

The effect of light intensity on photoelectric current. [p. 4-31]

	I	II	III	IV	V	VI	VII	VIII
	Hydrogen $_1$H							Helium $_2$He
1	•							•
core e^-	0							0
valance e^-	–1							–2
	Lithium $_3$Li	Beryllium $_4$Be	Boron $_5$B	Carbon $_6$C	Nitrogen $_7$N	Oxygen $_8$O	Fluorine $_9$F	Neon $_{10}$Ne
2	●	●	●	●	●	●	●	●
core e^-	–2	–2	–2	–2	–2	–2	–2	–2
valance e^-	–1	–2	–3	–4	–5	–6	–7	–8
	Sodium $_{11}$Na	Magnesium $_{12}$Mg	Aluminum $_{13}$Al	Silicon $_{14}$Si	Phosphorus $_{15}$P	Sulfur $_{16}$S	Chlorine $_{17}$Cl	Argon $_{18}$Ar
3	●	●	●	●	●	●	●	●
core e^-	–10	–10	–10	–10	–10	–10	–10	–10
valance e^-	–1	–2	–3	–4	–5	–6	–7	–8
	Potassium $_{19}$K	Calcium $_{20}$Ca						
4	●	●			Scale →| |← 100 pm			
core e^-	–18	–18						
valance e^-	–1	–2						

Figure 4.31

Electron shell model for the first 20 elements in the periodic table. [p. 4-56]

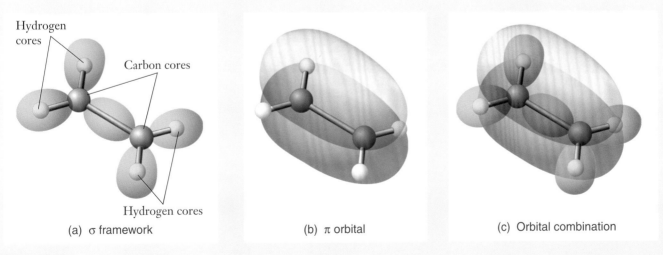

(a) σ framework   (b) π orbital   (c) Orbital combination

Figure 5.19

Molecular orbitals in ethene. The six atomic cores define a molecular plane. The σ framework is cut by this plane. The localized π orbital shown in (b) lies above and below the molecular plane and is cut by a plane perpendicular to the molecular plane. [p. 5-35]

Ethanol

Propanal

Propanone (acetone)

Ethanoic (acetic) acid

Methyl ethanoate (acetate)

Methyl amine

Propanamide

From Table 5.4

Examples of compounds with oxygen- and nitrogen-containing functional groups. [p. 5–67]

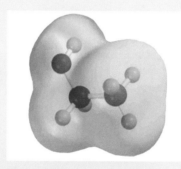

Figure 5.37

Computer-calculated electron density distribution in ethanol. [p. 5-69]

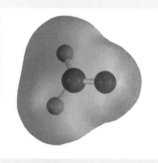

Figure 5.38

Computer-calculated electron distribution in methanal (formaldehyde). [p. 5-70]

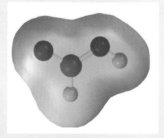

Figure 5.39

Computer-calculated electron distribution in methanoic acid (formic acid). [p. 5-71]

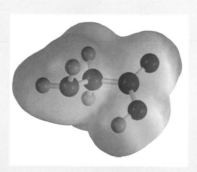

Figure 5.40

Computer-calculated electron distribution in glycine, an amino acid. [p. 5-72]

early Earth was hot. The simple compounds of these elements, such as methane (CH_4) and ammonia (NH_3), are gases or easily vaporized liquids that would have been driven off the newly forming hot planet. Sulfur, which is about a thousand times lower in abundance in the crust compared to the cosmos, also forms rather volatile compounds with hydrogen (H_2S) and other light elements and could also have been lost by vaporization.

> **3.59. Check This** *Where is the nickel?*
> Nickel is also much less abundant in the crust than we would expect, but it is not likely to have vaporized from the early Earth. Look at the structure of the Earth shown in Figure 3.24. Where is the "missing" nickel likely to be? Explain the reasoning behind your answer.

The age of the Earth

To estimate the age of the Earth, we take advantage of the properties of radioactive nuclei — their half lives and the products formed in the decay processes. For example, many rocks contain some uranium and the ^{238}U nucleus decays by a series of alpha and beta particle emissions to give the ^{206}Pb nucleus:

$$^{238}_{92}U^{92+} \rightarrow {}^{206}_{82}Pb^{82+} + 8\,{}^{4}_{2}He^{2+} + 6\,{}^{0}_{-1}e^- \tag{3.27}$$

The first step in this decay has a half life of 4.468×10^9 years and all the subsequent steps are at least 10^4 times faster, so the half life of the net process shown by equation (3.27) is the half life of its first step. Pb-206 and U-238 are always found together in rocks and other isotopes of lead are usually not present.

To find the age of a rock containing U-238 and Pb-206, we assume that, when the rock formed, all the metal started out as U-238. We also assume that all the Pb-206 was formed subsequently by the series of decays summarized by equation (3.27), and that no metal has been lost from the rock. If the sample originally contained a mole of U-238, 238 g, then after one half life, there would be one-half mole, 119 g, of U-238 left and one-half mole, 103 g, of Pb-206 would have been formed. The mass ratio of Pb-206 to U-238 in the sample would be:

$$\frac{\text{mass Pb-206}}{\text{mass U-238}} = \frac{103 \text{ g}}{119 \text{ g}} = 0.87$$

At the beginning, when no Pb-206 is present, this ratio is zero. The change in the mass ratio of Pb-206 to U-238 as a function of time after the beginning of the decay is shown in Figure 3.26. No rock with a mass ratio larger than about 0.85 has ever been found on Earth. The green lines in Figure 3.26 shows that this is the ratio we expect for a sample that is about 4.5×10^9

years old. Since the oldest known rock on Earth is 4.5 billion years old, the Earth must be at least this old, as shown in Figure 3.11.

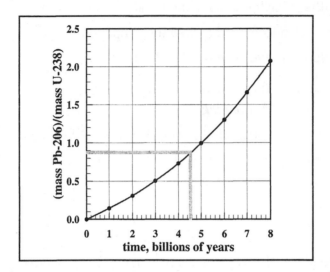

Figure 3.26. Pb-206/U-238 mass ratio as a function of sample age.
The green lines indicate the age of a rock with a Pb-206/U-238 mass ratio equal to 0.85.

3.60. **Check This** *Lead-uranium dating of a sample*

Some ancient fossils were found in rock that had a Pb-206/U-238 mass ratio of 0.46. About how old are the fossils?

The elements of life

When the early Earth had cooled enough for its solid crust to form, water began to collect on the surface. The water was forced out of the hot interior of the Earth and/or supplied by icy comets colliding with the newly forming planet. On a planet like Earth, which had both a relatively cool crust and a relatively strong gravitational field, liquid water could be retained on the surface where it formed oceans that contained an abundance of dissolved matter. The stage was set for life to emerge. The elements upon which life is based are not the most abundant on Earth. On the contrary, Figure 3.27 shows that there is little correlation between the elements that are most abundant in the Earth's crust and those most abundant in organisms (humans). Note that Figure 3.27 is not based on a comparison with the abundance of silicon. Because there is so little silicon in organisms, it is not a good standard in this case.

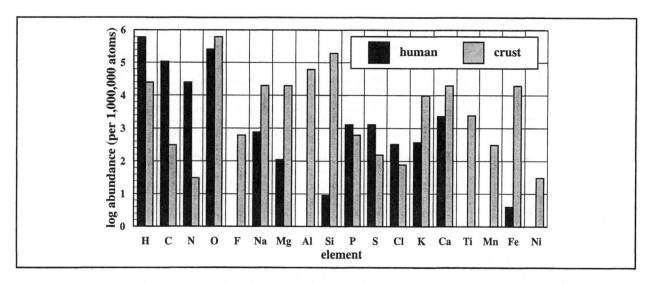

Figure 3.27. The abundance of selected elements in a human body and in the Earth's crust. The abundances are number of atoms out of every 1,000,000 atoms in the human or the crust.

About 99% of the atoms in living things are hydrogen, carbon, nitrogen, and oxygen. It is not surprising that hydrogen and oxygen predominate in humans, Figure 3.27, since water makes up about 70% of our mass. The fats and carbohydrates in our bodies are composed of carbon, hydrogen, and oxygen and all of our proteins contain these three elements plus nitrogen. Nucleic acids also contain these four elements. The properties of molecules constructed from these atoms must be so suitable for life that organisms evolve to use them, even when the atoms are scarce. We will examine some of these properties as we go on in the book.

3.61. **Check This** *What are the roles of other elements?*

(a) Figure 3.27 shows that calcium and phosphorus have about the same abundance in our bodies and are the next most abundant (along with sulfur) after hydrogen, carbon, nitrogen, and oxygen. What roles do calcium and phosphorus play that require them in such amounts? *Hint:* Review the opening paragraphs of Section 2.6 in Chapter 2.

(b) Which other elements are present to the extent of at least 50 atoms per one million atoms in our bodies? Explain how you determine which these are. Choose one of these elements and find out what its roles are in your body.

We end our history of the universe by noting that the elements of life are the most abundant elements in the cosmos (see Figures 3.7 and 3.25). They and their compounds are certain to be present in any planetary system and, under the right conditions, as on Earth, carbon-based life is likely to have evolved around many other stars we see twinkling like diamonds in the sky.

Origin of Atoms Chapter 3

Reflection and projection

It's easy to get caught up in the cosmic story and lose sight of some fundamental problems. From the beginning of the book, we have emphasized interactions of positive and negative electrical charge. In this chapter, for example, we said that repulsions between positive particles have to be overcome in order for nuclear fusion to occur. But you have learned that the binding energies holding protons and neutrons together in a tiny nucleus are enormous compared to chemical energies. What can be the nature of the forces acting within the nucleus? These forces are so strong that nuclei never change during chemical reactions; chemists leave the study of these forces to physicists with high energy particle accelerators.

The reverse problem arises when you consider the formation of atoms and molecules from nuclei and electrons. The nuclear model of the atom has the negative electrons held near the positive nucleus by their electrical attractions. Why aren't the electrons just attracted closer and closer until the atom collapses? The answer to this question governs all the chemical structures and chemical reactions around us and within us. In the next chapter we will examine the periodic properties of atoms, return to spectroscopy and a more detailed discussion of the properties of waves, and develop an atomic model that answers this question.

Section 3.8. Outcomes Review

We began the chapter by introducing spectroscopy as a method for learning about the identity and relative amounts of atoms in samples that emit light, including stars and other glowing objects in the universe. After reviewing the structure of atomic nuclei, we introduced the Big Bang theory for the formation of the universe and discussed the nuclear processes in stars that are responsible for the formation of the elements we find in the present-day universe. After practice in balancing nuclear reactions, we showed how the mass-to-energy conversion in nuclear reactions explains their enormous energy release. Nuclear binding energies determine whether a nucleus will undergo fusion or fission reactions and help explain the cosmic abundance of the elements. The elements formed in the first stars (and still being formed in present stars) are the matter from which planets, including the Earth, are formed and from which life can evolve.

Check your understanding of the ideas in the chapter by reviewing these expected outcomes of your study. You should be able to:

• describe a spectrograph and how it produces the spectrum from its light source [Section 3.1].

- distinguish between continuous and line spectra, identify an element from its line emission spectrum, and determine relative amounts of two elements by the brightness of their emission lines [Section 3.1].
- write the complete symbol and locate the subatomic particles in a diagram of any isotope of an elemental atom or ion [Section 3.2].
- explain cosmic condensation and nuclear fusion events as a function of the temperatures that make them possible [Section 3.3].
- write and balance reactions along the pathway for the formation of elements up to about ^{56}Fe in the first stars [Sections 3.3 and 3.4].
- write and balance reactions along the pathway for the formation of elements beyond ^{56}Fe in the first stars [Sections 3.3 and 3.4].
- describe the formation of planetary systems and the source of their elementary atoms [Sections 3.3 and 3.7].
- balance nuclear reactions (fusion, fission, and radioactive decay) by supplying the appropriate reactant or product nuclei and/or other particles [Section 3.4].
- use half lives to find the amount of a radioactive nucleus remaining after an elapsed time or to find the time elapsed since decay began [Sections 3.4 and 3.7].
- determine the energy changes for mass-to-energy conversions in nuclear fusion and fission reactions and *vice versa* [Section 3.5].
- determine nuclear binding energies from mass-to-energy conversion in formation of nuclei from protons and neutrons [Section 3.5].
- use nuclear binding energies to predict whether a nucleus is likely to undergo fusion or fission [Section 3.5].
- give the requirements for a nuclear chain reaction and diagram the differences between controlled and uncontrolled chain reactions [Section 3.5].
- use abundance data to estimate abundances of elements relative to one another in the universe, on Earth, and in organisms [Sections, 3.1, 3.6, and 3.7].
- use nuclear binding energies to explain trends in cosmic elemental abundances as well as deviations from the trends [Section 3.6].
- use the model for planetary formation to explain the structure of the Earth and the abundance of elements in the Earth's crust [Section 3.7].
- relate cosmic elemental abundances to possible evolution of carbon-based life [Section 3.7].

Origin of Atoms — Chapter 3

Section 3.9. EXTENSION — Isotopes: Age of the Universe and a Taste of Honey

Age of stars and the universe

One way to measure the age of the universe is to measure the age of stars. If our model, the Big Bang, is correct, the universe must be somewhat older than the oldest stars. If a star contains a long-lived radioactive isotope, its decay rate can be used to date the star, much as we discussed dating rocks in Section 3.7. In the year 2000, uranium was detected spectroscopically, Figure 3.28, in a star in a region of our Milky Way galaxy that is populated by relatively old stars. The uranium is assumed to be U-238 (half life = 4.468×10^9 yr).

The star, CS31082-001 in the

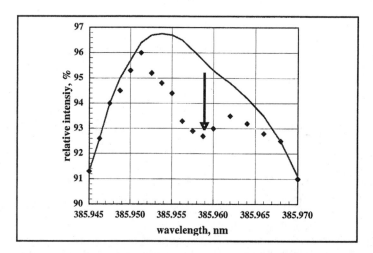

Figure 3.28. Absorption of light by uranium ions in star CS31082-001.
The red curve is the calculated emission from the star assuming no uranium is present. The points are the experimental data and the arrow marks the center of the uranium ion absorption line. Atomic absorption spectra are discussed in Chapter 4. Compare the shape of this absorption "line" with the shapes for the emission lines shown in Figure 3.6.

3.62. Consider This   *Why is the uranium in stars only the U-238 isotope?*
We have seen that uranium has two isotopes, U-238 and U-235 (half life = 7.1×10^8 yr). Both of these isotopes are formed in supernova explosions, Section 3.3. Why can we assume that all the uranium in an old star is U-238? Explain your reasoning clearly.

The amount of uranium, U-238, remaining in the star can be obtained from the data in Figure 3.28 (and the amounts of other elements from similar data at other wavelengths). In order to date the star, we also have to know how much U-238 the star contained when it formed. There is no way to know the absolute amount of uranium, or any other element, the star contained when it formed. However, astrophysicists can use the model for the evolution of the universe to calculate the *relative* amounts of the heavy elements formed in supernovae. We assume that these are the ratios of these elements that get incorporated into new stars. By comparing the ratio of one element to another in a present-day star to the theoretical ratio when the star was born, we can calculate the age of the star.

3.63. Worked Example *Star's age from its U/Ir ratio*

Our model for the evolution of the universe predicts that the elements uranium and iridium are formed in a ratio of about 0.050. The U/Ir (uranium to iridium) ratio determined for the star in Figure 3.28 is 0.0079. What is the age of the star?

Necessary information: We need to know the ratios given in the problem statement, the half life of U-238, 4.468×10^9 yr, and that iridium is not radioactive. We need Figure 3.14 or equation (3.18) to relate the fraction of U-238 left to the number of half lives that have elapsed.

Strategy: Determine the fraction of U-238 remaining in the star and use equation (3.18) to find out how many half lives have elapsed. The age of the star is this number of half lives times the half life of U-238.

Implementation: When the star formed, there were 50 U atoms for every 1000 Ir atoms and now there are only 7.9 U atoms for every 1000 Ir atoms. Thus, the fraction of U remaining is:

$$\text{fraction U remaining} = f_n = \frac{7.9 \text{ U}/1000 \text{ Ir}}{50 \text{ U}/1000 \text{ Ir}} = 0.158$$

Rearranging equation (3.18) to solve for the number of half lives gives:

$$n = \frac{\ln(f_n)}{-0.693} = \frac{\ln(0.158)}{-0.693} = \frac{-1.85}{-0.693} = 2.7 \text{ half lives}$$

$$\text{age of the star} = (2.7 \text{ half lives}) \times [4.468 \times 10^9 \text{ yr} \cdot (\text{half life})^{-1}] = 12 \times 10^9 \text{ yr}$$

Does the answer make sense? The age of the universe is estimated to be between 10 and 18 billion years. Our stellar age is in this range and suggests that the universe must be somewhat older than 12 billion years, since this star has to have been born after the first supernovae occurred. How do we know this star is younger than the first supernovae? Use this portion of a half-life decay curve to do another check on this calculation.

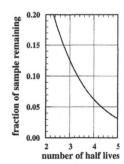

3.64. Check This *Star's age from its U/Os ratio*

Another isotope ratio that can be used to date this star is its U/Os (uranium to osmium) ratio. The present ratio is 0.0065 and the calculated ratio for the formation of these elements is 0.054.

> Osmium, like iridium, is a stable element. Use these data to calculate the age of the star and compare it with the age found in Worked Example 3.63. Do they reinforce one another? Explain.

A criticism of the ratio method for stellar dating relates to the fact that uranium is a far heavier atom than either iridium or osmium. Simple losses of elements into space from the star would favor the loss of the lighter, faster moving atoms and ions compared to the heavier ones. In order to do the calculations above, we have assumed that the only change in the ratios is due to radioactive decay of the U-238. To answer this criticism, we could use a ratio of isotopes that are quite close in mass. Thorium-232 meets this criterion and is present in the star we are considering. However, Th-232 is also radioactive and decays with a half life of 14.05×10^9 yr. Thus, to date the star, we have to account for the decay of both U-238 and Th-232.

We will use equation (3.18) to solve this problem, but will rewrite the number of half lives, n, as t/τ, where t is the time that has passed (which is what we are trying to calculate) and τ is the half life of the isotope of interest. We will also write explicit expressions for the fractions of isotopes remaining. In these fractions we use atomic symbols to represent the amount of the nuclei remaining and subscripts to represent the times. Thus, for U and Th in our star, equation (3.18) gives:

$$\ln(U_t/U_0) = -0.693 \cdot (t/\tau_U) \tag{3.28}$$

$$\ln(Th_t/Th_0) = -0.693 \cdot (t/\tau_{Th}) \tag{3.29}$$

Since t is the same time for both decays (they are in the same star), we can subtract equation (3.28) from equation (3.29) to get:

$$\ln(Th_t/Th_0) - \ln(U_t/U_0) = -0.693 \cdot t \cdot [(1/\tau_{Th}) - (1/\tau_U)] \tag{3.30}$$

Using the knowledge that $\ln(a/b) = \ln a - \ln b$, we can rewrite equation (3.30) as:

$$\ln(U_0/Th_0) - \ln(U_t/Th_t) = -0.693 \cdot t \cdot [(1/\tau_{Th}) - (1/\tau_U)] \tag{3.31}$$

> **3.65. Check This** *Show that equations (3.30) and (3.31) are identical*
> Rewrite the logarithms of the ratios in equations (3.30) and (3.31) as differences to show that the equations are identical.

Finally, solving equation (3.29) for t and rewriting the half life factor gives:

$$t = \frac{\ln(U_0/Th_0) - \ln(U_t/Th_t)}{0.693 \left(\dfrac{\tau_{Th} - \tau_U}{\tau_{Th} \cdot \tau_U} \right)} \tag{3.32}$$

Chapter 3 — Origin of Atoms

3.66. Worked Example *Star's age from its U/Th ratio*

At present, the star we have been analyzing has $\ln(U_t/Th_t) = -1.70$. One model calculation for the formation of U and Th gives $\ln(U_0/Th_0) = -0.59$. What is the age of the star?

Necessary information: We need the ratios in the problem statement and equation (3.32), as well as the half lives: $\tau_{Th} = 14.05 \times 10^9$ yr and $\tau_U = 4.468 \times 10^9$ yr.

Strategy: Substitute these known values into equation (3.32) to get the time that has elapsed since the star formed.

Implementation: Let's first calculate the denominator of equation (3.30):

$$0.693\left(\frac{\tau_{Th} - \tau_U}{\tau_{Th} \cdot \tau_U}\right) = 0.693\left(\frac{(14.05 \times 10^9\,\text{yr}) - (4.468 \times 10^9\,\text{yr})}{(14.05 \times 10^9\,\text{yr}) \cdot (4.468 \times 10^9\,\text{yr})}\right) = 0.105 \times 10^{-9}\ \text{yr}^{-1}$$

Then:

$$t = \text{age of the star} = \frac{\ln(U_0/Th_0) - \ln(U_t/Th_t)}{0.693\left(\dfrac{\tau_{Th} - \tau_U}{\tau_{Th} \cdot \tau_U}\right)} = \frac{(-0.59) - (-1.70)}{0.105 \times 10^{-9}\ \text{yr}^{-1}} = 11 \times 10^9\ \text{yr}$$

Does the answer make sense? Once again, the age of the star lies within the range of values for the age of the universe (which, of course, is determined from many different observations, including ones like this). This value is consistent with the previous calculations we have made, especially considering uncertainties in both the experimental and theoretical values used.

3.67. Check This *Star's age from its U/Th ratio*

(a) Another calculation (based on somewhat different assumptions about nuclear formation) for the formation of U and Th in supernovae gives $\ln(U_0/Th_0) = -0.23$. What does this value lead to for the age of the star we are studying? Is this age consistent with the others we have calculated? Explain why or why not.

(b) Considering all the values we have obtained for the age of this star, what would you report its age to be? Justify your response.

Stable isotopes and pure honey

Most of the atoms in the materials around and in us are composed of two or more **stable isotopes**, that is, isotopes that do not undergo radioactive decay. Several examples of the stable isotopes found in organisms are listed in Table 3.5. Although all the isotopes of a given atom react the same way chemically, they do so at slightly different rates. In any process that involves movement of atoms, such as a chemical reaction, a heavier isotope (or molecule containing the heavier atom) moves a little more slowly than a lighter isotope. Thus some discrimination between the isotopes can occur and can provide useful markers for the kind of process(es) that occurred to give the final result.

Table 3.5. A few of the stable isotopes in biological systems.

Element	Isotopes	Average Abundance, %
hydrogen	H-1	99.985
	H-2	0.015
carbon	C-12	98.892
	C-13	1.108
nitrogen	N-14	99.6337
	N-15	0.3663
oxygen	O-16	99.759
	O-17	0.0374
	O-18	0.2039

3.68. Check This *Isotopic combinations in molecules*

Molecules of carbon dioxide, CO_2, can have several different isotopic compositions. For example, one molecule might be (C-12)(O-16)(O-17) and another (C-12)(O-18)(O-18). How many different isotopic compositions can CO_2 have? Which do you think is most common? Second most common? Least common? Justify your choices.

One example of this discrimination between isotopes is found in photosynthesis by plants. Most plants produce sugars from carbon dioxide and water by one of two photosynthetic processes called the C_3 and C_4 pathways. The subscripts refer to the number of carbons in the product of the first step that incorporates a carbon atom from carbon dioxide. For both pathways, the percentage of C-13 that ends up in the sugars is less than the percentage of C-13 in the carbon dioxide (from the air) which the plant used to synthesize the sugars. However, the amount by which the percentage of C-13 is reduced is different for the two pathways. Even though the same sugars are formed in each pathway, they have different **isotopic signatures**, that is different ratios of the stable isotopes, $^{13}C/^{12}C$.

The $^{13}C/^{12}C$ ratio in a sugar sample is obtained by first burning the sample completely to convert all its carbon to carbon dioxide. A mass spectrometer, such as the one illustrated in

Figure 3.29, is then used to measure the ratio of CO_2 with C-13 to that with C-12. The results from such analyses are usually expressed as delta, δ, values in which the sample results are compared to a standard sample of known isotopic composition. You can think of the delta value as the number of atoms per 1000 that the heavy isotope in the sample differs from the heavy isotope in the standard. The expression for $\delta^{13}C$ in terms of the sample and standard isotopic ratios is:

$$\delta^{13}C = \left[\left(\frac{(^{13}C/^{12}C)_{sample}}{(^{13}C/^{12}C)_{standard}}\right) - 1\right] \cdot 1000 \qquad (3.33)$$

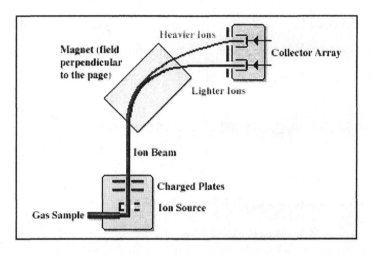

Figure 3.29. Schematic diagram of an isotope ratio mass spectrometer.
The gas sample is ionized by a beam of electrons in the ion source and then accelerated out of the source by electrically charged plates. The moving ion beam passes between the poles of a magnet which bends its path. The path an ion follows depends on its mass. The number of ions of each mass is measured separately and the numbers compared. The entire apparatus is in a vacuum chamber, so only the ions of interest are formed and detected.

3.69. **Check This**   *Determining a delta value*

If the standard has a $^{13}C/^{12}C$ ratio equal to that in Table 3.5 and a sample contains 1.099 percent C-13, what is $\delta^{13}C$ for the sample? Does the sign of $\delta^{13}C$ reflect direction of change of the sample from the standard? Why or why not? Explain your reasoning.

The usual standard for $\delta^{13}C$ is the carbonate from a particular limestone formation in South Carolina. Relative to this standard, almost all samples from living systems have less C-13, so their $\delta^{13}C$ values are negative, as you found in Check This 3.69. In particular, the $\delta^{13}C$ values for C_3 and C_4 plants are about -25 and -10, respectively. And here is where honey fits into this story. Most flowering plants from which bees collect nectar (sugar syrup) to make honey are C_3 plants, so the honey, which is mainly a collection of these sugars, has a $\delta^{13}C$ value of -25. On the other hand, corn and sugar cane are C₄ plants, so their sugars have a $\delta^{13}C$ value of -10. Honey is an expensive product, since it is produced in relatively small quantities. Corn syrup and cane sugar syrup are rather inexpensive products. It's difficult to taste the difference between pure honey and a mixture of honey with corn syrup or cane sugar syrup. Although it's not legal to sell a mixture like this as pure honey, there is a temptation to do so. Isotope ratios help food inspectors determine whether a product labeled "honey" has been adulterated with less expensive syrups.

3.70. Worked Example *The amount of honey in a "honey" sample*

An isotopic analysis of the sugars in a shipment of "honey" gave a $\delta^{13}C$ value of -14.3. The sample is obviously not pure honey. What percentage of the product is honey?

Necessary information: We make the assumption that $\delta^{13}C$ varies linearly with the proportion of the sample that is honey (or syrup), as shown in this graph.

Strategy: Read from the graph the composition corresponding to the measured $\delta^{13}C$.

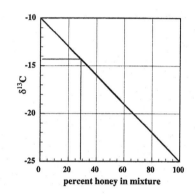

Implementation: The blue lines show how to read over from –14.3 to the line and then down to the percent composition, 29% honey.

Does the answer make sense? The measured $\delta^{13}C$ value is pretty close to that for pure C₄ sugars, so the percent of honey in the sample is low, as we found.

3.71. Check This *What is the legal limit for adulteration of honey?*

(a) If $\delta^{13}C$ varies linearly with the proportion of honey, the percent of honey in a sample is:

$$\% \text{ honey} = \left(\frac{\delta_{sample} - \delta_{syrup}}{\delta_{honey} - \delta_{syrup}} \right) \cdot 100\% \tag{3.34}$$

All the δ values in equation (3.34) are the $\delta^{13}C$ values for the subscripted substances. Show how this equation is derived and use it to check the answer in Worked Example 3.70. *Hint:* Write the equation of the line in slope-intercept form with the slope written in terms of the endpoints of the line and solve for the percent honey.

(b) For legal purposes, a difference of one unit in $\delta^{13}C$ from the value for pure honey is taken to represent an adulterated sample. To what percent adulteration does this difference correspond? Explain your reasoning clearly.

In this chapter, including this EXTENSION, we have explored several uses of isotopes, including household items, medicine, geology, astrophysics, and chemical analysis. These are important uses of isotopes and there are many others, a few of which we will meet in later chapters. The overarching message in this chapter is that all of these isotopes (except the very lightest that were a product of the Big Bang) were formed and are being formed in stars and supernovae. We can understand a great deal about our universe, our galaxy, our solar system, and our planet by understanding some of the principles that guide element formation, and isotopes help us read the story of their formation.

Index of Terms

A (mass number), 3-16
alpha particle, 3-25
atomic mass unit, 3-15
atomic number, 3-15
beta particle, 3-25
Big Bang Theory, 3-20
binding energy per nucleon, 3-42
chain reaction, 3-45
collision-induced fission, 3-45
condensation, 3-20
continuous spectrum, 3-8
coulombic repulsion, 3-21
critical mass, 3-46
daughter nucleus, 3-32
electron, 3-15
elementary charge, 3-15
gamma radiation, 3-25
half life, 3-32
ion, 3-19
isotopes, 3-17
isotopic signature, 3-62
line spectra, 3-8
mass number, 3-16
mass-energy equivalence, 3-37
N (number of neutrons in a nucleus), 3-16

neutrino, 3-25
neutron, 3-15
nuclear binding energy, 3-41
nuclear fission, 3-39
nuclear fusion, 3-21
nuclei, 3-15
nucleon, 3-42
nucleons, 3-16, 3-41
nucleus, 3-15
parent nucleus, 3-31
PET scan, 3-30
positron, 3-25
positron emission tomography, 3-30
proton, 3-15
radioactive decay, 3-31
radioisotope, 3-31
spectrograph, 3-8
spectroscope, 3-8
spectroscopy, 3-6
spectrum, 3-6
speed of light, 3-37
spontaneous nuclear fission, 3-44
stable isotopes, 3-61
supernova, 3-23
Z (atomic number), 3-15

Chapter 3 Problems

Section 3.1. Spectroscopy and the Composition of Stars and the Cosmos

3.1. Figure 3.3 shows the parts of a simple spectrograph and the result when the light from a hydrogen discharge (hydrogen atom) lamp enters and is diffracted. Is the diagram and diffraction consistent with the information from Figure 3.1? Explain why or why not.

3.2. Draw a diagram of a simple spectrograph, modeled after Figure 3.3, that uses a transmission diffraction grating, such as shown in Figure 3.2, instead of a prism to diffract the incoming light beam. Be sure the diffraction results you show on the photographic film are consistent with the information from Figure 2.2.

3.3. A spectrometer was used to analyze a source of light and the spectrum showed a series of emission lines. What can you conclude about the light source? Explain.

3.4. Flame tests can often be used to identify the metal ions in a compound. Robert W. Bunsen (German chemist and spectroscopist, 1811-1899) invented the bunsen burner in order to create a flame hot enough to cause light emission from metal ions in compounds that were placed in the flame. Two examples are shown here: the greenish flame from barium compounds and the scarlet flame from strontium compounds. How do the emissions from atoms discussed in Section 3.1 explain why flame tests work? Explain clearly.

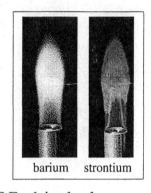
barium   strontium

3.5. More than 99% of all the atomic nuclei in the universe are hydrogen and helium. Are the data in Figure 3.7 consistent with this statement? Clearly explain why or why not.

3.6. As the previous problem states, almost all the nuclei in the universe are hydrogen and helium. Assume that almost all the mass of the universe is also hydrogen and helium. About what percentage of the mass of the universe is helium? Clearly explain the reasoning you use to get your answer.

3.7. How does the mass of iron in the universe compare to the mass of carbon? Explain your reasoning.

3.8. How does the mass of iron in the universe compare to the mass of all the other fourth period transition metals, scandium through zinc, combined?

Origin of Atoms   Chapter 3

Section 3.2. The Nuclear Atom

3.9. What do these ions, S^{2-}, Cl^-, K^+, and Ca^{2+}, have in common?

3.10. How does the mass of a proton compare to the mass of an electron?

3.11. What is an isotope?

3.12. Why are electrons not included when calculating the mass number of an isotope?

3.13. If an isotope of an element has 30 protons, 35 neutrons, 28 electrons:
  (a) What is the element?
  (b) Is this a anion or a cation? Explain.

3.14. What would be the mass, in grams, of 25 protons?

3.15. Write the atomic symbols for the following isotopes:
  (a) A = 19; Z = 40   (d) A = 13; Z = 28
  (b) A = 79; Z = 197   (e) A = 53; Z = 118
  (c) A = 54; Z = 118   (f) A = 83; Z = 189

3.16. How many protons, neutrons, and electrons do the following ions contain?
  (a) $^{58}Ni^+$ (d) $^{37}Cl^-$
  (b) $^{32}S^{2-}$ (e) $^{55}Mn^{7+}$
  (c) $^{65}Zn^{2+}$ (f) $^{56}Fe^{2+}$

3.17. Complete the following table. Some of the substances are ions and some are atoms.

Substance	# of protons	# of neutrons	# of electrons	Atomic number	Mass number	Nuclear Symbol
Rhodium		55				$^{100}_{45}Rh$
			23			$^{60}_{26}Fe^{3+}$
	88				225	$^{225}_{88}Ra$
	16	22	18			$^{38}_{16}S^{2-}$
			100			$^{248}_{100}Fm$

Chapter 3 Origin of Atoms

Section 3.3. Evolution of the Universe: Stars

3.18. The following questions deal with the kelvin temperature scale.

(a) Is 3 K, the present background temperature of the universe, hot or cold? Explain.

(b) The Earth is thought to have formed at a temperature just below 1000 K. Would the Earth, at that temperature, have had liquid water on its surface? Explain.

(c) If the average daytime high temperature for a city is 85 °F, what is the temperature in kelvin? *Hint:* A temperature of 32 °F is 273 K. A degree on the kelvin scale is 1.8 times the size of a degree on the fahrenheit (°F) scale.

3.19. According to the Big Bang theory stars, and eventually planets, formed when matter condensed upon cooling as the universe expanded. Give some examples of analogous phenomena showing condensation.

3.20. Put these reactions in order from the one that takes place at the lowest temperature to the one that requires the highest temperature. Explain the reasoning for each of your choices.

(a) $^{22}Ne^{10+} + {}^{22}Ne^{10+} \rightarrow {}^{43}K^{19+} + {}^{1}H^{+}$

(b) $^{13}C^{3+} + e^{-} \rightarrow {}^{13}C^{2+}$

(c) $^{12}C^{6+} + {}^{18}O^{8+} \rightarrow {}^{26}Mg^{12+} + {}^{4}He^{2+}$

(d) $^{12}C^{6+} + {}^{3}He^{2+} \rightarrow {}^{13}N^{7+} + {}^{2}H^{+}$

3.21. What kelvin temperature is necessary for nuclei to have sufficient kinetic energy to sustain nuclear fusion?

3.22. Where are the elements formed? Give examples of the processes by which elements can be formed.

Section 3.4. Nuclear Reactions

3.23. Complete the following nuclear reactions (nuclear charges have been omitted):

(a) $^{14}N + \underline{} \rightarrow {}^{17}O + p$

(b) $^{13}C + \text{neutron} \rightarrow \underline{} + \gamma$

(c) $^{1}H + {}^{1}H \rightarrow {}^{2}H + \underline{}$

(d) $^{20}Ne + \underline{} \rightarrow {}^{24}Mg + \gamma$

(e) $^{20}Ne + {}^{4}He \rightarrow \underline{} + {}^{16}O$

(f) $^{27}Al + {}^{2}H \rightarrow \underline{} + {}^{28}Al$

3.24. Complete the following nuclear reactions (nuclear charges have been omitted):

(a) $^{97}Tc \rightarrow {}^{97}Ru + \underline{}$

(b) $\underline{} + {}^{4}He \rightarrow {}^{243}Bk + n$

(c) $^{249}Cf + \underline{} \rightarrow {}^{263}Sg + 4n$

(d) $^{1}H + {}^{14}N \rightarrow \underline{} + {}^{4}He$

(e) $n + {}^{235}U \rightarrow \underline{} + {}^{94}Sr + 2n$

(f) $^{228}Ra \rightarrow \underline{} + {}^{228}Ac$

3.25. Write the balanced nuclear equation for the beta decay of ^{24}Na. Include both mass and atomic numbers.

3.26. The *p–n* reaction is a common nuclear reaction. In a *p–n* reaction, a proton reacts with a nucleus to produce a new nucleus and a neutron as products.

(a) The americium–241 used in smoke detectors, Investigate This 3.30, and Consider This 3.31, is extracted from spent nuclear reactor fuel rods. It is produced in the rods by a *p–n* reaction of plutonium. What isotope of plutonium is required? Write the balanced nuclear reaction that produces americium–241.

(b) A carbon isotope is produced by a *p–n* reaction of nitrogen–14 in the atmosphere. Cosmic rays (radiation and particles from the Sun) are the source of the protons. What isotope of carbon is produced? Write the balanced nuclear reaction for its production.

3.27. When ^{14}N captures a neutron, it decays to ^{3}H and another product. Write the balanced nuclear equation for this reaction.

3.28. Write equations to describe how the fusion of two ^{12}C^{6+} can lead to the formation of:
(a) ^{23}Na^{11+}
(b) ^{23}Mg^{12+}
(c) ^{20}Ne^{10+}
(d) ^{16}O^{8+}

3.29. Write equations to describe how the fusion of two ^{16}O^{8+} can lead to the formation of:
(a) ^{32}S^{16+}
(b) ^{31}P^{15+}
(c) ^{28}Si^{14+}
(d) ^{24}Mg^{12+}

3.30. Write equations to show how the three isotopes of Mg (^{24}Mg^{12+}, ^{26}Mg^{12+}, and ^{27}Mg^{12+}) are produced from the fusion of ^{12}C^{6+} and ^{16}O^{8+}.

3.31. Xenon-143 decays by a series of six successive beta particle emissions to a stable isotope. Write the series of decays and identify the stable isotope.

3.32. In 1996, the skeleton of an ancient hunter was found in the mud under the water near the shore of the Columbia river in Kennewick, Washington. A tiny sample from the skeleton (collagen in the bone) was analyzed for carbon-14 and found to contain 0.362 as much of this isotope as is present in living organisms. When did the "Kennewick Man" die?

3.33. At 8:15 a.m., a PET scan patient was injected with a compound containing fluorine 18. Assuming that none of the compound is excreted, what fraction of the F-18 remains in the patient's body at noon? Explain how you solve the problem. *Hint:* See Table 3.3.

3.34. Tiny quantities of iodine are essential for the proper functioning of our thyroid gland. A common treatment for patients with enlarged thyroid glands (hyperthyroidism) is ingestion of a compound, such as NaI, containing iodine-131. The iodine concentrates in the thyroid and beta particles produced by its decay kill the thyroid cells where it has accumulated.

(a) Why are only thyroid gland cells killed by the beta emission? *Hint:* See Consider This 3.31.

(b) Assuming that none of the I-131 is excreted, how long will it take for the radioactivity to decay to 10% of its initial level? *Hint:* See Table 3.3.

3.35. Rubidium-87 decays by beta emission with a half life of 4.9×10^{10} years.

(a) Write the balanced nuclear reaction equation for the decay of Rb-87.

(b) A rock sample from Greenland was found to have a Sr-87/Rb-87 mass ratio of 0.056. How old is the rock sample? Explain how you find the age and clearly state the assumptions you make.

Section 3.5. Nuclear Reaction Energies

3.36. What is binding energy?

3.37. What is the difference between fusion and fission? How might you predict whether a nucleus would undergo fission or fusion?

3.38. Calculate the binding energy per nucleon for:

(a) $^{20}Ne^{10+}$ (nuclear mass = 33.18913×10^{-24} g)

(b) $^{28}Si^{14+}$ (nuclear mass = 53.07632×10^{-24} g)

(c) Which nucleus is more stable? Explain your answer.

3.39. For the elemental nuclei, $^{6}Li^{3+}$ (nuclear mass = 9.98561×10^{-24} g) and $^{56}Fe^{26+}$ (nuclear mass = 92.8585×10^{-24} g), calculate the binding energy in kilojoules

(a) per nucleus

(b) per mole

(c) per nucleon

Origin of Atoms   **Chapter 3**

3.40. WEB Chap 3, Sect 3.5.1-4.

(a) Draw a picture of two nuclei that would undergo nuclear fusion, for example C-12 and He-4. Be sure to indicate the components of the nucleus as protons, neutrons, or electrons. Draw the resulting nucleus, after the nuclei have undergone fusion.

(b) Why would this type of fusion occur? Explain clearly.

(c) Why would these nuclei not undergo fission? Explain clearly.

3.41. (a) The mass of a Xe-142 nucleus is $235.63075 \times 10^{-24}$ g. What is its binding energy in kJ·mol^{-1}?

(b) Use your result from part (a) and the data in Figure 3.20 to calculate the energy released in this fission reaction:

$$^{235}U^{92+} + {}^1n^0 \rightarrow {}^{90}Sr^{38+} + {}^{142}Xe^{54+} + 4{}^1n^0$$

(c) How does your result in part (b) compare to the energy for the fission reaction in Figure 3.20? What might you conclude about the energies released in the fission reactions of U-235? Explain.

3.42. The energy change (in kJ·mol^{-1}) for one of the collision-induced nuclear fission reactions in Figure 3.21 is shown in Figure 3.20. The binding energies per nucleon for Kr 92, Rb 89, Cs 144, and Ba 141 are, respectively, -82.3, -83.7, -79.4, and -80.5×10^7 kJ·mol^{-1}.

(a) What are the energy changes for the other nuclear fission reactions in Figure 3.21?

(b) What might you conclude about the energies released in the fission reactions of U-235? Explain.

3.43. WEB Chap 3, Sect 3.4.3.

(a) Draw a nucleus of a carbon-12 atom. Clearly label the components of the nucleus as protons, neutrons, or electrons.

(b) Draw a picture of what this nucleus would look like if it separately underwent each of the main types of radioactive decay: alpha, beta, gamma and positron emission. Identify what the resulting elemental nucleus would be in each case.

(c) Which type of radiation do you think C-12 would be like likely to emit (if any). Give an example of an elemental nucleus that would be more likely than C-12 to undergo each type of radioactive decay.

3.44. The mass of a He-3 nucleus is 5.00642×10^{-24} g. Is this reaction endothermic or exothermic?

$$_2^3\text{He}^{2+} + {_2^3}\text{He}^{2+} \rightarrow {_2^4}\text{He}^{2+} + 2{_1^1}\text{H}^+$$

How much energy is required or released per gram of He-3 that reacts?

3.45. In March 2002, scientists from Oak Ridge National Laboratory reported that they had subjected a sample of acetone, $CH_3C(O)CH_3$, in which all the hydrogens had been replaced with deuterium, ^2H, to intense cavitation (formation and collapse of bubbles), and had detected the formation of neutrons and tritium, ^3H, in the sample. A tentative explanation for this observation was that fusion of deuterium nuclei, which can occur in two equally probable ways, was taking place at the high temperatures and pressures in the collapsing bubbles:

$$^2H + {^2H} \rightarrow {^3H} + p \qquad \text{(i)}$$
$$^2H + {^2H} \rightarrow {^3He} + n \qquad \text{(ii)}$$

(a) What are the energy changes for reactions (i) and (ii)? The masses of H-2, H-3, and He-3 nuclei are, respectively, 3.34357×10^{-24} g, 5.00736×10^{-24} g, and 5.00642×10^{-24} g.

(b) Which set of products is more stable? Explain the reasoning for your answer.

Section 3.6. Cosmic Elemental Abundance and Nuclear Stability

3.46. **(a)** The nuclear masses of helium-4 and beryllium-8 are 6.644655×10^{-24} g and 13.28949×10^{-24} g. What are the Δm and ΔE (in kJ·mol^{-1}) for this reaction:

$$^4\text{He}^{2+} + {^4}\text{He}^{2+} \rightarrow {^8}\text{Be}^{4+}$$

(b) Beryllium-8 decays by splitting into two alpha particle with a half life of about 7×10^{-17} s. Does your result in part (a) help you understand this very short lifetime? Explain why or why not.

(c) What is ΔE (in kJ·mol^{-1}) for this reaction:

$$^4\text{He}^{2+} + {^8}\text{Be}^{4+} \rightarrow {^{12}}\text{C}^{6+}$$

(d) The sum of the two reactions in this problem is reaction equation (3.7) in the text. What is ΔE for reaction (3.7)? Explain how you get your result.

(e) The text that accompanies reaction equation (3.7) says the fusion involves "almost-simultaneous collision of three helium-4 nuclei." How is the information in this problem related to this statement? Do your results in this problem help to explain the relative abundance of beryllium in the universe? Explain your responses.

3.47. A nuclear reaction sequence that occurs in some stars is:

$$^{12}C^{6+} \rightarrow {}^{13}N^{7+} \rightarrow {}^{13}C^{6+} \rightarrow {}^{14}N^{7+} \rightarrow {}^{15}O^{8+} \rightarrow {}^{15}N^{7+} \rightarrow ({}^{12}C^{6+} + {}^{4}He^{2+})$$

(a) Write the balanced nuclear reactions for each step of the sequence. *Hint:* Consider fusions involving protons, ^1H, and electron capture decays, as in Check This 3.39. Compare your reactions to those you proposed in Check This 3.56(b).

(b) Show that the *net* result of the series of reactions you wrote in part (a) is equivalent to nuclear reaction equation (3.6).

(c) The series of reactions you wrote in part (a) is usually referred to as the CNO (carbon-nitrogen-oxygen) cycle and carbon-12 is said to catalyze the formation of helium from hydrogen. Explain why carbon is assigned this role. *Hint:* A catalyst facilitates a reaction, but is not itself consumed in the net reaction.

Section 3.7. Formation of Planets: The Earth

3.48. (a) Since 70% of the mass of your body is water, we might say that you are mostly water, and the data in Figure 3.27 confirm that hydrogen and oxygen atoms are the most abundant in the human body. Is the ratio of these atoms consistent with the composition of water? Explain why or why not.

(b) Some might say that the crust of the Earth is mostly sand, silicon dioxide, SiO_2. Do the data in Figure 3.27 confirm this suggestion? Explain why or why not.

3.49. During the early part of the twentieth century, chemists spent a good deal of effort determining accurate values for relative atomic masses. When lead was studied, the metal from different ores had different atomic masses. From ores that also contained uranium, the relative atomic mass was close to 206 u, but from other ores, the value was close to 208 u.

(a) What would you suggest as the reason for the difference in the lead in different ores? Is your explanation consistent with the information in this problem and this chapter? Explain your answers clearly.

(b) The relative atomic mass of lead shown in the periodic table on the end papers of the book is 207.2 u. Clearly explain how you account for this value.

Section 3.9. EXTENSION — Isotopes: Age of the Universe and a Taste of Honey

3.50. Figure 3.29 shows a schematic diagram of an instrument called a mass spectrometer, which is named by analogy with the light spectrograph (or spectrometer) represented in Figure 3.3. A light spectrometer uses a prism or diffraction grating to disperse light into its component wavelengths. What does the mass spectrometer disperse? What is the part of the mass spectrometer that is responsible for the dispersion? Explain.

3.51. Since the temperature of the ocean surface has such a large impact on the Earth's climate, this is one of the variables whose history scientists wish to determine as they develop climate models. Most of the approaches to discovering this history make use of differences in stable isotope ratios. Fossil catfish from the shores of Peru were the focus of one such study. These catfish deposit calcium carbonate as otoliths, "ear stones," which grow throughout the life of the fish in layers (like the rings of a tree), so annual variations can be determined. The study examined the $\delta^{18}O$ in otoliths from contemporary fish and from 6000-6500 year-old fossil fish from a site on the northern coast of Peru and one from further south. The data are shown in this figure.

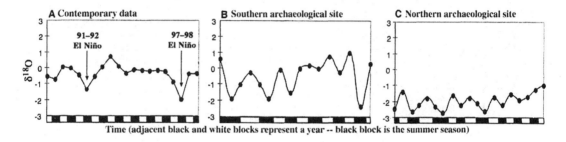

(a) During an El Niño year, the sea surface temperature (SST) of the eastern Pacific Ocean off the coast of South America gets warmer than usual. How does $\delta^{18}O$ vary with temperature in these calcium carbonate otoliths? Clearly explain the reasoning for your answer.

(b) The authors conclude that "the most plausible explanation for the archaeological $\delta^{18}O_{otolith}$ values is warmer summer SSTs at [site X] and nearly tropical conditions near [site Y] in the early mid-Holocene [the last 11000 years]." Which of the sites represented in the figure is site X and which site Y? Clearly explain the reasoning for your choices.

(c) Do the conditions at the sites you identified in part (b) make sense in terms of their geographic locations? Explain.

3.52. Potassium is relatively abundant, Figures 3.7 and 3.25, so geologists make extensive use of argon-potassium dating. Potassium-40 is radioactive and decays by two pathways: beta emission and electron capture. 10.7% of the decay occurs by the electron capture pathway. Argon-40 formed in the electron capture reaction is a gas and can escape from molten rock, but is trapped in the crystal lattice, if it is formed after the rock solidifies. Thus, Ar-40/K-40 dating indicates the age of the sample since it solidified.

(a) Write balanced nuclear reactions for the two modes of K-40 decay. *Hint:* See Check This 3.39.

(b) Assume that exactly 100 micrograms (μg) of K-40 are present in a sample when it solidifies. After one K-40 half life, 1.25×10^9 years, how many micrograms of Ar-40 will have been formed? What is the Ar-40/K-40 mass ratio at this time? *Hint:* 10.7% of the K-40 that reacted has become Ar-40.

(c) Calculate the Ar-40/K-40 mass ratio after two half lives have elapsed. *Hint:* The Ar-40 formed during the second half life adds to that formed during the first.

(d) The Ar-40/K-40 mass ratio as a function of the age of the rock (time since it solidified) is shown in this plot. Do your results from parts (b) and (c) fall on the curve? Explain why or why not?

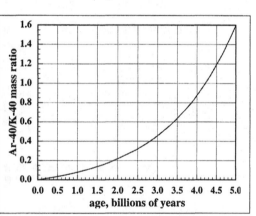

(e) A rock sample returned from the surface of our moon had an Ar-40/K-40 mass ratio of 1.05. What is the age of the sample? What does this result suggest about the age of the moon? Explain.

3.53. WEB Chap 3, Sect 3.5-9.

The plot of activity as a function of time for the radioactive decay of indium-113m on pages 7, 8, and 9 of this section of the Web Companion is low resolution and difficult to read. Another similar set of data for this decay are given here.

time, min	activity, counts per 30 s
0	68372
30	54852
60	45457
90	36901
120	29964
150	24570
185	18936
229	15020
240	12790

(a) Plot these data, as in the Web Companion, and use your plot to determine the half life for the decay. How does your value compare with the one from the Companion? *Hint:* One approach, which requires some mathematical manipulation, is to use a graphing calculator or computer plotting program to plot the data, find the equation of the exponential curve through the data, and use the equation of this curve to find the half life.

(b) Apply equation (3.28) to these data (with uranium replaced by indium) and plot the data to give a linear plot. Use the equation of the line to find the half life and compare your value with the one from the Web Companion.

Chapter 4. Structure of Atoms

Section 4.1. Periodicity and the Periodic Table 4-6
    Origin of the periodic table 4-6
    Filling gaps in the periodic table 4-8
    Ionization energy 4-9

Section 4.2. Atomic Emission and Absorption Spectra 4-11
    Spectrum of the sun: elemental analysis by absorption of light 4-11

Reflection and projection   *4-13*

Section 4.3. Light as a Wave 4-14
    Wave nomenclature 4-14
    Superimposed waves and diffraction 4-16
    Diffraction of light 4-18
    Electromagnetic waves 4-19
    Electromagnetic spectrum 4-21
    Source of electromagnetic radiation 4-23

Reflection and projection   *4-24*

Section 4.4. Light as a Particle: The Photoelectric Effect 4-24
    Emission from glowing objects 4-25
    Planck's quantum hypothesis 4-26
    The photoelectric effect 4-28
    The dual nature of light 4-31

Reflection and projection   *4-32*

Section 4.5. The Quantum Model of Atoms 4-33
    Quantized electron energies in atoms 4-33
    Electron energy levels in atoms 4-34

Section 4.6. If a Wave Can Be a Particle, Can a Particle Be a Wave? 4-38
    Wavelength of a moving particle 4-38
    Experimental evidence for electron waves 4-39

Reflection and projection   *4-41*

Section 4.7. The Wave Model of Electrons in Atoms 4-41
Waves and atomic emissions 4-41
Standing electron waves 4-42
Probability picture of electrons in atoms 4-43
Orbitals 4-44

Section 4.8. Energies of Electrons in Atoms: Why Atoms Don't Collapse 4-45
Kinetic energy of an electron 4-45
Potential energy of a nucleus and an electron 4-46
Total energy of an atom 4-47
Why atoms don't collapse 4-49
Ionization energy for the hydrogen atom 4-49

Reflection and projection *4-50*

Section 4.9. Multi-electron Atoms: Electron Spin 4-50
The electron wave model for helium 4-51
The electron wave model for lithium: a puzzle 4-52
Magnetic properties of gaseous atoms 4-52
Electron spin 4-53
Pauli exclusion principle 4-53
Exclusion principle applied to lithium: puzzle solved 4-54

Section 4.10. Periodicity and Electron Shells 4-54
Patterns in the first ionization energies of the elements 4-54
Electron shell model for atoms 4-55
Atomic radii 4-57
Electronegativity 4-58
A closer look at the electron shell model 4-59
Electron spin and the shell model 4-62

Reflection and projection *4-63*

Section 4.11. Wave Equations and Atomic Orbitals 4-64
Schrödinger wave equation 4-64
Wave equation solutions: orbitals for one-electron atoms 4-65
Electron configurations 4-67
Energy levels in multielectron atoms 4-68
Electron configurations for multielectron atoms 4-69

Chapter 4 — Structure of Atoms

Section 4.12. Outcomes Review 4-72

Section 4.13. EXTENSION — Energies of a Spherical Electron Wave 4-74

    Kinetic energy of an electron wave 4-74

    Potential energy of an electron wave 4-76

    Total energy of an atom 4-76

Structure of Atoms Chapter 4

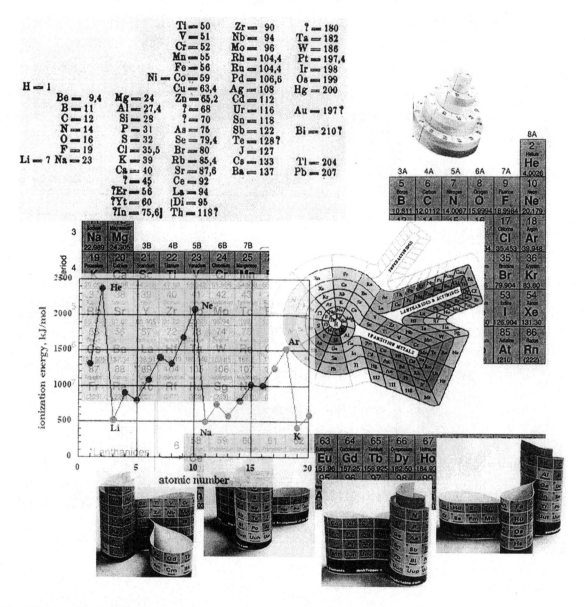

The first periodic table (upper left) based on the repeating properties of the elements as a function of relative atomic mass was devised by Dmitri Mendeleev in 1869. Since then, many other versions of the periodic table have been developed to help visualize the conceptual basis for periodicity. The repeating pattern of the ionization energies of elemental atoms, shown in the superimposed graph, is one of the strongest clues to the conceptual basis for periodicity, the electronic shell structure of atoms.

Chapter 4. Structure of Atoms

> [I]t is possible to take advantage of the laws discovered by chemistry [the law of periodicity] without being able to explain their causes.
>
> *Dmitri I. Mendeleev (1836-1907)*

In Chapter 3, you saw that the atoms in our bodies and all the matter that surrounds us result from the Big Bang and billions of years of nucleosynthesis in stars. You also saw that the observed abundances of elemental atoms in the Universe are a consequence of these processes in combination with the relative stabilities of atomic nuclei. Now we will leave nuclear transformations and focus our attention on the arrangements of electrons in atoms.

These arrangements are reflected in the structure of the periodic table of the elements, several versions of which are shown on the facing page. The periodic table is a succinct device that helps us remember and/or predict trends and similarities in chemical and physical properties of the elements; this understanding of periodicity is what we hope you will attain from this chapter. Our present interpretation of periodicity is based on the structure of atoms (the behavior of electrons in atoms), so, in this chapter, we develop a model of atomic structure.

However, the first periodic tables were devised and used before electrons and nuclei had been discovered and only about 60 elements were known. The quotation above from Dmitri Mendeleev (Russian Chemist, 1836-1907), who devised one of the very first periodic tables, reminds us that the empirical correlations upon which these tables were based were useful, even though the basis of the correlations was not understood. It is the search for this understanding that drove a good deal of the scientific effort described in this chapter.

We will begin the chapter by considering the kinds of correlations that led to the development of the first periodic tables and then use a more modern correlation, the ionization energies of atoms, to connect to the atomic shell model introduced in Chapter 1. The spectra of atoms that are so useful in determining the occurrence and abundance of elements in stars and throughout the universe are also useful in determining atomic structure, so we will also

> **Personal Tutor**
> Using Tables and Graphs will be important for the analyses in this chapter.

return to spectroscopy to help develop our atomic models. As you study this chapter, we ask you to do two things: (1) look at experimental data and try to see in them *evidence* for the shell model and (2) look at experimental data and try to interpret them as a *result* of the shell model.

Structure of Atoms Chapter 4

Section 4.1. Periodicity and the Periodic Table

4.1. Consider This *Are there patterns in the molar volumes of the elements?*

Work together in small groups on this investigation and then discuss your findings with the whole class. You can use the densities of solid and liquid elements and their relative atomic masses to calculate their volume per mole. Data for the elements known in 1869 are shown here.

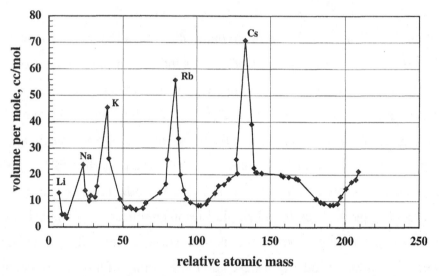

Figure 4.1. Molar volume of the elements (1869) as a function of atomic mass.
The data points are connected to make trends easier to see.

(a) How is the molar volume a measure of atomic volume?

(b) Describe any patterns in atomic size you see in Figure 4.1. Are these patterns represented in the periodic tables shown in the chapter opening illustration? Explain why or why not.

(c) Radium, an **alkaline earth metal** (Be, Mg, Ca, Sr, Ba, and Ra) with a relative atomic mass of 226, was unknown in 1869. What do you predict (from the data in Figure 4.1) for the molar volume of radium? Explain how you arrive at your answer. Use a reference handbook to find density data to check your prediction.

(d) Describe how **alkali metal** (Li, Na, K, Rb, and Cs) atoms vary in size with atomic mass. How might you use our present-day model of the nuclear atom to interpret this variation?

Origin of the periodic table

The most powerful organizing concept in chemistry is **periodicity** — the pattern of repeating chemical properties — which is represented by the **periodic table**. Periodicity was conceived or discovered independently in 1869 by Mendeleev and Lothar Meyer (German chemist, 1830-

1895). Their work was guided, in part, by the availability of a consistent set of relative atomic masses proposed in 1858 by Stanislao Cannizzaro (Italian chemist, 1826-1910). These masses were accepted by all scientists, because they resolved many problems among different atomic mass scales and were based on a wealth of experimental evidence that had built up over the previous century.

Mendeleev and Meyer put the known elements in order by relative atomic mass and noticed repeating patterns in their properties. Meyer, for example, calculated the ratio of elemental density to relative atomic mass, plotted the data, as in Figure 4.1, and noted the obvious peaks for the low density alkali metals, with the denser elements in troughs between them. Another physical property he considered was boiling point, Figure 4.2. Meyer based his periodic table on patterns like the ones you see in these figures.

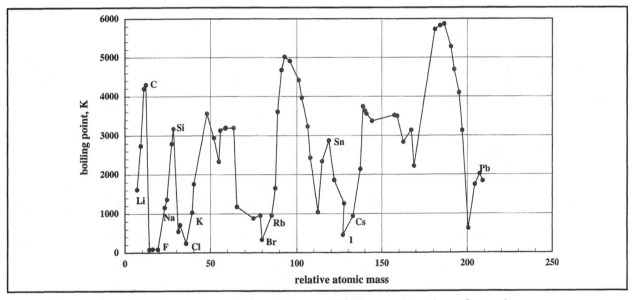

Figure 4.2. Boiling points of the elements (1869) as a function of atomic mass.
These are modern data. Meyer and Mendeleev did not know the actual values for the very low and very high boiling points. The data points are connected to make trends easier to see.

4.2. **Consider This** *How do you interpret patterns in the boiling point data?*

(a) Describe the pattern(s) you see in the boiling point data as a function of relative atomic mass in Figure 4.2. Note that there are gaps in the data because not all the elements had yet been discovered. Where does the pattern seem to be interrupted by missing data? Explain what you think is missing.

(b) In 1869, the **noble gases** had not yet been discovered. Their relative atomic masses and boiling points are tabulated here. If these data are added to the plot in Figure 4.2, are periodic patterns more evident or less evident? Explain your answer.

noble gas	He	Ne	Ar	Kr	Xe	Rn
atomic mass, u	4	20	40	84	131	222
boiling point, K	4.2	27.1	87.3	119.5	165.1	211.3

Filling gaps in the periodic table

In contrast to Meyer, Mendeleev focused more on chemical properties, such as reactivity and binary compounds formed by the elements. The alkali metals (elements in the first column of the periodic table), for example, all react vigorously (in some cases violently) with water to form basic solutions, as shown in Figure 4.3. They also all form one-to-one halide salts, such as NaCl, KI, and CsBr. In order to make elements with similar chemical properties align in families, Mendeleev took the bold step of leaving gaps in his periodic table for undiscovered elements. (In his first periodic table, shown in the chapter opening illustration, the families are in horizontal rows. Within two years he had rotated his table so that families were in columns, as they have remained in many versions of the table.)

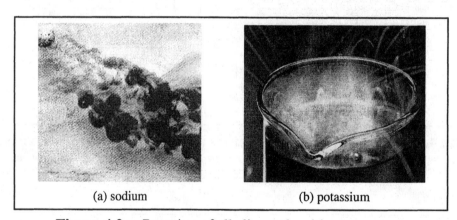

(a) sodium   (b) potassium

Figure 4.3. Reaction of alkali metals with water.
The water contains phenolphthalein, an acid-base indicator that turns red in the presence of base.

Based on the periodic properties of the known elements, Mendeleev predicted the properties of several undiscovered elements. Guided by these predictions, other scientists soon experimentally searched for and found several of the "missing" elements, thus confirming the value and power of the concept of periodicity. All of these developments occurred before the discovery of any subatomic particles and before a model for the structure of an atom had been developed. The organization was based on the macroscopic properties of the elements.

> **4.3. Consider This** *How do you predict the properties of an undiscovered element?*
>
> In Mendeleev's periodic table, there are missing elements (denoted by question marks in the chapter opening illustration) at atomic masses of 45, 68, 70, and 180. The middle two of these four elements are the ones we now know as gallium, Ga (atomic mass 69.7), and germanium, Ge (atomic mass 72.6), which, as shown by Mendeleev, are in the boron and carbon families, respectively. From the data and periodic patterns in Figures 4.1 and 4.2, predict the molar volumes and boiling points of these elements. Clearly explain how you make your predictions. Use a reference handbook to check your predictions.

Ionization energy

With the discovery of electrons and as a result of experimentation on ionization of gaseous elemental ions, scientists were able to measure properties of atoms themselves and find out whether these properties also exhibited periodicity. Recall that, when we discussed the formation of ionic compounds (Chapter 2, Section 2.4), we introduced the idea that atoms can lose and gain electrons to form, respectively, positive and negative ions. For example, the ionization of a gaseous sodium atom can be represented by this reaction equation:

$$Na(g) \rightarrow Na^+(g) + e^-(g) \qquad (\Delta E_{ionization} = 496 \text{ kJ·mol}^{-1}) \qquad (4.1)$$

The energy required to remove an electron from a gaseous atom (or ion), $\Delta E_{ionization}$, is called its **ionization energy**; the ionization energy for sodium atoms is 496 kJ·mol^{-1}. The ionization energies for gaseous atoms of the first twenty elements are plotted in Figure 4.4 as a function of their atomic numbers. Atomic numbers now replace relative atomic mass as the independent variable, because we now know that the identity of an element depends on the number of protons in its nucleus, not on the mass of its atoms.

Structure of Atoms

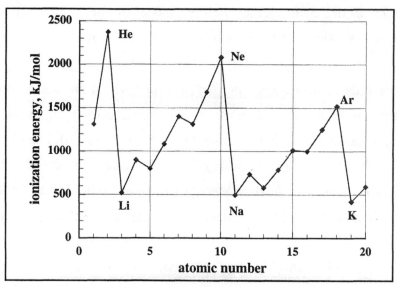

Figure 4.4. Ionization energies for the first twenty elements.
The data points are connected to make trends easier to see.

4.4. Consider This   *How do you interpret patterns in the ionization energy data?*

(a) Describe the pattern(s) you see in the ionization energy data as a function of atomic number in Figure 4.4.

(b) Do the major features of your pattern correlate with the shell model of atoms introduced in Chapter 1, Section 1.2, especially Figure 1.7? If so, what do you think is the physical basis for the correlation? Explain your reasoning clearly.

As you see, an atomic property, the ionization energy of the atoms, shows striking periodicity. The periodicity in macroscopic properties that Mendeleev and Meyer represented in their periodic tables must be a reflection of periodicity in the properties of the atoms themselves. In our nuclear model of the atom, the electrons make up the bulk of the volume of an atom and are responsible for most of its interactions with external forces. When we look at Figure 4.4, we can see that the three major peaks seem to correspond to the shells in our electron shell model from Chapter 1. However, the smaller steps between the peaks are not accounted for by simple shells, so we will probably need to refine our model.

In the remainder of the chapter, we will explore how electrons behave in atoms and further develop our model of atomic structure. Many of the clues that guide our exploration and refinement come from the study of atomic spectra, which we introduced in the previous chapter and take up again in the next section.

Chapter 4   Structure of Atoms

Section 4.2. Atomic Emission and Absorption Spectra

4.5. Investigate This   *Is the visible spectrum affected by substances in the light beam?*

Use the set-up described in Investigate This 3.2 to produce a visible spectrum on the projection screen. Cover about half the length of the slit with a flat-bottomed Petri dish containing a solution of potassium permanganate, $KMnO_4$. Write a description comparing and contrasting the spectrum you observe from light that has passed through the solution with the spectrum from the light that has not.

4.6. Consider This   *Why is a spectrum affected by substances in the light beam?*

(a) In Investigate This 4.5, what is the correlation, if any, between the spectra you observe and the color of the potassium permanganate solution? How do you explain this correlation?

(b) Solutions of nickel chloride, $NiCl_2$, are green. If a solution of nickel chloride is substituted for the potassium permanganate solution, what will the spectra look like? Check your prediction experimentally.

Spectrum of the sun: elemental analysis by absorption of light

In Chapter 3, we discussed analysis of stars, including our sun, by emission spectroscopy. By the middle of the 19th century, spectrographs had been improved to the point that another feature of the sun's spectrum was observed. Superimposed on the background of the visible spectrum are hundreds of dark lines, some of which you can see here in Figure 4.5. The dark lines meant that these wavelengths were missing in the spectrum from the sun. Scientists noticed that many of the wavelengths of these dark lines corresponded to the emission wavelength lines of particular elements. Compare this observation with the experimental results from Chapter 3, shown in Figures 3.4 and 3.5 and your analysis in Check This 3.7.

Figure 4.5.   A few of the most prominent dark lines in the solar spectrum.

This phenomenon of missing wavelengths can be studied in laboratories on earth. Two spectroscopic experiments and their results are shown schematically in Figure 4.6. In the experiment represented in Figure 4.6(a), the visible emission from a sodium atom discharge lamp, like the ones you used in Investigate This 3.4, is examined in a spectrograph and the typical atomic line emission is observed. In the experiment represented in Figure 4.6(b), the continuous emission from a lamp that emits white light passes through a sample of hot gaseous sodium atoms (which are not emitting light) before it is examined in the spectrograph.

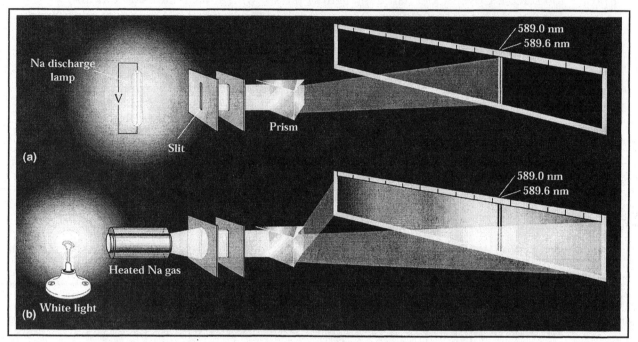

Figure 4.6. Two spectroscopic experiments with sodium atoms.

4.7. Consider This *How do you interpret the results shown in Figure 4.6?*

Work in small groups to compare and contrast the spectra that are obtained in the two experiments represented in Figure 4.6. What is your interpretation of the reasons for the similarities and differences between the spectra?

To explain the dark lines in the solar spectrum, Gustav Kirchhoff (German physicist, 1824-87) proposed that atoms in a layer or cloud around the sun absorbed the sun's continuous emission and accounted for the missing wavelengths. Whatever process produces the emission from elemental atoms is reversible. The atoms can absorb light of the same wavelength and produce dark lines, an **absorption spectrum** [Figure 4.6(b)], where light is missing in the spectrum from a continuous source. Thus, we can use absorption spectra to analyze the composition of the cooler regions around stars and emission spectra to analyze the composition of their hotter surfaces. We also use absorption spectra to analyze the clouds of atoms and molecules between the stars and our telescopes.

4.8. **Consider This** *Does light absorption explain the results in Investigate This 4.5?*

(a) Why do some substances appear colored? When we look at the light passing through a colored solution, what happens to the wavelengths of light that do not reach our eyes?

(b) Instead of absorbing light at discrete wavelengths as atoms do, substances in solution usually absorb light over a range of wavelengths. Are your observations in Investigate This 4.5 an example of an absorption spectrum? Explain why or why not.

4.9. **Check This** *Elemental identification by absorption spectroscopy*

An experiment similar to the one represented in Figure 4.6(b) gave this result:

700   600   500   400

Use the emission spectral data in Figure 3.5 to identify the elemental atomic gas in the heated container. *Hint:* An atomic absorption spectrum does not necessarily absorb at *all* the wavelengths that are present in an emission spectrum of the same element.

The emission and absorption spectra of elemental atoms provide clues about their structures and properties. Emission occurs from atoms that have been given a lot of energy by a spark or a flame and absorption occurs from less energetic atoms at a lower temperature. This probably means that the emission and absorption processes in Figure 4.6 represent changes in the energy of the atoms. Perhaps an atom with extra energy can lose energy by emitting light of a certain wavelength and, conversely, perhaps an unenergetic atom can gain energy by absorbing light of the same wavelength. In order to connect wavelengths of light and atomic energies, we need to know more about the nature of light, which we will examine in the next two sections.

Reflection and projection

The recognition by Mendeleev and Meyer that the properties of the elements repeated as a function of the atomic mass, was an enormously valuable breakthrough for chemistry. This periodicity was symbolized by them and is still symbolized today by the periodic table of the elements. Keep in mind, however, that it is the periodicity underlying the table, not the table itself, that is so important for seeing correlations among and making predictions about the properties of the elements. And, as Mendeleev said, the periodicity of the elemental properties is useful, even if we do not know why the properties are periodic. But we are always curious about the "whys," because we are not content with correlations; we want more fundamental understanding of atomic structure.

We recalled from Chapter 3 that a fundamental difference between the emission spectra from gaseous atoms and the emissions from other materials is that only certain wavelengths of light are emitted by energized atoms. And we have seen here that gaseous atoms also absorb only certain wavelengths of light, all of which are also emitted by the energized atoms. It seems likely that these discrete emissions and absorptions might provide a clue to the structure of atoms, especially the arrangement of electrons around the atomic nucleus. However, in order to analyze these spectra in more detail and to relate them to the energies in atoms, we need to know more about the nature of light, the topic of the next two sections.

Section 4.3. Light as a Wave

4.10. Investigate This   *What are the characteristics of waves?*

Do this as a class investigation and work in small groups to discuss and analyze the results. Use a small ripple tank to observe water wave patterns. Set up the tank so that parallel waves strike a barrier containing a single gap. Observe the wave pattern before the barrier and beyond the barrier and make a sketch of them.

4.11. Consider This   *How do you characterize water waves?*

In Investigate This 4.10, how are the waves beyond the barrier the same as the ones before they strike the barrier? How are they different? Do your sketches show these similarities and differences? Why or why not?

Wave nomenclature

Water waves are familiar to almost everyone because the undulations on the surface of a lake, pond, or the ocean are so easily seen. Light also exists as waves. All waves, including water and light waves, are characterized by oscillations described by a sine curve, as shown in Figure 4.7. The distance from crest to crest, the **wavelength**, λ (lower-case Greek "lambda"), and the displacement of a crest or valley from the zero (null) level, the **amplitude**, are characteristic properties of a wave. The places where the amplitude is zero are called **nodes**.

WEB Chap 4, Sect 4.2.2
Try interactive animations of waves and wave properties.

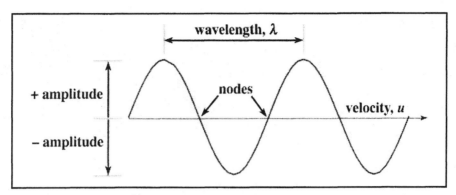

Figure 4.7. Components used to describe a wave.
Amplitude has value and sign as the wave oscillates with the same amplitude above and below its zero (null) value.

The **velocity** of the wave, u, is the distance the wave moves in one second. If you were able to position yourself at some point and count the number of crests that passed that point each second, you would have measured the **frequency** of the wave, ν (lower-case Greek "nu"). Frequency has units of $1/\text{second}$ (s^{-1}). The relationship between velocity, wavelength, and frequency is:

$$u = \lambda \cdot \nu \tag{4.2}$$

4.12. Worked Example *Wavelength of water waves*

Suppose you are standing on a 10-meter long dock to which a rowboat is tied. Watching the waves from the lake washing past the dock, you note that the rowboat bobs up and down three times during the 5 seconds it takes one wave crest to travel from the outer end to the shore end of the dock. What is the wavelength of the waves?

Necessary information: We need the wave velocity from the problem statement, and equation (4.2) relating the frequency and velocity to wavelength.

Strategy: Substitute the frequency and wave velocity into equation (4.2) to get wavelength.

Calculations: Each time the boat bobs up a wave crest has raised it. This occurs 3 times in 5 seconds, so the waves have a frequency:

$$\nu = 3/(5\text{ s}) = 0.6 \text{ s}^{-1}$$

It takes one wave crest 5 seconds to travel 10 meters, so the waves travel with a velocity:

$$u = (10 \text{ m})/(5 \text{ s}) = 2 \text{ m}\cdot\text{s}^{-1}$$

Rearrange equation (4.2) to solve for the wavelength, λ, and substitute for ν and u:

$$\lambda = u/\nu = (2 \text{ m}\cdot\text{s}^{-1})/(0.6 \text{ s}^{-1}) = 3 \text{ m} \text{ (one significant figure consistent with data)}$$

Does the answer make sense? Since the boat bobs three times while one wave crest travels 10 m, there must be a wave crest about every 3 m, as we have found.

4.13. Check This *Wave properties*

(a) This figure is a representation of waves in a ripple tank, as in Investigate This 4.10. The figure shows the tank at one instant in time. Lighter areas are wave crests and darker areas are valleys. If the distance from crest to crest of the ripples is 8.7 mm and each point on the surface crests 1.87 times per second, what is the velocity of travel of the ripples?

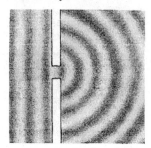

(b) WEB Chap 4, Sect 4.2.2. View the animations of the three different waves and answer the four questions. Assume that the grid lines (light blue lines) are spaced one centimeter apart. What are the amplitudes, wavelengths, frequencies, and velocities of each of the three waves? Is equation (4.2) satisfied by each wave? Show why or why not.

4.14. Investigate This *How do water waves interact?*

Use the ripple tank again, but replace the single-gap barrier with one that has two gaps. Observe the wave pattern beyond the barrier and make a sketch of it. In your group, agree on a way to describe the pattern(s) you observe.

4.15. Consider This *What patterns do interacting water waves create?*

(a) In Investigate This 4.14, are the water waves starting out from each of the two gaps in the barrier the same as those starting out from the single gap in Investigate This 4.10? Explain.

(b) When the water waves from each of the two gaps overlap (interact), how would you describe the observed pattern(s)? As a class, develop a common description of the pattern(s).

Superimposed waves and diffraction

The patterns illustrated in Figure 4.8, and which you saw in Investigation 4.14, when two waves overlap, are called **diffraction patterns**. The processes of wave **reinforcement** and **cancellation** that created those diffraction patterns are caused by **superimposition** (layering or adding together) of the two waves. Where the waves intersect, the single wave that results is the sum of the combined waves.

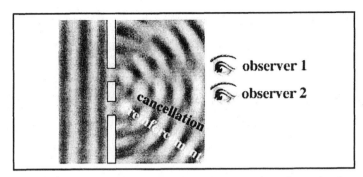

Figure 4.8. Diffraction of overlapping water waves in a ripple tank.
This shows the tank at one instant in time. Lighter areas are wave crests and darker areas are valleys. Position of the observers refers to Check This 4.16.

For two identical waves, the two extreme cases of superimposition are illustrated in Figure 4.9. In Figure 4.9(a), the two waves are exactly matched to each other. They reinforce each other to give a wave with twice the amplitude as either of the original two. Reinforcement leads to the brighter crests and darker valleys in Figure 4.8; one ray of reinforced waves is labeled. In Figure 4.9(b), the two waves are exactly unmatched; the highest crest of one wave corresponds to the deepest valley of the other. The result is that the two waves exactly cancel each other out. In Figure 4.8, the narrow regions without waves (one of which is labeled) are where the waves have cancelled one another. Waves of all kinds, including light waves, undergo this process of superimposition and diffraction.

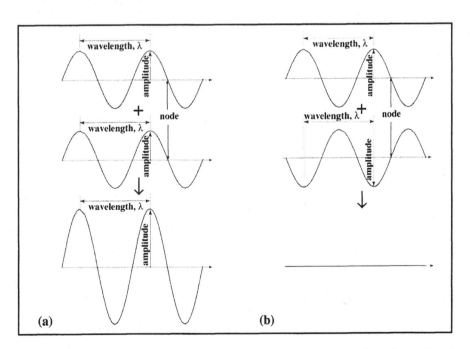

Figure 4.9. Superimposition of waves of the same wavelength and amplitude.
Nodes are matched with one another in both cases, but in (a) the crests are matched with one another and in (b) crests are matched with valleys.

Structure of Atoms Chapter 4

4.16. Check This *Observed results of wave superimposition*

(a) Suppose you are observer 1 in Figure 4.8 and are watching the level of the water at this position. Sketch a graph that shows the variation of the water level with time. How does your sketch correlate with the superimpositions shown in Figure 4.9? Explain.

(b) Suppose you are observer 2. Make a sketch of the variation of the water level with time and answer the same question as in part (a).

4.17. Investigate This *How is light affected by passing through narrow openings?*

(a) Hold your first and second fingers straight and together. Somewhere along the length of the fingers, most often near the palm, there will be a place where they don't quite touch. Close one eye. Hold your fingers close to your open eye and look through the gap between your fingers at a source of light (like a ceiling light) or a bright, light-colored wall. Use the thumb and fingers of your other hand to squeeze the gap shut. Describe any pattern of light and dark areas you see as the gap gets narrower. How small does the gap have to be to see a pattern?

(b) Do this as class investigation and work in small groups to discuss and analyze the results. Observe the spot of light from a laser pointer on a light-colored wall or projection screen. Place a transmission diffraction grating in the laser beam. (A transmission diffraction grating, see Figure 3.2 and Investigate This 3.4, is a series of very narrow, closely spaced transparent slits in an opaque film.) Observe and record how the spot of light from the laser is affected by the grating.

4.18. Consider This *How do the properties of water waves and light compare?*

What are the similarities between your observations in Investigation This 4.14 and Investigation This 4.17? Can you explain the observations in Investigate This 4.17 as a result of superimposition of waves? If so, what is observed when light waves reinforce one another? when they cancel one another?

Diffraction of light

In 1803, Thomas Young (English physicist and mathematician, 1773-1829) did experiments similar to yours in Investigate This 4.17(b). When he passed light of a single wavelength through slits in a barrier and observed the light on a parallel screen beyond the barrier, he saw, as in Figure 4.10, bright and dark regions that did not correspond directly to the slits. Young concluded that light moves through space as waves, and that, after passing through the slits, the waves reinforce and cancel to form a diffraction pattern. Where the beams overlap, the bright regions result from wave crests from both slits reaching the screen at exactly the same time; the dark regions result from wave crests from one slit reaching the screen at exactly the same time as wave valleys from the other slit.

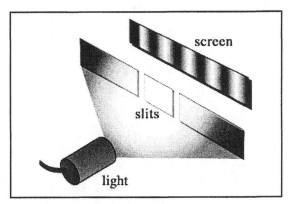

Figure 4.10. An illustration of Thomas Young's light diffraction experiment.
Light passing through the slits produces the diffraction pattern of alternating light and dark areas on the screen.

Imagine that the water waves in Figure 4.8 are light waves and the right hand side of the illustration is a screen. In Check This 4.16, you showed that the pattern on the screen denotes places where no wave crests or valleys reach the screen and other places with strong crests and valleys. Light waves (radiation) of different wavelengths are diffracted at different angles the slits, so white light is dispersed into its spectrum of colors, as you saw in Chapter 3, Figure 3.2.

4.19. Consider This   How are Figures 4.8 and 4.10 related?
WEB Chap 4, Sect 4.2.1. Explain in your own words how Figures 4.8 and 4.10 are related and use this page in the Web Resource to check your response.

Electromagnetic waves

Young's diffraction experiments were convincing proof of the wave nature of light. However, a basic question for 19th century scientists was, *"What is it that oscillates when light waves move through space?"* For water, the answer is easy; the surface of the water moves up and down. For light, the answer is not at all obvious. The answer developed by James Clerk Maxwell (Scottish physicist, 1831-79) is the **electromagnetic wave theory** that we still use to model many interactions of radiation (including light) with matter. Let's try to connect what we know about electric and magnetic phenomena to waves.

You know from experience and the activities in Chapter 1, Section 1.2, that an electrically charged object can attract (or repel) other charged objects from a distance. You also know from experience that a magnet can attract iron objects from a distance. To explain how electric and magnetic attractions can act at a distance, scientists developed the concepts of electric and magnetic fields. These fields exert forces on objects that are not in contact with the source of the fields. Magnetic fields are easy to visualize with iron filings, as shown in Figure 4.11. Moving electrical charges, an electric current, also create magnetic fields; that is how electromagnets and electric motors work. And, conversely, magnetic fields attract and repel moving charges. Electric and magnetic phenomena are closely linked.

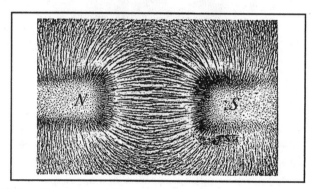

Figure 4.11. Using iron filings to visualize a magnetic field.
Iron filings scattered on a stiff piece of paper align with the magnetic field of two bar magnets under the paper with their opposite poles facing one another as indicated by the *N* and *S* designations.

Electromagnetic waves (also called **electromagnetic radiation**), including visible light, are a combination of an oscillating electric field (an electric field that changes amplitude periodically

WEB Chap 4, Sect 4.2.3-4
View animations of Figure 4.12 and the oscillating electric and magnetic fields.

like the waves represented in Figures 4.7 and 4.9) and, perpendicular to it, an oscillating magnetic field. Their wavelengths are identical, and the crests and valleys occur at exactly the same places along the direction of travel of the wave as illustrated in Figure 4.12.

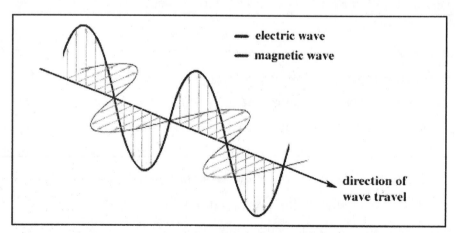

Figure 4.12. Oscillating electric and magnetic fields in electromagnetic waves.
The light gray arrows indicate the direction and magnitude of the wave amplitudes.

Electromagnetic spectrum

By the end of the 19th century, almost all the phenomena associated with light had been successfully explained by assuming an **electromagnetic spectrum** that is composed of electromagnetic radiation of different wavelengths, as shown in Figure 4.13. The wave nature of light seemed firmly established. Wave properties explained, for example, how prisms and diffraction gratings disperse white light into the spectrum of wavelengths (colors). Wave properties also provided mathematical descriptions of light waves. Light waves can't be observed in the same way as water waves on the surface of a pond, but their observable effects, such as diffraction, are explained by the mathematical model.

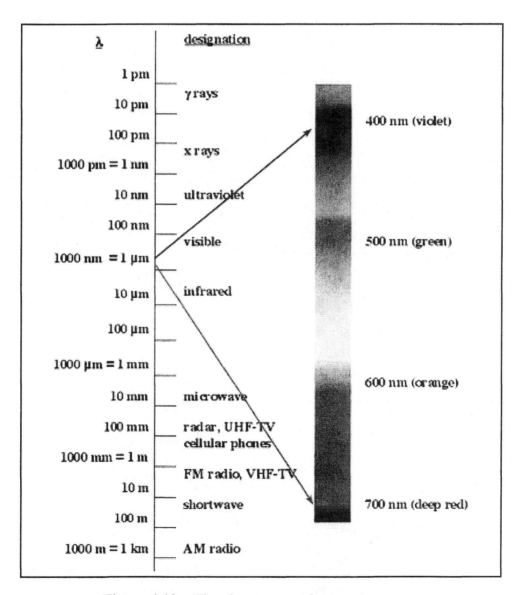

Figure 4.13. The electromagnetic spectrum
The spectrum is shown on a logarithmic wavelength, λ, scale. The visible region of the spectrum, a tiny part of the entire spectrum is expanded.

Structure of Atoms Chapter 4

Figure 4.13 shows that the wavelengths of electromagnetic radiation vary from less than a picometer (1 pm = 10^{-12} m) to many meters. Visible light wavelengths (all the wavelengths in white light) are from about 400 nm, violet, to about 700 nm, deep red. An electromagnetic wave moves through a vacuum with a velocity of 3.00×10^8 m·s^{-1}. This is the **speed of light**, c, that we used in Einstein's mass-energy equivalence relationship in Chapter 3, Section 3.5.

4.20. Check This   *Electromagnetic radiation*

WEB Chap 4, Sect 4.2.2. Explain why these animations of three waves do not represent electromagnetic radiation. *Hint:* Which property of the waves would have to be the same, if all the representations were electromagnetic radiation?

4.21. Worked Example   *Frequency of gamma, γ, radiation*

Calculate the frequency of gamma radiation, an electromagnetic wave, that has a wavelength of 2.45 pm.

Necessary information: We need equation (4.2), the wavelength, λ, of this electromagnetic wave, which is given in the problem statement, and the speed, c, that a light wave travels.

Strategy: For electromagnetic waves, the wave velocity is c, so equation (4.2) becomes:

$$c = \lambda \cdot \nu \qquad (4.3)$$

Substitute the wavelength and speed of light into equation (4.3) and solve for ν.

Implementation: Since the length unit for the speed of light is meters, the wavelength has to be converted to meters, 1 m = 10^{12} pm:

$$\text{wavelength of the gamma ray in meters} = \lambda = (2.45 \text{ pm}) \cdot \left(\frac{1 \text{ m}}{10^{12} \text{ pm}}\right) = 2.45 \times 10^{-12} \text{ m}$$

$$\text{frequency of the gamma ray} = \nu = \frac{c}{\lambda} = \frac{3.00 \times 10^8 \text{ m·s}^{-1}}{2.45 \times 10^{-12} \text{ m}} = 1.22 \times 10^{20} \text{ s}^{-1}$$

Does the answer make sense? Equations (4.2) and (4.3) show that frequency and wavelength are inversely proportional: a wave with a short (small) wavelength has a high (large) frequency. This is the result we observe here and again in Check This 4.22.

4.22. Check This   *Frequencies and wavelengths of electromagnetic radiation*

(a) Calculate the frequency of green light that has a wavelength of 515 nm.

(b) Find the frequency of a wave with a wavelength of 21.11 cm. In what region of the electromagnetic spectrum do you find waves of this frequency?

(c) Based on the results from Worked Example 4.21 and parts (a) and (b) here, how would you describe the relationship of the frequency to the wavelength of electromagnetic radiation?

(d) In what region of the electromagnetic spectrum, Figure 4.13, would you find light with a frequency of 3.45×10^{13} s^{-1}?

Source of electromagnetic radiation

In Investigate This 3.4, Chapter 3, you observed light from a light bulb and looked at its spectrum. How does the hot filament in the light bulb produce electromagnetic radiation? What causes the oscillating electric and magnetic fields that make up electromagnetic radiation? As we said above, moving electrical charges create magnetic fields. Motion of charges in matter is the source of electromagnetic radiation produced by matter. As we go on, we will point out specific examples of radiation sources. One of the simplest to think about is a television or radio transmission antenna. Essentially, the antenna is a length of wire through which an electric current (moving electrons) passes back and forth, that is, oscillates. First one end of the wire is negative and then the other. The frequency of the emitted electromagnetic waves depends upon the frequency of the electronic oscillation. Different television stations use different frequencies; you tune your receiver (select a channel) to choose the one you wish to see and hear. Sometimes the frequency of electromagnetic waves, especially radio and television frequencies, is given in **hertz** (Hz), 1 Hz = 1 s^{-1}.

4.23. Check This *Electromagnetic wavelength detection and emission*

(a) Electromagnetic radiation detectors interact best with wavelengths that are approximately the same size as the detector. The television antennas (several lengths of metal tubing, not satellite dishes) you see on the roofs of houses and other buildings are designed to receive VHF television signals. UHF antennas, on the other hand, are short loops of wire that often are attached directly to the input on the back of the television set. Use the information in Figure 4.13 to explain why these antennas are so different.

(b) FM radio frequencies are in the range 88 —108 MHz (megahertz, 1 MHz = 10^6 Hz). What is the wavelength of emission (transmission) from your favorite FM station?

Emissions of electromagnetic waves from objects in the universe have been detected in all regions of the electromagnetic spectrum. At one extreme, bursts of gamma radiation have been

Structure of Atoms — Chapter 4

detected by specially designed satellites. The sources of some of these bursts appear to be near the edge of the universe; how they are created is still a mystery, but they may come from events that occurred shortly after the Big Bang that we discussed in the previous chapter. At the other end of the spectrum, the Earth is continuously bathed in microwave radiation that is responsible for about one percent of the interference ("snow") on a television set that gets its signal from an antenna. This microwave background radiation is the "glow" left over from the Big Bang. Its existence and variations throughout the universe are major pieces of evidence supporting the Big Bang model.

Reflection and projection

In this section, we have found that light acts like a wave and shares the properties of other more easily observable waves, like water waves. In particular, we can characterize light by its wavelength, frequency, and speed of travel, which are simply related as: $c = \lambda \cdot \nu$.

We further found that light is an electromagnetic wave and that the electromagnetic spectrum of wavelengths extends over an enormous range from the longest radio waves to the shortest gamma waves, with visible light somewhere in between and accounting for only a tiny fraction of the entire range. This information about light was all known by the end of the 19th century and electromagnetic radiation seemed to be a well-understood phenomenon. It came as a surprise to scientists to find that they only knew half the story. Both parts of the story are required to help us unravel the mysteries of atomic structure, so we will continue by examining what happened at the beginning of the 20th century.

Section 4.4. Light as a Particle: The Photoelectric Effect

4.24. Investigate This *Are the temperature and color of a hot filament related?*
Do this as a class investigation and work in small groups to discuss and analyze the results. Plug a lamp with a clear glass light bulb into a variable voltage transformer. Start at zero voltage and increase the voltage slowly until the filament just begins to glow. Record the color of the filament. Hold your hand close to the light bulb and note how warm it feels. *CAUTION*: Don't touch the glass; it could be hot enough to burn you. **Turn up the voltage until the filament is glowing brightly, make the same observations, and answer the same questions.**

4.25. Consider This *How are the temperature and color of a hot filament related?*

In Investigate This 4.24, what, if anything, is different between the observations when the filament just begins to glow and when it is glowing brightly? What, if anything, is the same? How might you explain these observations?

Emission from glowing objects

The explanations of dispersion and diffraction phenomena are successes of the wave theory of light. Near the end of the 19th century, however, scientists were uncomfortably aware that the wave theory could not explain some experimental results. The observations you made in Investigate This 4.24 are an example of a phenomenon that the wave theory could not explain. The filament in a light bulb is heated by the electric current flowing through it and emits the light you see. You found that a filament at a lower temperature glows red and at a higher temperature glows white, that is, emits all visible wavelengths. The wave theory did not predict this temperature dependence of light emission. Figure 3.14 shows the prediction of the wave theory of light and the experimental results at two temperatures.

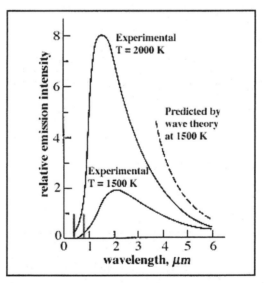

Figure 4.14. Intensity of emission from a glowing object.
The emission intensity at any wavelength is a function of the temperature of the object. The visible region of the electromagnetic spectrum is between the red (0.7 μm = 700 nm) and blue (0.4 μm = 400 nm) lines on the figure.

Structure of Atoms Chapter 4

4.26. Consider This   *How are Investigate This 4.24 and Figure 4.14 related?*

Which experimental curve in Figure 4.14 corresponds to the bulb with the just-glowing filament in Investigate This 4.24? Explain clearly how your observations of color and warmth of the bulb justify this response.

Planck's quantum hypothesis

Max Planck (German physicist, 1858-1947), found that he could explain the emission from glowing objects if he assumed that the oscillating charges that emitted the radiation could have only certain energies

$$E = n \cdot h \cdot \nu \tag{4.4}$$

Here n is any integer, ν is the frequency of the oscillator in s^{-1}, and h is a proportionality constant with units of J·s. Thus, in this model, the oscillator energies are **quantized**; the only energies allowed are multiples of the energy **quantum**, $h \cdot \nu$. Further, Planck had to assume that when an oscillator emitted or absorbed energy, it did so by changing from one energy level, with $n = n_1$, to another energy level, with $n = n_2$, so that the change in energy (the energy emitted or absorbed) is:

> "Quantized," "quantum," and the more familiar "quantitative" are derived from Latin, *quantus* = how much.

$$\Delta E = n_2 \cdot h \cdot \nu - n_1 \cdot h \cdot \nu = \Delta n \cdot h \cdot \nu \tag{4.5}$$

Planck chose the numeric value of the proportionality constant, h, to make the predictions from his model fit the experimental data in Figure 4.14. The constant has been named **Planck's constant** and its modern value is $h = 6.6256 \times 10^{-34}$ J·s.

4.27. Consider This   *Are emission and absorption of energy by oscillators related?*

(a) If an oscillator with four quanta (plural of quantum) of energy changes to one with two quanta of energy, is energy emitted or absorbed? Use equation (4.5) to explain your response.

(b) If the oscillator in part (a) goes from having to two quanta of energy to having four quanta, is energy emitted or absorbed? How is the energy change in this change related to the one in part (a)? Explain.

(c) Do you see any connection between your results in parts (a) and (b) and the experimental observations represented in Figure 4.6? Explain your response.

With Δn = 1, equation (4.5) gives the energy, E, associated with light (electromagnetic radiation) of frequency ν. This relationship can also be written in terms of the wavelength of the light, by combining equation (4.5) (with Δn = 1) and equation (4.3):

$$E = \frac{h \cdot c}{\lambda} \tag{4.6}$$

4.28. Check This *Energy of electromagnetic radiation*

(a) Carry out the combination of equations (4.3) and (4.5) (with Δn = 1) to get equation (4.6).

(b) Which are more energetic, γ rays or microwaves? Use equation (4.6) and the information in Figure 4.13 to answer this question. Does the energy of the electromagnetic radiation increase or decrease as you go from the bottom to the top of the scale in Figure 4.13?

(c) What is the energy (in J) of visible light with a wavelength of 414.5 nm? What color is this light? Explain your response.

Planck's assumption (hypothesis) said that the oscillators in a glowing object could have only certain energies that depend on their frequency. Hotter objects have more energy, so there could be more oscillators with higher energies. This explained why the higher temperature curve in Figure 4.14 has its maximum at a shorter wavelength (higher frequency) of emitted radiation. Since the total energy in the object is limited, the number of high energy oscillators is limited. This explained why the emission curves finally drop toward zero intensity at shorter wavelengths. Scientists had previously thought that the energy of oscillators was associated with the amplitude of the emission, not its wavelength. Planck's hypothesis was met with a great deal of skepticism, because it seemed ridiculous to think that oscillator energies were quantized.

4.29. Investigate This *What is the effect of light on silver chloride?*

Do this as a class investigation and work in small groups to discuss and analyze the results. Use a rectangle of absorbent paper, four colored plastic filters (red, green, blue, and colorless), 0.1 M aqueous solutions of silver nitrate, $AgNO_3$, and sodium chloride, $NaCl$, and a bright lamp. Near one of the long edges, label the paper with the letters "R," "G", "B," and "C," about equally spaced apart. A few centimeters beneath each label, place a drop of the silver nitrate solution. Below each drop of silver nitrate, place a drop of sodium chloride solution so that the solutions overlap as they spread on the paper. Recall from Chapter 2 that silver ion and chloride ion react to form solid silver chloride:

$$Ag^+(aq) + Cl^-(aq) \rightarrow AgCl(s)$$

Structure of Atoms — Chapter 4

Record the appearance of the paper. Cover each pair of spots with the filter of the color corresponding to the label on the paper, as shown in the photograph. Shine a bright light on the covered paper, taking care that each pair of spots gets the same amount of illumination. After five minutes, remove the light and filters, examine the paper, and record its appearance.

4.30. Consider This   *Do all colors of light affect silver chloride the same way?*

Silver halides are affected by light, which is why photographic film contains silver halide crystals. In damp silver chloride precipitates, we can write the reaction caused by light as:

$$\left(Ag^+ \; :\!\ddot{\underset{..}{Cl}}\!:^- \right)_{(s)} \xrightarrow{\text{light energy}} Ag\cdot_{(s)} + :\!\ddot{\underset{..}{Cl}}\cdot_{(aq)}$$
silver metal   chlorine atom

(4.7)

The chlorine atoms formed in this reaction can react with one another to form chlorine molecules, Cl_2, and/or with water to form chloride ions and oxygen gas, O_2. The silver forms tiny specks of silver metal that darkens the paper, as shown here.

(a) Do your results from Investigate This 4.29 provide any evidence that reaction (4.7) has occurred? Explain your response.

(b) If reaction (4.7) occurs, is it the same or different for light of different colors? If it is different, what color (wavelength) light has the greatest effect? Which the least? Can you interpret these observations using the wave theory of light? Why or why not?

The photoelectric effect

Five years after Planck's quantum hypothesis, his ideas were used by Albert Einstein to explain another experimental result that the wave theory had failed to explain: In the **photoelectric effect**, when light shines on a metal surface electrons can be knocked out of the surface. The wave theory predicted that, if the light beam were bright enough (had a large enough amplitude or intensity), any frequency (wavelength) of light could knock an electron out.

> The photoelectric effect is the principle behind the "electric eyes" that control such things as outdoor lighting and elevator doors. The presence of light falling on a photoelectric surface causes an electric current that triggers lights to come on or go off, for example.

However, experimental results illustrated for low frequency light in Figure 4.15 show that no current flows in this circuit, — that is, no electrons are knocked out —no matter how bright the light beam. Low frequencies of light cannot knock electrons out of the metal. The experimental results illustrated in Figure 4.16 show that when a higher frequency light is used, current flows. Electrons can be knocked out of the metal by high frequency light. The results for a number of different frequencies are summarized in Figure 4.17.

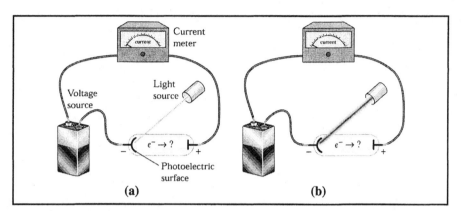

Figure 4.15. Photoelectric experiments with low frequency light of different intensities.
Red represents low frequency (long wavelength — Figure 4.13) light. The thickness of the light beam shows its intensity (brightness): (a) is low intensity and (b) is high intensity.

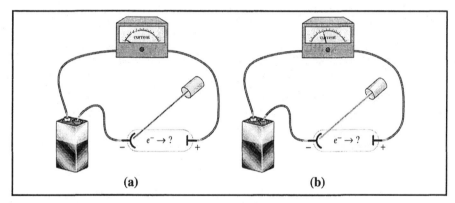

Figure 4.16. Photoelectric experiments with (a) low and (b) high frequency light.
Blue, (b), represents higher frequency (shorter wavelength — Figure 4.13) light than red, (a).

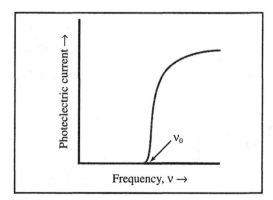

Figure 4.17. Photoelectric current as a function of frequency of the light.
The intensity (brightness) of the light is the same at all frequencies.

Below a minimum frequency, v_0, no electrons are knocked out of the metal. Einstein assumed that it must take a certain minimum amount of energy to eject an electron from a particular metal surface. If Planck was right about the relationship between energy and frequency, equation (4.5), then a certain minimum frequency would be required to eject electrons. The Planck quantum hypothesis explained the data in Figure 4.17.

In addition, other photoelectric experiments showed that the maximum energy of the ejected electrons depended on the frequency of the light, as shown in Figure 4.18. Einstein reasoned that the maximum energy an ejected electron could have would be the energy provided by the light minus the energy required to eject the electron:

$$E_{electron} = E_{light} - E_{ejection} \tag{4.8}$$

He wrote the two energies on the right in terms of frequencies, v, and Planck's proportionality constant, h:

$$E_{electron} = h \cdot v - h \cdot v_0 = h \cdot (v - v_0) \tag{4.9}$$

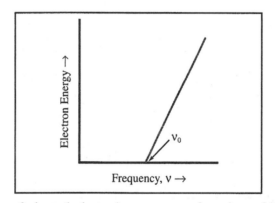

Figure 4.18. Energy of ejected photoelectrons as a function of light frequency.

When Einstein analyzed the experimental data quantitatively, he found that the numerical value of the proportionality constant, h, in equation (4.9) was the same value Planck had used to explain the experimental results for emission from glowing objects. Thus, Einstein showed that light is quantized and comes in discrete energy packets, $E = h \cdot v$. He used the term **photon** to designate an energy packet of light.

4.31. **Consider This** *How does photoelectric current depend on light intensity?*

(a) Figure 4.19 shows the effect of light intensity on the photoelectric current for light with a frequency $v > v_0$. How does the model based on photons explain the results? In the photon model, what does "intensity" mean?

(b) What does the photon model predict for the energy of the ejected electrons in Figure 4.19(b) compared to the electrons in Figure 4.19(a)? Explain your reasoning.

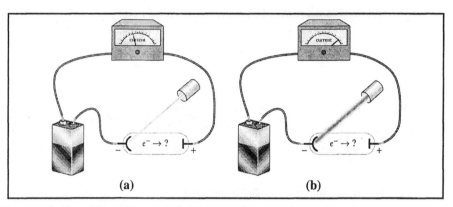

Figure 4.19. The effect of light intensity on photoelectric current.

4.32. Consider This *How is the darkening of silver chloride related to photons?*

(a) Can you explain the results from Investigate This 4.29 using the photon ("energy packet") model of light? Explain why or why not. What assumption(s) do you have to make about reaction (4.7).

(b) Reaction (4.7) requires an energy of about 4×10^{-19} J for each (Ag^+Cl^-) that reacts. Are your results consistent with this energy requirement? Explain why or why not.

4.33. Check This *Energy of photons*

(a) What is the energy of a photon of green light, wavelength 515 nm? What is the energy of a mole of these photons?

(b) What is the energy of a photon of wavelength 21.11 cm, an emission from hydrogen atoms that we will discuss later.

(c) What is the energy of a photon of the frequency emitted by your favorite FM radio station [Check This 4.23(b)]? What is the energy of a mole of these photons?

The dual nature of light

Since the same hypothesis and proportionality constant explained two entirely different phenomena, most scientists soon accepted Planck's original hypothesis, even though it seemed strange then and still does. We use the wave model to interpret and explain some electromagnetic phenomena. We were able to understand the diffraction of light (Section 4.3), for example, from a glowing object or a laser, in terms of a wave model. In this section, however, we found that the temperature dependence of the intensity of emission from a glowing object cannot be explained by the wave model, but requires a photon (quantum) model of light. The photon model also enables us to interpret and explain the photoelectric effect and the wavelength dependence of a chemical reaction that requires energy from light. There is no analog in the world of objects that are large enough to see and touch that helps us understand the dual nature of light. The justification for accepting this dual nature is that it works to give correct predictions of observable results.

4.34. **Check This**   *Wave and photon phenomena*

For each of these examples, tell whether the wave theory or photon (quantum) theory of light best explains the phenomenon and give your reason(s).

(a) the play of colors from the surface of a CD

(b) the color of a neon light

(c) getting sunburned

(d) a rainbow

Reflection and projection

We have found that the electromagnetic wave model of light successfully explains many phenomena, including diffraction. However, the wave model fails to explain other phenomena, such as the wavelength dependence of the intensity of emission from hot, glowing solids and the photoelectric effect. In order to explain the latter phenomena, we need to assume that electromagnetic oscillators can emit energy at only certain frequencies. The oscillators are quantized; the energy of the emitted light is directly proportional to the frequency of the oscillator. Further, in some experiments, light acts as though it has particle-like properties; the energies of the photons (light particles) are directly proportional to the frequency of the light. The proportionality constant, Planck's constant, is the same for both cases.

The dual nature of light leads to the question: Does matter also have a dual nature, both particle and wave? We will take up this question in Section 4.6, but first we will look at how the concept of energy quantization was used to develop a quantum model of atoms that explained line emissions of light from energized atoms.

Section 4.5. The Quantum Model of Atoms

We began our discussion of the properties of light, because we thought the emission and absorption of light by elemental atoms would provide clues about their structures and properties. Recall, from Figure 4.6, that emission occurs from atoms that have been given a lot of energy by a spark or a flame and absorption occurs from less energetic atoms at a lower temperature. This probably means that the emission and absorption processes in the figure represent changes in the energy of the atoms. An atom can lose energy by emitting light with a wavelength corresponding to the amount of energy lost. The wavelength of the emitted light is given by the Planck radiation law, equation (4.6). Conversely, an atom can absorb light energy and change from a lower to a higher energy. These are the processes illustrated in Figure 4.20.

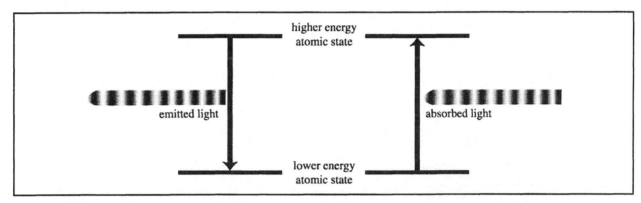

Figure 4.20. Atomic emission and absorption of light of the same wavelength.
Light is emitted when an atom goes from a higher energy to lower energy state. When the energy of the atom is boosted from the lower to the higher energy state, light of this same wavelength is absorbed by the atom.

4.35. **Consider This** *What property of atoms causes line emission (and absorption)?*

Assume that emission and absorption spectra of atoms are the result of their electrons giving off or taking up energy.

(a) If the electrons can have any energy (consistent with being part of the atom), what would their emission spectra look like? Explain.

(b) If the electrons can have only certain energies, what would their emission spectra look like? Explain.

Quantized electron energies in atoms

When gaseous atoms and ions are heated to a high temperature, or when an electric spark is passed through the gas, much of the energy input is transferred to the electrons in the atoms. When the electrons give up that extra energy, it is released as electromagnetic radiation in the visible, infrared, and ultraviolet regions. Solids, such as the tungsten filament in an incandescent light, also emit light when they are heated. The enormous difference, as you observed in Investigate This 3.4, is that the emission from the tungsten filament consists of *all* the wavelengths in a wide range, whereas only a few wavelengths of light are emitted by energetic electrons in gaseous atoms and ions. Since Planck's hypothesis links energy and wavelength, this observation must mean that energetic electrons in atoms shed excess energy by emitting only certain amounts of energy.

Early in the 20th century, after the nuclear model of the atom and Planck's quantum hypothesis had been developed, Niels Bohr (Danish physicist, 1885-1962) applied these ideas to the structure of atoms. Bohr used his **quantum model** of atomic structure to calculate the energies (wavelengths) that would be emitted by an energetic hydrogen atom. The experimental

> **WEB Chap 4, Sect 4.4.1**
> View an animation of the H atom emission to give the data in Figure 4.21.

values for the wavelengths of hydrogen atom emissions in the visible region of the spectrum are shown in Figure 4.21; Bohr's calculations gave exactly these same values. This agreement of a model and an experiment was convincing evidence that *electron energies in an atom are quantized*, which is why emissions from atoms are limited to a few particular energies.

Figure 4.21. The visible emission spectrum from atomic hydrogen.

Electron energy levels in atoms

Bohr's achievement was enormous; it suggested that the quantization of electron energies in atoms is a fundamental property of the electron. Figure 4.22 is a representation of the quantized energies of the electron that Bohr calculated for the hydrogen atom. Each line, or level, in the figure indicates the relative energy of an electron, just like the other energy diagrams we have been using. The lowest energy is chosen as the starting point (E_1) and the energy levels of the electron continue to increase until the electron has so much energy that it can no longer be held by the attraction of the nucleus. When an electron receives this much energy, it separates from the original atom; the atom has been ionized. The loss of the electron leaves the ion with one unit of net positive charge. In multi-electron atoms, more electrons can be lost in the same way, with each loss increasing the net positive charge of the ion. Notice that the spacing between the energy levels gets closer and closer as the electron energy increases toward the ionization limit.

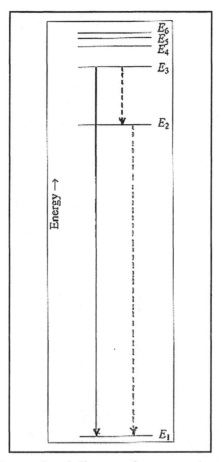

Figure 4.22. Energy level diagram for a one-electron atom.
The electron energy levels are for an H atom or an ion such as He^+, Li^{2+}, and so on. The arrows show pathways for an electron to lose energy and return from the excited state, E_3, to the ground state, E_1. The loss of energy can occur in one step, solid arrow, or two steps, dashed arrows.

4.36. Check This *Experiment with a one-electron atom simulation*

WEB Chap 4, Sect 4.4.2. Follow the directions and drag the pointer to different energy levels. Describe what occurs to the "atom." Drag the pointer to an intermediate point between energy levels. What happens? Is it possible for the electron to have an energy intermediate between two of the energy levels? Explain.

The lowest total energy of an atom is called its **ground state**, level E_1 in Figure 4.22. When an electron in the atom gains the energy to be in any one of the higher energy levels, say E_3, the atom is said to be in an **excited state** and the electron has been excited. To return to its ground-state energy, the electron has to get rid of the excess energy. There are two ways back to the ground state for this electron, as shown by the arrows in Figure 4.22. It can emit a photon with an energy equal to $E_3 - E_1$, the solid blue arrow, or it can emit one photon with energy of $E_3 - E_2$

Structure of Atoms — Chapter 4

and then a second one having an energy of $E_2 - E_1$, the dotted red and green arrows, respectively. In either case, only specific energies can be emitted and only the corresponding wavelengths of light will be observed. This is how the quantum model of the electron accounts for the observation that energized atoms or ions emit light of only a few wavelengths.

4.37. Worked Example *Energy level differences in the hydrogen atom*

The emission wavelengths corresponding to the $E_3 - E_1$ and $E_2 - E_1$ energy differences for hydrogen, Figure 4.22, are at 102.57 and 121.57 nm, respectively. What are these energy differences in joules? in joules per mole?

Necessary Information: We need the Planck relationship, equation 4.6, Planck's constant, 6.6256×10^{-34} J·s, and Avogadro's number, 6.022×10^{23} mol^{-1}.

Strategy: The energy of an emission is equal to the difference in energy between the higher and lower energy states. We use the Planck relationship to relate the wavelength of an emission to the difference in energies that produce the emission. To get the energy in J·mol^{-1}, we multiply the energy for a single atom by Avogadro's number.

Implementation: Wavelengths are subscripted to show which energy levels are involved.

$$E_3 - E_1 = \frac{h \cdot c}{\lambda_{3-1}} = \frac{(6.626 \times 10^{-34} \text{ J·s})(3.00 \times 10^8 \text{ m·s}^{-1})}{102.57 \times 10^{-9} \text{ m}} = = 1.938 \times 10^{-18} \text{ J}$$

$$E_3 - E_1 \text{ (per mole)} = (1.938 \times 10^{-18} \text{ J}) \cdot (6.022 \times 10^{23} \text{ mol}^{-1}) = 1.167 \times 10^6 \text{ J·mol}^{-1}$$

$$E_2 - E_1 = \frac{h \cdot c}{\lambda_{2-1}} = \frac{(6.626 \times 10^{-34} \text{ J·s})(3.00 \times 10^8 \text{ m·s}^{-1})}{121.57 \times 10^{-9} \text{ m}} = = 1.635 \times 10^{-18} \text{ J}$$

$$E_2 - E_1 \text{ (per mole)} = (1.938 \times 10^{-18} \text{ J}) \cdot (6.022 \times 10^{23} \text{ mol}^{-1}) = 0.985 \times 10^6 \text{ J·mol}^{-1}$$

Do the results make sense? We see that, when the results are expressed in terms of a mole of atoms, the energies are substantial. For hydrogen, the difference in energy between the ground and first excited state is 985 kJ·mol^{-1}. This explains why so much energy is required to excite the atoms in a discharge tube to emit light.

4.38. Check This *Energy level differences in the hydrogen atom*

(a) Use the results in Worked Example 4.37 to find $E_3 - E_2$ (in joules) for the hydrogen atom.

(b) What is the emission wavelength corresponding to $E_3 - E_2$, the red arrow in Figure 4.22? Compare your answer with the emission wavelengths shown in Figure 4.21.

(c) WEB Chap 4, Sect 4.4.3. Drag the pointer to the blue line. Explain in your own words what the animation represents in terms of what is happening to the electron and the associated energy change(s). Explain where the photon of light comes from and why it is emitted rather than absorbed.

A basic idea to take from our discussion is that, when high-energy electrons in an atom or ion return to lower energy levels, the emitted energy takes the form of light. A light wave is produced from the difference in energy between the two energy levels. Conversely, the photons in a light wave can transfer energy to an electron in an atom *only* when the photon's energy exactly equals the *difference* in energy between two energy levels of the electron. This explains why the emission and absorption processes, Figures 4.6 and 4.20, emit and absorb light of exactly the same wavelength(s).

4.39. Check This   *Predicting emission and absorption wavelengths*

(a) The emissions from excited hydrogen atoms in Figure 4.21 are emissions from the E_3, E_4, E_5, and E_6 levels to return to the E_2 level. What is the $E_4 - E_3$ emission energy? At what wavelength would this emission occur? In what region of the electromagnetic spectrum (Figure 4.13) would this wavelength be found? (This emission wavelength was calculated, as you are doing, before it was found experimentally exactly where it was predicted to be.)

(b) If you were trying to detect hydrogen atoms by their absorption of energy to go from the ground state to the first excited state, what region of the electromagnetic spectrum would you need to use as your light source?

Due to limitations of the model, the only atomic energy levels the Bohr quantum model could calculate exactly were those for hydrogen and other one-electron ions like He^+ and Li^{2+}. Even more dismaying to chemists was the fact that the model could not be applied to molecules. Nonetheless, the quantum model served a useful purpose in interpreting the spectral lines from atoms in gas-discharge tubes and it lead scientists to search for other models that would retain the quantization of electron energies without the shortcomings of the Bohr quantum model. That search brings us back to waves.

Structure of Atoms Chapter 4

Section 4.6. If a Wave Can Be a Particle, Can a Particle Be a Wave?

Wavelength of a moving particle

Once the work of Planck, Einstein, and others had convinced scientists that light exhibits the properties of both waves and particles, it didn't take long to develop a wave model for the behavior of particles. In 1924, Louis de Broglie (French physicist, 1892–1987) postulated (with no experimental justification) that the behavior of moving electrons can be described as though they have wave properties. His model showed that the wavelength of a moving, wave-like particle is:

$$\lambda = \frac{h}{m \cdot u} \tag{4.10}$$

This wavelength, often called the **de Broglie wavelength**, is inversely proportional to the mass, m, and velocity, u, of the particle. The proportionality constant, h, in equation (4.10) is Planck's constant, which crops up in almost every equation dealing with atomic-level phenomena. Equation (4.10) is the bridge that relates the wave nature of electrons to the particle nature of electrons. The product, $m \cdot u$, in the denominator of equation (4.10) is the **momentum** of the particle:

$$\text{momentum} = m \cdot u \tag{4.11}$$

Momentum measures the amount of "push" a moving object can exert on another object: the larger the mass, m, and faster the motion, u (velocity), the larger the momentum.

4.40. Worked Example   *de Broglie wavelength of an electron*

Calculate the de Broglie wavelength of an electron traveling at 9.4×10^5 m·s^{-1}, a velocity that is easy to attain. If the electron acts like an electromagnetic wave, to what region of the spectrum, Figure 4.13, does this wavelength correspond?

Necessary information: We need equation (4.10), Planck's constant, the velocity of the electron from the problem statement, and, from Table 3.1, its mass, 9.1×10^{-28} g.

Strategy: We substitute the mass and velocity into equation (410), using the correct units for mass and velocity to be compatible with Planck's constant. The energy unit in Planck's constant is the joule ($= \text{kg} \cdot \text{m}^2 \cdot \text{s}^{-2}$).

Implementation:

$$\lambda = \frac{h}{m \cdot u} = \frac{6.63 \times 10^{-34} \text{ J} \cdot \text{s}}{(9.1 \times 10^{-31} \text{ kg}) \cdot (9.4 \times 10^5 \text{ m} \cdot \text{s}^{-1})} = 7.8 \times 10^{-10} \text{ m} = 0.78 \text{ nm} = 780 \text{ pm}$$

Figure 4.13 shows that electromagnetic waves in the region 100–1000 pm are in the x-ray region of the spectrum.

Chapter 4 — Structure of Atoms

Does this answer make sense? The calculated wavelength of the electron is in the x-ray range, so we might expect electrons moving at this velocity to have properties like x-rays, which are used extensively by scientists to study the structure of crystals. To figure out whether this de Broglie wavelength makes sense, we have to know what happens if these electrons are substituted for x-rays in a crystal structure experiment, which is the topic of the next paragraph.

4.41. Check This *de Broglie wavelengths*

(a) What is the de Broglie wavelength for an electron moving at one-tenth the speed of light? To what region of the electromagnetic spectrum does this wavelength correspond?

(b) Some baseball pitchers can throw a 145 g baseball 100 miles per hour (about 44 m·s^{-1}). What is the de Broglie wavelength for this fastball?

(c) Thermal neutrons (such as those in nuclear reactors) are neutrons moving at the speed they would have if they were atoms in a gas at room temperature, about 1 km·s^{-1}. What is the de Broglie wavelength of thermal neutrons? To what region of the electromagnetic spectrum does this wavelength correspond?

Experimental evidence for electron waves

If moving electrons act like waves, they should show the phenomenon of diffraction, as in Figures 4.8 and 4.10. Shortly after de Broglie proposed the electron wave model, two American physicists, C. J. Davisson and L. H. Germer, interpreted the results of one of their experiments as being due to the diffraction of electrons by atoms in a crystal. The diffraction of x-rays by crystals, as in Figure 4.23(a), was a well-known phenomenon. Note that the diffraction pattern from a crystal, is more complicated than the pattern you observed from a grating in Investigate This 4.17(b). This is because the "gratings" in a crystal are the spacings between atoms and there are several different spacings that depend upon the geometric arrangement of the atoms in the crystal. Information from x-ray diffraction had been used to analyze the structure and spacings in crystals since the early part of the 20th century.

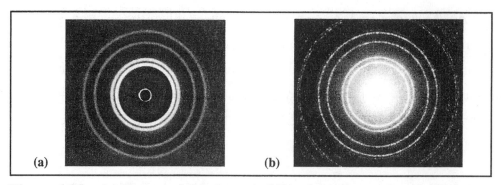

Figure 4.23. (a) X-ray and (b) electron diffraction patterns for Al foil.
The foil is made up of an enormous number of tiny crystals that are responsible for the diffraction. The pair of intense inner rings and pair of less intense outer rings are identical in the two patterns. A much weaker further pair of rings in the electron diffraction pattern are too weak to see in the x-ray pattern.

When electrons are accelerated to known velocities (about the same as in Worked Example 4.40) in an electric field and directed through a very thin metal foil, the electrons are diffracted, as shown in Figure 4.23(b). The patterns for x-ray diffraction and electron diffraction from the same sample are identical. The wavelength of the electron waves can be measured from the spacings of the diffraction pattern. The measured wavelength and the calculated wavelength from the de Broglie model, equation (4.10) are the same; our result in Worked Example 4.40 makes sense.

All diffraction methods, including spectroscopy, are related. Diffraction occurs when the wavelength of the waves being diffracted is comparable in size to the "grating" that is responsible for diffracting the waves. Since moving electrons and x-rays of the same wavelength are diffracted by crystals, the spacing between atoms in the crystals must be about 100–1000 pm (see Figure 4.13). The wavelength of a fast-pitched baseball, as in Check This 4.41(b), is so short that there is no physical structure small enough to detect its wave properties. We can safely ignore the wave properties of massive objects like baseballs and automobiles. Heavy atomic- or subatomic-size particles are a different matter. Your calculation in Check This 4.41(c) shows that thermal neutrons have wavelengths comparable to x-rays, that is, a size that can be diffracted by the atoms in crystals. Neutron diffraction is a powerful technique that is extensively used to explore the structures of molecules, especially biological molecules like proteins, in crystals.

The results from this section and the previous one upset simple differentiations between light and matter. We see that both light and matter have wave-like and particle-like properties. The property that is observed depends on the phenomenon being investigated. Now we will use these new understandings to investigate atoms in more detail.

Chapter 4 — Structure of Atoms

Reflection and projection

The first quarter of the 20th century was both exciting and perplexing for scientists. It began with the demonstration that the well-established electromagnetic model of light was not the complete story and that light has particulate (quantum) properties as well as wave properties. It ended with the demonstration that matter, at the atomic level, also has wave properties, as shown by electron diffraction, as well as its familiar particulate properties. In the years between, Bohr and others developed a quantum model of the atom that had spectacular success interpreting atomic line spectra. The quantum model of the atom says that electrons in atoms can have only certain energies and that the loss or gain of energy by the atom can only occur by emission or absorption of light that corresponds to the difference in energy between two of these energy levels.

However, problems with the quantum model, especially in applying it to multielectron atoms and molecules, were evident. A fundamental question about atomic structure, which we raised at the end of Chapter 3, is: Why don't atoms collapse? Electrons are negatively charged and are attracted by the positively charged nucleus. What keeps electrons from being drawn closer and closer to their nuclei until all atoms collapse to the size of their nuclei? The quantum model could not explain why they were not. The advent of the de Broglie wave model for electrons provided an exciting new way to model atoms that also explained their energy quantization and got beyond the quantum model problems. In particular, the wave model leads to an explanation for why atoms don't collapse.

Section 4.7. The Wave Model of Electrons in Atoms

Waves and atomic emissions

Some waves, like water and light waves, are propagated through space; these are **traveling waves**. Other waves, **standing waves**, are held in place; they are constrained in some way. Vibrating guitar strings are an example of standing waves. In a guitar, the waves cannot extend beyond the posts that anchor each end of the string. The amplitude of such a wave is the maximum displacement of the string from its rest position. A node must always occur at each post. The wavelength is determined in the same way as for traveling waves—the distance between wave crests. Figure 4.24 illustrates typical standing waves in guitar strings. Figure 4.24(a) represents time exposure photographs, while Figure 4.24(b) represents instantaneous exposures.

Structure of Atoms   Chapter 4

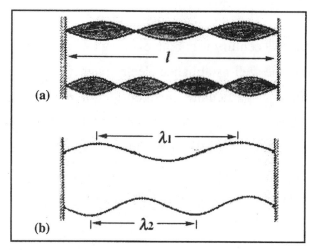

Figure 4.24. Standing waves in guitar strings of length l.
(a) represents time exposure photographs that blur the strings. (b) represents instantaneous exposures when the waves are at maximum amplitude.

4.42. Check This *Standing waves on a string*

Not just any wavelengths can be accommodated on the strings in Figure 4.24. The wavelengths that "fit" on a string of length l are:

$$\lambda = \frac{2l}{n}, \text{ where } n = 1, 2, 3,\ldots \tag{4.12}$$

(a) What is n for each of the waves in the figure? If $l = 0.75$ m, what are the wavelengths of the waves? How many nodes are there in each wave?

(b) What are n and λ for the longest wavelength on a 0.75 m string? Sketch this wave, using the representations in Figure 4.24 as models.

Standing electron waves

Scientists studying atomic emission spectra in the 19th century had found mathematical expressions that described the patterns of the emission wavelengths they observed, such as those from hydrogen in Figure 4.21. There are striking similarities between these mathematical expressions for atomic emission patterns and the mathematical expressions for standing waves, like the simple series for waves on a string, equation (4.12). These similarities were part of the evidence that led scientists to apply the de Broglie model of matter waves (for the electron) to construct a wave mechanical model of atomic structure.

The electron is held in the atom by its attraction to the positive nucleus, much like a guitar string is attached to the body of the instrument. Using **wave mechanics**, the physics that provides mathematical descriptions of wave motion, we describe the electron in an atom as a standing wave, and, like the standing waves of a guitar string, the waves that are associated with an electron can have only certain wavelengths. Planck's relationship, equation (4.6), specifies that an electron wave of a particular wavelength would have an energy that is inversely proportional to the wavelength. Thus, the electron (in an atom or ion) described as a wave can have only certain energies and no others. Quantization of electron energy is a natural consequence of the electron wave model.

Probability picture of electrons in atoms

The Bohr quantum model provided a picture of an atom as a nucleus with electrons in orbit about the nucleus. It would be nice to have a comparable picture of the motion of electrons that behave as waves. The problem with electrons behaving like waves, rather than like baseballs or marbles or planets, is that we can never tell exactly where the electrons are or which way they are going at a particular instant. This is because of the uncertainty principle, another property of atomic-size systems that, like waves, has no analog in the macroscopic world of baseballs and marbles. The uncertainty principle was formulated by Werner Heisenberg (German physicist, 1901–1976) within a year of de Broglie's electron-wave hypothesis. Qualitatively, the **uncertainty principle** states that it is impossible to measure simultaneously both the exact location and the exact momentum of an electron. Recall that the momentum of the electron is a part of the de Broglie model, equation (4.10), which closely ties the wave model to the uncertainty principle.

The best we can do for finding an electron in an atom, ion, or molecule is to determine its *probability* of being in a particular location when it is described by one of the energy states (levels) like those shown in Figure 4.22. The location of an electron near an atomic nucleus is described by a probability distribution, much as are the holes in a dartboard that represent attempts to hit the bull's-eye. The fraction of holes in each target ring, Figure 4.25(a), is the probability of a dart having landed in that ring. The likelihood of finding the electron in any particular location near the nucleus is represented in Figure 4.25(b) by the density of dots (darkness) around the nucleus (at the very center of the distribution). The probability of finding the electron decreases as distance from the nucleus increases.

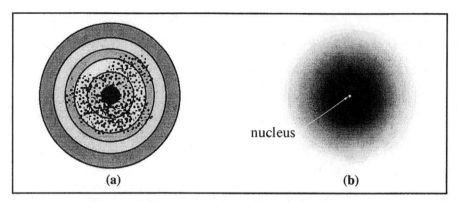

Figure 4.25. Probability distributions.
Probability of (a) hitting a dartboard bull's-eye and (b) finding an electron around a nucleus. The "density of dots" in a region represents the probability of (a) having hit that region with a dart or (b) finding the electron.

4.43. Check This *Probability in three dimensions*
What three-dimensional geometric figure would you choose to represent the probability distribution shown two dimensionally in Figure 4.25(b)? Explain your choice.

The de Broglie wave model and Heisenberg's uncertainty principle ended scientists' attempts to describe the *path* of an electron moving from one place to another in an atom, as Bohr had done in his quantum model. Use Figure 4.24 as an analogy to visualize a standing electron wave in an atom. The instantaneous exposures of the vibrating string in Figure 4.24(b) are analogous to electrons in defined orbits, which we have had to give up, because we can't find the electron at any particular instant in time. But, we can get the electron distribution over time, as illustrated in Figure 4.25(b), and this is analogous to the blurred time exposures of the vibrating string in Figure 4.24(a).

Orbitals

We need a three-dimensional model for this blurred electron wave. The simplest model is a sphere, which is probably the geometry you chose in Check This 4.43. A sphere is simple because it requires only two variables to describe it: the location of its center and its radius. We will visualize a standing electron wave in an atom as a sphere that encloses the volume where the electron is most likely to be found. The wavelength, λ, of a spherical electron wave is proportional to the radius of the sphere, R:

$$\text{wavelength of a spherical electron wave} = \lambda \propto R = \text{radius of the sphere} \qquad (4.13)$$

By the time the wave model was developed, the idea of electron *orbits* in atoms had become so widely used that we now use the word **orbital** to name electron waves (probability distributions) in atoms, like the one represented in Figure 4.25(b). Simpler pictures of orbitals, as in Figure 4.26(a), represent the volume in space within which there is roughly a 95% probability of finding an electron that has a particular energy. The spherical surface in Figure 4.27(b) shows how the simpler surfaces are related to the probability distribution. Recall from the Planck relationship, equation (4.6), that the energy of a wave is inversely proportional to its wavelength; a more constrained (smaller) electron wave has a higher energy. Before you can interpret and use orbital pictures, we need to discuss what determines electron orbital energies. This discussion will also help us understand why atoms don't collapse.

> **WEB Chap 4, Sect 4.6.1**
> View this animation of an orbital in both representations; compare with Check This 4.43.

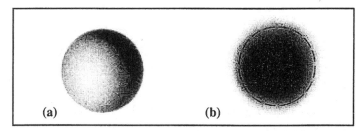

Figure 4.26. An atomic orbital; a standing spherical electron wave around a proton.
(a) Orbital represented as a smooth sphere. (b) The dashed line is shown to enclose the volume within which there is 95% probability of finding the electron. The proton is at the center of the sphere.

Section 4.8. Energies of Electrons in Atoms: Why Atoms Don't Collapse

We've said a lot about electron energy levels without saying what we mean by energy in this context. In a one-electron atom (or ion), which we have been using as a model, we are concerned with the kinetic energy of the moving electron and the potential energy of attraction between the positive nucleus and the negative electron. The motion of the nucleus is usually not considered in atomic models, because the nuclei are so slow moving relative to the electrons that their effect on the energy is insignificant.

Kinetic energy of an electron

We cannot pick energy up, feel it, turn it around to look at the back side, check its density, or throw it at the wall in the same way we can matter. Nonetheless, you can tell the difference between a thrown baseball moving at 44 m·s^{-1} and a baseball at rest beside the pitcher's mound. The idea that an object in motion has a property that the object at rest does not have is easy enough to accept; that property is **kinetic energy**. Kinetic energy (KE) is expressed quantitatively as a function of mass (m) and velocity (u):

$$KE = \frac{m \cdot u^2}{2} \qquad (4.14)$$

Structure of Atoms Chapter 4

4.44. Worked Example *Kinetic energy of electrons in the diffraction experiment*

What is the kinetic energy of an electron traveling at 9.4×10^5 m·s^{-1}, as in Worked Example 4.40? What would be the kinetic energy of a mole of these electrons?

Necessary Information: We need equation (4.14), the mass of an electron, 9.11×10^{-31} kg, from Table 3.1, the velocity from the problem statement, and Avogadro's number, 6.022×10^{23} mol^{-1}.

Strategy: We substitute our known values for mass and velocity in equation (4.14) to get the kinetic energy for one electron and then multiply by Avogadro's number to get the kinetic energy for a mole of them.

Implementation:

$$KE \text{ (per electron)} = \frac{m \cdot u^2}{2} = \frac{(9.11 \times 10^{-31} \text{ kg})(9.4 \times 10^5 \text{ m·s}^{-1})^2}{2} = 4.02 \times 10^{-19} \text{ J}$$

$$KE \text{ (per mole)} = (4.02 \times 10^{-19} \text{ J}) \cdot (6.022 \times 10^{23} \text{ mol}^{-1}) = 2.4 \times 10^5 \text{ J·mol}^{-1} = 240 \text{ kJ·mol}^{-1}$$

Does the answer make sense? A mole of electrons has a tiny mass, 5.5×10^{-7} kg, but, when traveling at these speeds, has a substantial kinetic energy. The energy is comparable to the energies of electrons in hydrogen atoms that we found in Worked Example 4.37. This makes sense, since we imagine the electrons in atoms are moving quite rapidly about the nucleus.

4.45. Check This *Kinetic energy of more massive objects*

(a) Calculate the kinetic energy of the baseball in Check This 4.41(b).

(b) Calculate the kinetic energy of a mole of the thermal neutrons in Check This 4.41(c).

(c) Compare the results in parts (a) and (b) with the kinetic energy of the electrons in Worked Example 4.44.

Potential energy of a nucleus and an electron

The kinetic energy associated with motion is obvious to almost everyone. It is somewhat less obvious that two objects at rest, but physically in different locations, also differ in their energy content. A textbook placed on a high shelf has significantly more gravitational **potential energy** —*PE*, the energy resulting from position— than does a textbook on the floor. If the textbook on the shelf is knocked off, it will move under the influence of gravity to rest on the floor. The book on the floor, on the other hand, will not move up to the shelf without an input of energy.

The potential energy we are interested in is the potential energy of coulombic attraction between an electron and a proton. As you saw in equation (2.3), Section 2.4, the potential energy of coulombic attraction between two charges (Q_1 and Q_2) that are a distance r apart is:

$$PE \propto \frac{Q_1 Q_2}{r} \qquad (4.15)$$

For a proton and an electron, we write the unit charges as +1 and –1, so the potential energy is negative:

$$PE \propto \frac{(+1)(-1)}{r} = -\frac{1}{r} \qquad (4.16)$$

When the nucleus and electron are far apart, (when r is large), $PE \approx 0$. As they get closer to each other, the potential energy goes farther and farther below zero. *The closer the nucleus and electron are (the smaller the atom), the more negative their potential energy.* This is the direction the energy must go to form a stable atom.

4.46. Consider This *What is the effect of nuclear charge on the potential energy?*
For the same distance of separation, how does the coulombic attraction of an electron and a 2+ nucleus compare to the attraction of an electron and 1+ nucleus? Which species, H or He$^+$, would you expect to be smaller? Why?

Total energy of an atom

For any system, the **total energy**, E, is the sum of its kinetic and potential energies:

$$E = KE + PE \qquad (4.17)$$

A stable atom is one that has a lower total energy than the total energy associated with its nucleus and electrons when they are separated from each other. Scientists use an energy scale that defines the energy of the separated nucleus and electrons as zero. To determine the size and other properties of our wave mechanical atom, we need to find the conditions where the *total energy of the nucleus and electron wave, E, is most negative*.

The kinetic energy and the potential energy of a spherical wave are both functions of the wave radius, R. These can be calculated, and one way to do this is given in the EXTENSION to this chapter. At this point, we need to know only the results of those calculations:

$$E = KE + PE \propto \frac{1}{R^2} - \frac{1}{R} \qquad (4.18)$$

The momentum of a matter wave, equation (4.11), is inversely related to the wavelength of the wave, equation (4.10), and the kinetic energy of a particle is directly related to the square of its momentum. These relationships give the $1/R^2$ dependence of KE. The average distance of the electron from the nucleus is related to the size of the electron wave, and the potential energy, equation (4.16), is inversely related to this distance. These relationships give the $1/R$ dependence of PE. Figure 4.27 shows the two energies, KE and PE, and their sum, the total energy, E, plotted as a function of R.

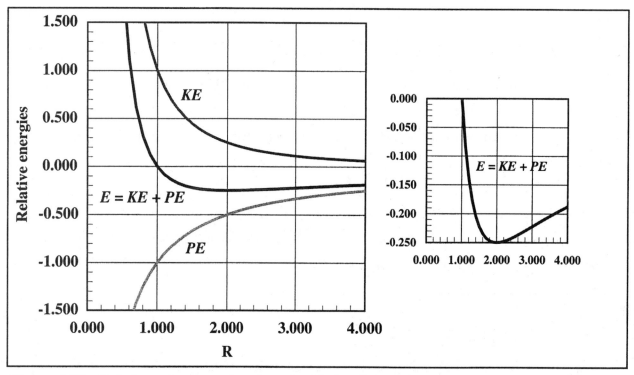

Figure 4.27. Relative energies as function of orbital radius, R, for a wave mechanical atom. Energies are for a spherical electron wave attracted to a positive nucleus. The smaller plot expands the scale for the energy range from –0.250 to 0 to show the minimum in total energy more clearly.

4.47. Consider This   *How can you interpret Figure 4.27?*

WEB   Chap 4, Sect 4.7.1. Carry out the directed actions on this interactive animation. Explain in your own words what happens as the distance between the electron and proton and the size of the electron wave decreases in terms of:

(a) the potential energy. What causes the potential energy to change? What would happen if the potential energy were the only energy in the system?

(b) the kinetic energy. What causes the kinetic energy to change? What would happen if the kinetic energy were the only energy in the system?

(c) the total energy. When is the system most stable?

Why atoms don't collapse

Figure 4.27 illustrates how the kinetic and potential energies of the atom vary with the size of the electron orbital (the size of the atom). The kinetic energy is always positive and increases as the orbital radius decreases. The potential energy of attraction is always negative and gets more negative as the radius of the orbital decreases. The smaller the orbital (atom), the more favorable (lower) the potential energy. This is the attraction that would, by itself, cause atoms to collapse. The kinetic energy of an electron orbital increases as the orbital gets smaller, so the kinetic energy counteracts the orbital contraction.

At some radius, the sum of the kinetic and potential energies, the total energy, goes to a minimum value. If the atom were to get any smaller, its total energy would increase and it would not be stable relative to its energy minimum. Thus, *the size of a one-electron atom is determined by a balance between the negative potential energy of attraction and the positive kinetic energy of the electron wave*. It is the kinetic energy of the electron wave that prevents the atom from collapsing.

Ionization energy for the hydrogen atom

Our wave mechanical model can be made quantitative and we can calculate the energy minimum in Figure 4.27, and the size of the orbital at this point. The energy minimum is, –1312 kJ·mol^{-1}, at an orbital radius near 80 pm. The experimental value for the ionization energy of the hydrogen atom, the amount of energy required to separate the electron and proton, is 1312 kJ·mol^{-1}, that is:

$$H(g) \rightarrow H^+(g) + e^- \quad (\Delta E_{ionization} = 1312 \text{ kJ·mol}^{-1}) \quad (4.19)$$

We can't determine the exact size of a hydrogen atom because the electron probability distribution has no sharp boundary. The electron probability envelope resembles a ball of cotton more than it does a baseball. On the other hand, we know from many different experiments that atoms have radii that are near 100 pm, so we are certainly in the right ballpark. Thus the wave model is accepted by scientists because it both predicts and explains experimentally observed properties of electrons and atoms, such as electron diffraction, quantized energies, atomic size, and ionization energy.

4.48. Check This *Ionization energy and nuclear charge*

Which species has the higher ionization energy, H or He$^+$? Why? Use what you learned in Consider This 4.46 to help answer these questions.

Structure of Atoms Chapter 4

Reflection and projection

The properties of waves have brought us to the wave model of the atom. The formation of a stable atom corresponds to reaching the lowest possible value for the total energy of the atom. The total energy of a one-electron atom is the sum of the kinetic and potential energies of the electron orbital attracted to the positively-charged nucleus. The kinetic and potential energies are related to the size of the electron orbital (the size of the atom). At some orbital size, the balance between the always-positive kinetic energy and the always-negative potential energy of attraction produces a minimum in the total energy. This is the stable atom. The fundamental concepts to remember are: **atoms (electrons and nucleus) are held together by mutual attractions of unlike charges and their sizes are limited by the kinetic energies of the electron waves (orbitals)**.

As we go on to atoms with more than one electron, the number of interactions that must be included in the energy calculations grows. More electrons mean more possibilities for changes in energy levels and much more complicated emission and absorption spectra. Analysis of these spectra requires a specialized knowledge of atomic structure that is unnecessary for our basic understanding of the properties of atoms. In the next section we will examine some other atomic properties to see what patterns we can find in the data that can provide the clues we need to develop our wave mechanical model further.

Section 4.9. Multi-electron Atoms: Electron Spin

4.49. Consider This   *What patterns do you find in atomic ionization energies?*

(a) Successive ionization energies for the lightest three elements are given in Table 4.1. The ionization energy for the H atom is the energy required for reaction (4.19). For the helium atom with two electrons, the ionization energies in Table 4.1 are for these reactions:

$$He(g) \rightarrow He^+(g) + e^- \qquad (\Delta E_{ionization} = 2373 \text{ kJ·mol}^{-1}) \qquad (4.20)$$

$$He^+(g) \rightarrow He^{2+}(g) + e^- \qquad (\Delta E_{ionization} = 5248 \text{ kJ·mol}^{-1}) \qquad (4.21)$$

What are the reactions that correspond to the three entries for the lithium atom, Li, in the table?

(b) The first ionization energy for Li is very low compared to the first ionization energies for H and He. The first ionization energy for Li is also very low compared to its second ionization energy. What might account for this very low first ionization energy for Li? In other words, why is the electron held so loosely to its nucleus?

Table 4.1. Successive ionization energies in kJ·mol^{-1} for H, He, and Li atoms.

	1st electron	2nd electron	3rd electron
H	1312		
He	2373	5248	
Li	520	7300	11808

The electron wave model for helium

Helium atoms have two electrons and two protons in their nuclei, so the electrons are attracted by a +2 charge. Both electron waves can surround the nucleus and form a single, spherical electron orbital with a –2 charge. The electron waves can be thought of as reinforcing one another, just as the two waves described in Figure 4.9(a) reinforce one another. The potential energy for the helium atom is more complicated than the potential energy for the hydrogen atom. For hydrogen, we had to consider only the attraction of a proton and an electron. For helium, we must consider the attractions and repulsions among three particles. Each of the two electrons is attracted to the positive (+2) nucleus. At the same time, the two negative electrons repel each other. Without doing any calculations, you can estimate some of the effects of these differences from the hydrogen atom; in fact, you already began doing this in Check This 4.48.

The higher nuclear charge will attract each electron more strongly and draw it closer to the nucleus. Counteracting the nuclear attraction will be the repulsion between the electrons. This repulsion is not as strong, however, since each electron has only a single negative charge. Also counteracting the negative potential energy of attraction is the kinetic energy of the electron orbital, which gets more positive as the orbital gets smaller. On balance, we can predict that the larger nuclear attraction should make the helium atom smaller than the hydrogen atom and should hold the electrons more tightly. The experimental values for the ionization energies in Table 4.1 confirm that the electrons are held more tightly.

Simple calculations for the helium atom, like those for hydrogen discussed in Section 4.8, give an atomic radius about half that of hydrogen and a total minimum energy, *E*, of about –6700 kJ·mol^{-1}. Thus, 6700 kJ·mol^{-1} is the energy required to remove both electrons:

$$He_{(g)} \rightarrow He^{2+}_{(g)} + 2e^- \qquad (4.22)$$

Reaction equation (4.20) is the sum of reaction equations (4.20) and (4.21). Experimentally, the total energy required is the sum of the two ionization energies for helium, 7621 kJ·mol^{-1}, from Table 4.1. The calculated value and the experimental value do not agree exactly, but the calculation is only off by a little more than 10%. These results—a smaller electron cloud radius and a lower energy minimum for helium than for hydrogen—are what we predicted above. Recall from our previous discussions of the periodic table that atoms get smaller as you go from left to right across a period. You have just successfully crossed the first period.

The electron wave model for lithium: a puzzle

Each successive element adds another unit of charge to its atomic nuclei, and another accompanying electron wave. Lithium is the next element and, based on the discussion for helium, we might expect its three electrons to be held even more tightly by the increased nuclear charge. The total ionization energy for lithium atoms, from Table 4.1, is 19628 kJ·mol^{-1}. This is the energy required for the sum of the reactions you wrote for lithium in Consider This 4.49(a):

$$Li(g) \rightarrow Li^{3+}(g) + 3e^- \tag{4.23}$$

Our expectation is confirmed; much more energy is required to remove all the electrons from lithium than from helium. However, lithium's very low first ionization energy—even lower than that for hydrogen, which has a nuclear charge of only one—is puzzling. The second and third ionization energies are quite large, so it is clear that these electrons are tightly held. The problem is with the first electron, which seems more weakly held than expected. Either our wave model is no good or it needs some modification.

Magnetic properties of gaseous atoms.

We have been using the wave model because it explains the quantization of energy levels in atoms and accounts for the wave properties of electrons. Whenever a model that seems to be working fails in some way, the first approach is usually to try and fix it. Almost always, this means looking at other properties of the system that is giving the problem — the lithium atom, in this case — which brings us back to spectra. While studying atomic spectra, scientists often observed changes when the atoms were placed in a magnetic field. They also found that the paths of some atoms streaming through an inhomogeneous magnetic field (a field with curved field lines, as at the top and bottom of Figure 4.11) were unaffected by the field, while the paths of other atoms were split into two streams, as illustrated in Figure 4.29.

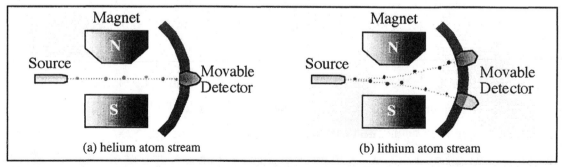

Figure 4.28. Effect of an inhomogeneous magnetic field on streams of atoms.

A stream of helium atoms, Figure 4.28(a), is an example of atoms that are not deflected in the magnetic field. They pass straight through. On the other hand, lithium atoms, Figure 4.28(b), are split into two streams by the magnetic field. Half the lithium atoms are deflected toward one pole of the magnet and half are deflected toward the other. The *lithium atoms act as though they are themselves tiny magnets*, half of which are oriented in one direction and half in the other.

Electron spin

It's a magnetic property of the electrons that is responsible for the behavior of the atoms observed in these experiments. Electrons can be thought of as acting like tiny magnets; the name we use to characterize this magnetic property is **spin**. Although the name suggests images of spinning tops or figure skaters, these are probably not accurate images of electron spin. Electron spin can take two orientations in space, just like a bar magnet can be oriented with the north pole up or the north pole down, Figure 4.29. When it is important to show the spin of an electron, we use an arrow, as shown in the figure.

Figure 4.29. Electron spin. The analogy of electron spin to the orientation of a magnet.

Pauli exclusion principle

Electron spin is essential to our discussion of electrons in atoms because *two (and no more than two) electrons can be described by the same spatial orbital only if they have opposite spin.* The two electrons have the same energy, and are identical in every respect except for their spins. A corollary of this statement is that *two electrons of the same spin cannot be described by the same electron orbital*. This concept, developed by Wolfgang Pauli (Austrian-born physicist, 1900-1958), is called the **Pauli exclusion principle**. It is an essential tool that enables us to describe energy levels and regions of space occupied by electrons in multi-electron atoms.

Protons also have spin and magnetic properties (different size than electrons). Proton spin is the basis of powerful techniques for probing the structure of matter, including humans via MRI, magnetic resonance imaging.

The proton magnet and electron magnet affect one another. In the hydrogen atom, the magnets can be paired or not, as shown in this energy diagram. The nuclear magnet is represented by the heavy blue arrow and the electron by the red arrow. You know that magnets resist being brought together with the same poles facing one another. This is the higher energy state.

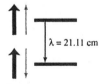

The emission from H atoms that have been excited by collisions with other H atoms is detected by radio astronomy and is a basis for the estimates of the amount of hydrogen in the universe. Also see Check This 4.22 and 4.33

For helium, the exclusion principle tells us that the two electron waves can occupy exactly the same space if, and only if, the two electrons have opposite spins. Two electrons with opposite spin are called **spin-paired electrons**. The magnetism of spin-paired electrons cancels out, and produces no net magnetic properties in the atom or molecule where they are located. Thus, helium atoms with their spin-paired electrons pass unaffected between the poles of a magnet. We often symbolize spin-paired electrons as a pair of oppositely oriented arrows: (↑↓).

Exclusion principle applied to lithium: puzzle solved

For lithium, the two lowest-energy electron waves have paired spins, occupy exactly the same space, and surround the nucleus. The third electron is attracted to the Li$^+$ ion that results (the +3 nucleus plus the two –1 electrons). However, because it has the same spin as one or the other of the two electrons that are already there, it cannot share exactly the same space with them. The orbital that describes it is at a larger average distance from the nucleus than the other two. As a result, its potential energy of attraction to the nucleus is less negative, an unfavorable effect. At the same time, because the size of the orbital is larger, the kinetic energy of the third electron is less positive, a favorable effect. The net effect is a total energy that is less negative (a low ionization energy). In terms of the energy levels for the atom (see Figure 4.22), this electron will be at a higher energy than the first two electrons. Since the third electron in lithium is not spin-paired with another electron, the atom acts like a magnet, which explains its behavior in the magnetic field experiments, Figure 4.28. Taking spin and the exclusion principle into account solves the puzzle of why lithium's first ionization energy is so low, but more data are needed to tell you whether the solution is general and can be applied to elements with even more electrons.

Section 4.10. Periodicity and Electron Shells

Patterns in the first ionization energies of the elements

Since it is ionization energies that signaled a problem for our simplest atomic model, let us take up where we left off in Section 4.1 and further examine elemental ionization energies. Figure 4.4 plotted the first ionization energies for the first 20 elements. Taking a somewhat different look at the first ionization energies for gaseous atoms of all the elements, Figure 4.30 plots the *negative* of the first ionization energies for gaseous atoms as a function of their atomic numbers (nuclear charge). The negative value of the ionization energy is the energy that is given off when an electron is added to an atom that is missing one electron. For example:

> WEB Chap 4, Sect 4.9.1
> View and manipulate a three-dimensional representation of the ionization energies.

$$C^+(g) + e^-(g) \rightarrow C(g) \quad (\Delta E = -1086 \text{ kJ·mol}^{-1}) \quad (4.24)$$

The more negative the energy, the more stable the atom relative to its monocation and electron.

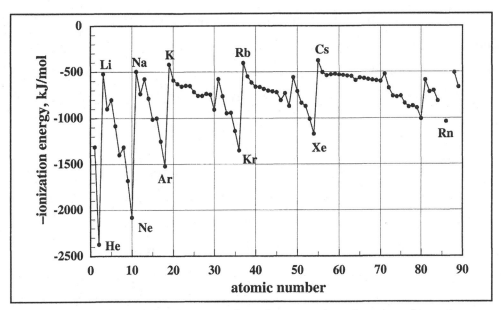

Figure 4.30. The first ionization energies of atoms plotted as negative values.
The lower the energy the more stable the atom, X(g), relative to its ion and an electron, X$^+$(g) + e$^-$(g). The data points are connected to make trends easier to see. Values for some the heavier elements are unknown.

4.50. Consider This   *How do you interpret patterns in these ionization energy data?*

(a) Atoms from which family (group) of the periodic table are most stable with respect to loss of an electron? least stable? What is the relationship of these families in the common version of the periodic table that is shown on the inside front cover of the book?

(b) As you go from left to right across the plot in Figure 4.30, the nuclear charge is increasing. Higher nuclear charge should have greater attraction for the electrons. Are there places in plot that show this effect? If so, where are they?

(c) As you go from left to right across the plot, does the attraction for the last electron generally increase, decrease, or stay about the same? Is your answer consistent with the increase in nuclear charge across the plot? If not, what factor(s) might explain why it is not?

Electron shell model for atoms

In Section 4.9, we said that the third electron in lithium is not as close to the nucleus as the first two, because its spin excludes it from having the same spatial orbital (wave) as the first two. Its orbital is, on average, further from the nucleus. As a consequence of being further from the nucleus, this electron is not held as tightly and, as we see in Figure 4.30, not as much energy is released when it forms the Li atom by joining with the Li$^+$ ion. As the nuclear charge increases going from lithium to neon, the energy released upon formation of the atoms generally increases.

Structure of Atoms   Chapter 4

This is just what we expect from the increasing attraction by the nuclei for electron orbitals that form a second shell around the nucleus.

In going from neon to sodium, there is again a large change in the energy of attraction with the last electron in sodium being even more weakly held than the last electron in lithium. If our model is correct, this means that the spin of the 11th electron excludes it from having the same spatial orbital as any of the 10 electrons already present in Na$^+$. The ionization energy data suggest that the orbital that describes the 11th electron is, on average, even further from the nucleus and begins a third shell. This is the essence of the **shell model** for atoms — electrons behave as if they are being added in layers (shells) as atomic numbers increase. Within a shell, all the electrons are about the same distance from the nucleus. Between shells, there is a difference in the average distance from the nucleus. These concepts are summarized in Figure 4.31, an abbreviated periodic table with a shell structure shown for the first 20 elements. This model does raise a puzzling question: how can the second and third shells accommodate eight electrons and still not violate the exclusion principle? We'll return to this question in Section 4.11.

	I	II	III	IV	V	VI	VII	VIII
1	Hydrogen $_1$H •							Helium $_2$He •
core e^- valance e^-	0 −1							0 −2
2	Lithium $_3$Li	Beryllium $_4$Be	Boron $_5$B	Carbon $_6$C	Nitrogen $_7$N	Oxygen $_8$O	Fluorine $_9$F	Neon $_{10}$Ne
core e^- valance e^-	−2 −1	−2 −2	−2 −3	−2 −4	−2 −5	−2 −6	−2 −7	−2 −8
3	Sodium $_{11}$Na	Magnesium $_{12}$Mg	Aluminum $_{13}$Al	Silicon $_{14}$Si	Phosphorus $_{15}$P	Sulfur $_{16}$S	Chlorine $_{17}$Cl	Argon $_{18}$Ar
core e^- valance e^-	−10 −1	−10 −2	−10 −3	−10 −4	−10 −5	−10 −6	−10 −7	−10 −8
4	Potassium $_{19}$K	Calcium $_{20}$Ca			Scale →| |← 100 pm			
core e^- valance e^-	−18 −1	−18 −2						

Figure 4.31. Electron shell model for the first 20 elements in the periodic table.

Atomic radii

Our atomic model provides an indication of the sizes, or at least the relative sizes, of atoms. We have seen that the size of an atom is a fuzzy concept, because the electron probability distribution, Figures 4.25(b) and 4.26(b), has no sharp boundary. We have tried to indicate this fuzziness in Figure 4.31 as well. Experimental measurements of atomic size are not straightforward and simple. One approach to determining the size of atoms is to measure the distance between nuclei in covalent compounds and assign part of the distance as the radius of one atom and the rest as the radius of the other atom. (We will say more about bond distances in the next chapter.) For bonds between the same two elemental atoms, **homonuclear bonds** (*homo* = same), as in F_2 or I_2, the atomic radius would be just half the bond length. For bonds between different elemental atoms, we could start with one radius from the homonuclear bonds and assign the rest of the bond distance as the radius of the other atom.

4.51. **Check This** *Atomic radii for carbon and chlorine*

The distance between the two nuclei in the chlorine molecule, Cl_2, is 198 pm. Chlorine forms a compound with carbon, CCl_4, tetrachloromethane (carbon tetrachloride), in which the distances between the carbon and chlorine nuclei are all the same, 178 pm. What is the atomic radius of chlorine? What is the atomic radius of carbon? *Note*: Distances between nuclei are determined by several spectroscopic and diffraction techniques.

If we analyze a large number of compounds, by the method you used in Check This 4.51, a reasonably consistent set of atomic radii can be derived. Someone else, using a different choice of compounds, might get a somewhat different set of atomic radii. You will find different atomic radii in different books. For our purpose here, minor variations are unimportant. We are interested in the trends and patterns for the atomic radii shown in Figures 4.31 and 4.32. Figure 4.32 shows the relative size of atomic radii derived from analyses of compounds containing each element (or, in a few cases, calculations based on the wave model of atoms).

WEB Chap 4, Sect 4.9.2
View and manipulate a three-dimensional representation of the atomic radii.

Structure of Atoms Chapter 4

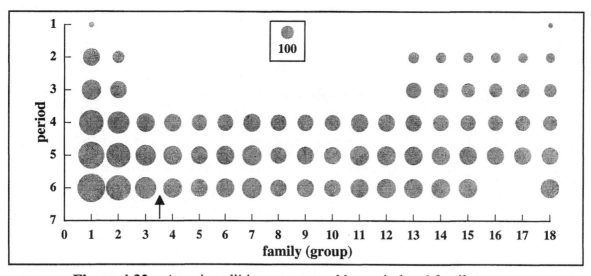

Figure 4.32. Atomic radii in pm arranged by period and family.
The 14 lanthanide elements (at the arrow) and the period 7 elements are not shown. For reference, a circle representing an atomic radius of 100 pm is shown.

4.52. Consider This *What are the patterns in atomic size?*

(a) In Figure 4.32, how would you describe the trend in atomic radii across a period? Can you think of a reason for this trend? Explain.

(b) How would you describe the trend in atomic radii down a family (group or column)? Can you think of a reason for this trend? Explain.

(c) Does our atomic model provide a consistent explanation for these patterns? That is, is your reasoning in part (b) consistent with your reasoning in part (a)? Explain why or why not?

(d) Compare any patterns in Figure 4.32 with those in Figure 4.1. Explain why there should (or should not) be a correlation between these two plots.

Electronegativity

Electronegativity, which we have met in the previous chapters, is a measure of the attraction that an atom (atomic core) in a molecule has for the bonding electrons it shares with another atom. We have warned that electronegativity is a concept defined by scientists and not a measurable property of an atom like its ionization energy. Linus Pauling developed his set of electronegativities based on interrelationships of the energies released in bond formation for a large number of molecules. Other scientists have developed other sets of electronegativities based on different models and data. The trends in all these values of electronegativities are the same; Figure 4.33 shows Pauling electronegativities.

> **WEB Chap 4, Sect 4.9.3**
> View and manipulate a three-dimensional representation of the electronegativities.

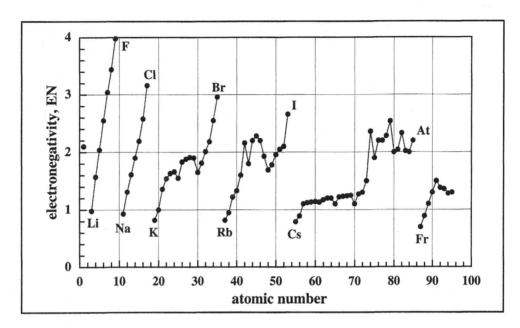

Figure 4.33. Pauling electronegativities plotted as a function of atomic number. The data points are connected to make trends easier to see. The gaps are the missing data points for the noble gases.

4.53. Consider This *What are the patterns in electronegativity?*

(a) What patterns do you observe for electronegativities in Figure 4.33? How do they vary within periods and families of the periodic table?

(b) Does Figure 4.33 provide any further evidence for an electron shell model for atoms? If so, explain clearly how you interpret the plot as evidence for the shell model. If not, explain clearly how you interpret the plot as evidence against the shell model.

(c) What relationships can you see among the plots in Figures 4.30, 4.32, and 4.33? How do you interpret these relationships

A closer look at the electron shell model

The electron shell model of atoms is based on the kinds of experimental evidence that you have been examining from the beginning of this chapter. This experimental evidence is reinforced by the wave model for atoms, which permits us to interpret the periodicity of atomic properties in terms of electron waves that are attracted as close as possible to the atomic nuclei. Restrictions on how close the waves come to the nucleus are imposed by their kinetic energies and by the exclusion principle. We interpret the shell structure as a result of the balance between the attractions and restrictions. As we said in the introduction to this chapter, we have asked you to do two things: (1) look at experimental data and try to see in them *evidence* for the shell model

Structure of Atoms   Chapter 4

and (2) look at experimental data and try to interpret them as a *result* of the shell model. We hope you have become comfortable working both ways.

> **4.54. Consider This** *What are some further properties of electron shells in atoms?*
>
> **(a)** How many shells are represented in Figures 4.30, 4.32, and 4.33? How many electrons are in each shell?
>
> **(b)** How does the spiral periodic table shown in the center of the chapter opening illustration, and in Figure 4.34, reflect the patterns shown by the data in Figures 4.30, 4.32, and 4.33? Explain with reference to specific elements or groups of elements from the periodic table and the data. Does this periodic table and/or the data provide any evidence for "structure" (subshells) within the electron shells? If so, what is the evidence?

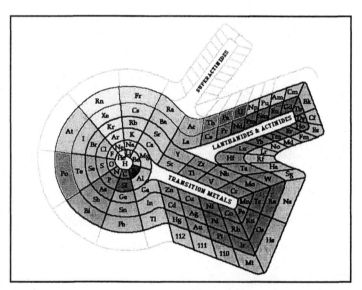

Figure 4.34. A spiral version of the periodic table of the elements.

The data in Figures 4.30 and 4.33 show that not all the electron shells are the same. In the first place, they do not all accommodate the same number of electrons. In Section 4.9, we learned why the first shell has only two electrons. For the succeeding shells, if you count from an alkali metal to the following noble gas, K to Kr, for example, the number of elements in periods 2 through 6 is 8, 8, 18, 18, and 32. (Period 7 is not complete; the heavier elements have yet to be synthesized.) The patterns of ionization energies and electronegativities for periods 2 and 3 are very similar, but in periods 4 and 5, the pattern is interrupted by a series of 10 elements all of which have quite similar values for these properties. Period 6 shows this same interruption with yet another longer interruption by a series of 14 elements. It appears as though we might have to amend the shell model to a subshell model, where the subshells can accommodate orbitals containing totals of 8, 10, and 14 electrons. The arms coming off the spiral form of the periodic table in Figure 4.34 help you visualize this structure.

4.55. Check This *Number of electron subshells in each shell*

For each of periods 2 through 6, which subshells (the 8-, 10-, and/or 14-electron subshells) are used to build the series of elements across the period? Show how the spiral periodic table illustrates your answers.

In Figures 4.4 and 4.30, we can also see that within the 8-electron shells (or subshells) in periods 2 and 3, the first ionization energies seem to show a stair step pattern of two, three, and three elements. We might, however be seeing some effect of the increase in coulombic attraction as the nuclear charge increases from one elemental atom to the next. To check this possibility, we can look at successive ionization energies for an element, as we did in the previous section for He and Li to learn more about their structures. The energy required for each successive ionization has been measured for many elemental atoms and some data for the first 20 elements are given in Table 4.2.

4.56. Consider This *How can successive ionization energies be interpreted?*

(a) What relationships are there between the numbers of core and valence electrons shown in Figure 4.31 and the data in Table 4.2? Explain your relationships with specific examples.

(b) For which atoms does Table 4.2 give the ionization energies for all the electrons in complete 8-electron shells (subshells)? Explain how you make your choices.

(c) Plot the successive ionization energies in complete 8-electron shells (subshells) for two or three of the atoms you chose in part (b). If available, use a spreadsheet program or graphing calculator to do this analysis. Do your plots show evidence for the two and three element steps that are apparent in Figures 4.4 and 4.30? If so, what is the evidence?

Structure of Atoms

Table 4.2. Successive ionization energies, in kJ·mol⁻¹, for the first 20 elements.
The column number is the number of the electron removed. Column 1, for example, is the first ionization energies that have been plotted in Figures 4.4 and 4.31. The shading calls attention to the ionization energies before a large jump (more than a factor of 3) for the next electron lost.

atom	1	2	3	4	5	6	7	8	9	10
$_1$H	1312									
$_2$He	2373	5248								
$_3$Li	520	7300	11808							
$_4$Be	899	1757	14850	20992						
$_5$B	801	2430	3660	25000	32800					
$_6$C	1086	2350	4620	6220	38000	47232				
$_7$N	1400	2860	4580	7500	9400	53000	64400			
$_8$O	1314	3390	5300	7470	11000	13000	71300	84000		
$_9$F	1680	3370	6050	8400	11000	15200	17900	92000	106000	
$_{10}$Ne	2080	3950	6120	9370	12200	15000	20000	23100	115000	131000
$_{11}$Na	496	4560	6900	9540	13400	16600	20100	25500	28900	141000
$_{12}$Mg	738	1450	7730	10500	13600	18000	21700	25700	31600	35500
$_{13}$Al	578	1820	2750	11600	14800	18400	23300	27500	31900	38500
$_{14}$Si	786	1580	3230	4360	16000	20000	23800	29200	33900	38700
$_{15}$P	1012	1904	2910	4960	6240	21000	25400	29900	35900	40960
$_{16}$S	1000	2250	3360	4660	6990	8500	27100	31700	36600	48100
$_{17}$Cl	1251	2297	3820	5160	6540	9300	11000	33600	38600	43960
$_{18}$Ar	1521	2666	3900	5770	7240	8800	12000	13800	40800	46200
$_{19}$K	419	3052	4410	5900	8000	9600	11300	14900	17000	48600
$_{20}$Ca	590	1145	4900	6500	8100	11000	12300	14200	18200	20380

Electron spin and the shell model

Although the steps you find in Consider This 4.56(c) are not as pronounced as those in Figures 4.4 and 4.30, there is definite evidence that the successive ionization energies do not increase smoothly, but show two and three element steps. Another piece of evidence for more structure within the shells comes from spectroscopic studies of atoms in magnetic fields that tell us how many unpaired electron spins the atoms have. The details of these studies are not important for us here, but the results shown in Table 4.3 for the second period elements are important for our understanding of electrons in atoms.

Table 4.3. Number of unpaired electron spins for ground-state, second-period atoms.

$_3$Li	$_4$Be	$_5$B	$_6$C	$_7$N	$_8$O	$_9$F	$_{10}$Ne
1	0	1	2	3	2	1	0

Chapter 4 — Structure of Atoms

We know that any atom with an odd number of electrons, such as Li or B, must have at least one electron spin unpaired, and we might have anticipated that any atom with an even number of electrons would have them all spin paired. For Be with two electrons in the second shell, the electron spins are paired, as we anticipate. But C with two further electrons in the second shell has two unpaired electrons and N with three further electrons has three unpaired electrons. It appears as though the third, fourth, and fifth electrons added in the second shell are unpaired.

4.57. Check This *Electron spin pairing in second period elements*
As three more electrons are added in O, F, and Ne, how can spin pairing explain the results shown in Table 4.3?

We could once again amend the shell-subshell model to account for these experimental results, but that is not a fruitful approach. The value of the shell model is its simplicity. As you have seen, a great deal of the data on periodic properties of the elements can be explained and/or predicted from the shell model of the atom which is based on a simple application of the wave nature of electrons. The more we tinker with the model, the further we get from the basic concepts of electron attraction by the nucleus and the restrictions imposed by kinetic energy and electron spin on the average distance of the electron waves from the nucleus. To explain more subtle effects, such as electron-spin pairing (or lack of pairing), we need to return to the wave model of the electron in atoms and see how it is used to describe the waves in more detail.

Reflection and projection

The wave mechanical model of the atom that accounted for the kinetic energy of an electron wave and the potential energy of attraction between electrons and the nucleus had to be modified to take into account electron spin and the exclusion principle. Only two spin-paired electrons can be described by the same electron wave (orbital). Thus, electrons in an atom, after the first two, which enclose the nucleus in a spherical electron wave, are, on average, further from the nucleus and form successive shells of electrons. The consequences of this structure are the periodic properties of the elements that repeat as each new shell is built up from one element to the next across a period of the periodic table.

You examined several properties of atoms—ionization energies, electronegativities, and sizes—and found patterns in these properties that varied periodically. Each of these variations can be used as evidence for the atomic model or explained as a result of the atomic model. The model balances increasing electron attraction by increasing nuclear charge with restrictions on how close the electrons are to the nucleus. For example, the size of the atoms decreases across a period as the nuclear charge increases and attracts the electrons from the valence shell closer. Conversely, the size of the atoms increases as the nuclear charge increases in families (groups) of elements because a new shell, further from the nucleus, is added in going from period to period.

Even within the patterns that are readily explained by the shell model, there are some variations that are not explained and there are other properties, such as the number of unpaired electron spins in atoms that also cannot be accounted for. To account for such observations, we need to look in more detail at electron waves in atoms.

Section 4.11. Wave Equations and Atomic Orbitals

We introduced the idea of a wave model for electrons in Section 4.7 with the analogy to standing waves on a guitar string, Figure 4.24. The standing waves on the acoustic body of the guitar, as it is played, are more complex, Figure 4.36. As a consequence, the nodes of these standing waves and the wave equations describing the waves are more complex than the nodes on the strings of the instrument and equation (4.11) in Check This 4.37. For a three-dimensional electron wave, our only pictorial representation has been the spherical probability distribution in Figures 4.25, 4.26, and 4.27. Solutions to the wave equations for electron waves also yield more complex shapes than spheres.

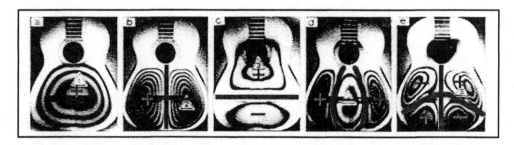

Figure 4.35. Standing waves on the acoustic body of a guitar.
The nodes are shown by red lines and curves. Each node traces a part of the surface that is not moving. At any instant the surfaces on opposite sides of a node are moving in opposite directions (toward you, +, or away from you, –), just like a guitar string on either side of one of its nodes, Figure 4.24.

Schrödinger wave equation

In Sections 4.8 and 4.9, you saw how the balance of kinetic and potential energies for an electron wave determines the size of the wave and its ionization energy. Soon after de Broglie introduced the wave model, scientists developed mathematical descriptions that accounted for the kinetic and potential energies of electron waves. The equations can be solved to give the total energy *and* a probability picture of electron waves. One of these equations is the **Schrödinger wave equation** developed by Erwin Schrödinger (Austrian-born physicist, 1887-1961):

$$-\frac{h}{4\pi m}\nabla^2\psi + V\psi = E\psi \qquad (4.25)$$

$$\underset{\text{energy}}{\text{kinetic}} + \underset{\text{energy}}{\text{potential}} = \underset{\text{energy}}{\text{total}}$$

The two terms on the left side of the wave equation give the kinetic and potential energies as a function of the **electron wave function,** ψ (Greek lower case "psi"), a mathematical description of the wave. Their sum gives the total

> The Schrödinger wave equation is a differential equation, that requires calculus to solve. $\nabla^2\psi$ is a second derivative with respect to the spatial coordinates. Many mechanical systems in physics are described by wave equations like this. This is why the wave equation approach to describing atomic level systems is called wave mechanics.

energy on the right. If you continue your study of chemistry, you will learn, in later courses, how to find wave functions that are solutions to the Schrödinger wave equation. For this discussion, we will present the solutions in a pictorial form and relate them to the ionization energy data in Figures 4.4 and 4.30 and Table 4.2.

Wave equation solutions: orbitals for one-electron atoms

The wave function, ψ, depends on the spatial coordinates x, y, and z (the directions in a three-dimensional coordinate system). Thus, it is possible to calculate the value of ψ at any point in space. In wave mechanics, the square of the electron wave description, ψ^2, calculated for a given location, is proportional to the probability of finding the electron in that location. Plots of these probabilities in the region around the atomic nucleus show the shape of the electron wave (orbital) and its nodes. The lowest energy solutions to the wave equation for a one-electron atom or ion are shown in Figure 4.36. The probabilities may be represented by the density of dots on a page, as in Figure 4.25(b), but more often simpler boundary surface pictures, as in Figure 4.26(a), are used to represent orbitals. Boundary surfaces are drawn to enclose a high

> **WEB Chap 4, Sects 4.6.1 & 4.10.1**
> View lowest energy orbitals rotated in three dimensions.

percentage of the electron probability (density), 95-99%, and with a shape that captures the shape of the probability distribution (orbital).

Structure of Atoms — Chapter 4

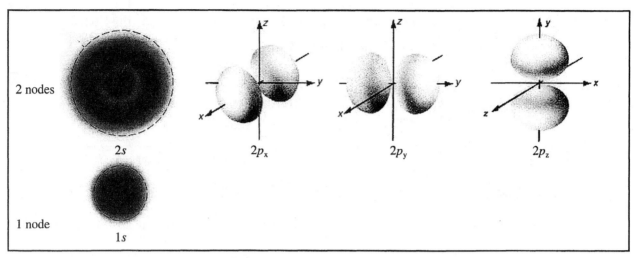

Figure 4.36. Atomic orbitals for a one-electron atom.
The s orbitals are shown as cross-sections of dot density diagrams; the p orbitals as boundary surface diagrams. All the orbitals have a node at infinity which is not represented in the drawings. Orbitals in the second energy level all have a second node which is represented.

Orbitals are named, $1s$, $2s$, $2p_x$, and so on to identify their energy, shape, and orientation in space. The numerical value that begins each name is the **principal quantum number, n**, and designates the energy level, Figure 4.22, for the electron described by that orbital. In a one-electron atom or ion, all the orbitals with the same principal quantum number, n, have the same total energy. Orbitals with the same energy are called **degenerate**. The lower case letter that follows the numeral designates the three-dimensional shape of the orbital. The letters chosen are related to spectroscopic observations on the atoms. You can remember "s" because these orbitals are *s*pherical. The *p* orbitals always come as a set of three that are mutually *p*erpendicular to each other and can be visualized as oriented along the three axes of a three-dimensional coordinate system, as shown in Figure 4.36. All three have the same energy and shape, so a subscript is used to indicate their orientation in space with respect to one another.

> "Quantum" refers to the quantized energies of the electrons in atoms. "**Quantum number**" refers to a particular energy level. The names of the orbitals were originally used for energies in the Bohr model and then were transferred to the corresponding wave mechanical solutions.

> There is no negative meaning to the term "degenerate energy levels." It simply denotes two or more solutions to the wave equation that have the same total energy.

4.58. Consider This *How are orbital nodes described?*

(a) What would the three-dimensional boundary surfaces for a $1s$ and $2s$ orbital look like? Could you tell them apart? Why or why not?

(b) For the orbitals in Figure 4.36, the nodes represented are surfaces. What is the shape of the nodal surface for the $2s$ orbital? For a $2p$ orbital? Give the reasoning for your answers.

Chapter 4 — Structure of Atoms

It is very important to remember that an orbital is a description of an electron wave; *orbitals do not have any existence separate from the electron.* For *convenience*, scientists often speak about "electrons in an orbital," "adding an electron to an orbital," "half-filled orbitals," or even "empty orbitals." All of these phrases imply that orbitals are like boxes into which electrons can be placed. What is meant is that a *calculated* energy level (and corresponding spatial probability distribution) exists that could be used to describe an electron wave. If no actual electron is described by the calculation, then the orbital is "empty." If an actual electron is described by the calculation, then the electron is "in" the calculated orbital. Because these phrases are so convenient, we will use them, but always keep in mind that *orbitals are not boxes*.

4.59. Consider This *How are principal quantum numbers, nodes, and orbitals related?*

(a) What is the relationship between the number of nodes and the principal quantum number, n, in Figure 4.36? Sketch a picture of a 3s orbital, using the 2s representation in the figure as a model. Clearly explain the reasoning you use to decide how to make your sketch.

(b) All of the orbitals with n = 1 and n = 2 are shown in Figure 4.36. For n = 3, there are nine energy-degenerate spatial orbitals: one 3s, three 3p, and five 3d orbitals. Do you see any evidence for *d* orbitals in the structure of any of the periodic tables in the chapter opening illustration and/or in Figures 4.30 or 4.33? If so, clearly explain the evidence.

(c) What is the relationship between n and the number of energy-degenerate spatial orbitals? How many energy-degenerate spatial orbitals are there with n = 4? Explain how you reach your conclusion.

Electron configurations

The 1s orbital is lowest in energy. A ground-state hydrogen atom has one electron described by a 1s orbital. The **electron configuration** of this hydrogen atom is $1s^1$. The superscript shows how many electrons in the atom are described by the orbital. A ground-state helium atom has two spin-paired electrons that can be described as superimposed 1s spatial orbitals; the electron configuration is $1s^2$. The energy of the 1s orbitals in helium is lower than the 1s orbital in hydrogen because of the increased nuclear charge. This description of the electron waves in hydrogen and helium is the same as the one in Sections 4.8 and 4.9.

You can imagine continuing this process of adding one electron at a time (with the appropriate change in nucleus) to build the rest of the elements. For atoms with three or more electrons, two electrons can be 1s electrons. Because of the exclusion principle, the rest of the electrons have to be in orbitals with n = 2 or larger. The four orbitals with n = 2 can accommodate eight electrons if there are two spin-paired electrons described by each spatial orbital. These eight electrons would account for the second shell of electrons in our shell model and a repeat of this build up for the 3s and 3p orbitals would account for the third shell of eight electrons.

But this reasoning based on the energy degeneracy of orbitals with the same principal quantum number is just the shell model again and does not help to explain the substructure of the shells that is evident in the ionization energies. Furthermore, as you found in Consider This 4.58(b), the n = 3 orbitals are nine-fold degenerate in a one-electron atom, so we might expect the third shell to accommodate 18 electrons, but there are only eight elements, Na through Ne, in period three. Only when we get to period four are there 18 elements. What's wrong with the model?

Energy levels in multielectron atoms

The orbitals in Figure 4.36, which are solutions to the wave equation (4.25) for a one-electron atom, are not correct for multielectron atoms. The wave equation cannot be solved exactly for an atom with two or more electrons. The calculations for multielectron atoms are all based on approximations that try to account for electron-electron repulsions and the fact that electrons cannot be told apart from one another. Simple pictorial representations of electron waves disappear in these calculations. The names of the orbitals survive to designate the energy levels derived from the calculations; these energy levels for multielectron atoms or ions are shown in Figure 4.37. The energies form a pattern like our shell structure with n = 1, 2, ... shells. No energy scale is given in the figure, because the actual energies of the levels vary from element to element (depending on the nuclear charge and total number of electrons in the atom or ion).

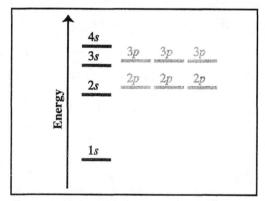

Figure 4.37. Energy levels for multielectron atoms and ions.
Relative energies are shown for the ten lowest-energy levels for multielectron atoms and ions.

Chapter 4 — Structure of Atoms

4.60. Worked Example — *The electron configuration for lithium*

What is the electron configuration for the ground state of the lithium atom?

Necessary information: We need to know the relative energies in Figure 4.37 and that the lithium atom has three electrons.

Strategy: Add the electrons, one at a time to the orbitals represented by the energy levels in Figure 4.37 in a way that makes the energy as low as possible and obeys the exclusion principle.

Implementation: Add the electrons to the energy level scheme in Figure 4.37:

1st electron   2nd electron   3rd electron

The first two electrons can spin pair at the lowest energy, $1s$, but the third electron has to go to the next higher energy, $2s$. The electron configuration is $1s^2 2s^1$.

Does the answer make sense? This configuration has the energy and spin properties necessary to explain the behavior of the lithium atom that we discussed in Section 4.9.

4.61. Check This — *The electron configuration for boron*

What is the electron configuration for the ground state of the boron atom? Explain.

Electron configurations for multielectron atoms

Figure 4.37 shows that the energy degeneracy among all the orbitals with the same principal quantum number no longer holds for multielectron atoms. There is still energy degeneracy among the orbitals that you can picture as having the same shape, but different orientations in space. You can use the pictures in Figure 4.36 as a reminder of the number of orbitals of the same shape and use the energy level diagram, Figure 4.37, as the basis for writing electron configurations for atoms and ions with more than one electron.

Atoms, in their ground states, have the lowest total energy possible. Thus, to build atoms and figure out their electron configurations, you choose from Figure 4.27 the lowest energy level possible for each added electron. We have done this for the first ten elements in Table 4.4. For the first five elements, the electron configurations are easy to predict; the last electron added goes to the lowest energy level that is not already occupied by two spin-paired electrons.

Structure of Atoms

Table 4.4. Electron configurations for the first ten elemental atoms.

The configurations are correlated with the first ionization energies. Steps of two-orbital, three-orbital, and three-orbital progressions in adding electrons and the corresponding ionization energies are highlighted in increasing shades of gray.

atom	electron configuration		first ionization energy, kJ·mol^{-1}
H	$1s^1$		1312
He	$1s^2$	$[He]^2$	2373
Li	$1s^2 2s^1$	$[He]^2 2s^1$	520
Be	$1s^2 2s^2$	$[He]^2 2s^2$	899
B	$1s^2 2s^2 2p^1$	$[He]^2 2s^2 2p^1$	801
C	$1s^2 2s^2 2p^1 2p^1$	$[He]^2 2s^2 2p^1 2p^1$	1086
N	$1s^2 2s^2 2p^1 2p^1 2p^1$	$[He]^2 2s^2 2p^1 2p^1 2p^1$	1400
O	$1s^2 2s^2 2p^2 2p^1 2p^1$	$[He]^2 2s^2 2p^2 2p^1 2p^1$	1314
F	$1s^2 2s^2 2p^2 2p^2 2p^1$	$[He]^2 2s^2 2p^2 2p^2 2p^1$	1680
Ne	$1s^2 2s^2 2p^2 2p^2 2p^2$	$[He]^2 2s^2 2p^2 2p^2 2p^2$	2080
		$[He]^2 2s^2 2p^6 = [Ne]^{10}$	

For carbon, the sixth electron could be paired with the first 2p electron or be in one of the other degenerate 2p orbitals. If two electrons are described by 2p orbitals, the electrons are farther apart and repel each other less when they are in different orbitals. Less electron-electron repulsion lowers the energy and makes the atom more stable. Thus, the three electrons added going from boron to nitrogen, are in separate 2p orbitals. An electron in one 2p orbital has the same energy as an electron in another of the 2p orbitals in the same atom. The increasing nuclear charge accounts for the steady increase in ionization energy from boron to nitrogen. The next electron added, to give oxygen, must occupy the same 2p orbital as an electron already present, since that is the lowest energy level available. The increased electron-electron repulsion raises the energy of the atom and makes it a bit easier to ionize. Thus, the first ionization energy of oxygen is a little lower than that of nitrogen, even though the nuclear charge has increased.

4.62. Consider This *How are electron configurations related to ionization energies and electron spin?*

(a) Write the electron configurations for the elemental atoms from Na through Ar. Use your electron configurations to make correlations between the pattern of first ionization energies in Figure 4.4 and 4.30 and the energy levels in Figure 4.37.

(b) Consider the successive ionization energies for the neon atom and ions in Table 4.2. What is the electron configuration for the atom or ion that loses an electron at each step? Is there any evidence in the pattern of ionization energies that supports the relative energy levels in Figure 4.37? Explain why or why not. *Hint*: This problem is related to what you did in Consider This 4.56(c).

(c) What is the relationship between the electron configurations in Table 4.4 and the number of unpaired electrons in these atoms, Table 4.3? Try to state the relationship as a general rule (for these cases).

(d) Use the electron configurations you wrote in part (a) and the rule you derived in part (c) to predict the number of unpaired electrons in the ground states of the third period elements. Do you find family relationships between your predictions and the data in Table 4.3? Did you expect such relationships? Explain why or why not.

Shell structure of atoms. The energy levels shown in Figure 4.38 for electrons with the same principal quantum number are similar in energy, even if not degenerate. This energy level grouping describes an electron-shell structure for atoms that also has some substructure in the shells. We can think of the orbitals discussed in this section as a refinement of the shell structure model that accounts for the variations in numbers of electrons in a shell and for the observed steps in ionization energies in periods two and three. When all the electrons are accounted for in the solutions to the wave equation for *any* isolated atom or ion, the electron probability distribution is spherical. The pictorial representation of isolated atoms we have used in Figure 4.31 and elsewhere is based on the shell structure and this spherical distribution.

Studying the properties of atoms has lead us to many insights about the nature of the physical world at this very tiny scale. However, you seldom encounter atoms in any everyday application. Helium-filled balloons, mercury and sodium vapor lights (used for street lighting), and neon signs are among the few common examples. None of these applications entails chemical changes. Essentially everything you meet involves molecules, not atoms. Recall that chemists were disappointed with the Bohr quantum model, because it couldn't be applied to molecules. The wave mechanical model of electrons does not suffer this drawback. Wave mechanics focuses on finding electron waves that balance coulombic attraction with coulombic repulsion and kinetic energy to give the lowest possible total energy. In principle, there is no restriction on the number of nuclei that can be included in the calculations. We developed the wave mechanical model for the simpler case of atoms so that we can apply it to building molecules in the next chapter.

# Structure of Atoms	Chapter 4

Section 4.12. Outcomes Review

Many properties of the elements vary in a periodic way as they increase in atomic mass (or atomic number, in the present view) and these variations are codified in periodic tables that were first devised by Mendeleev and Mayer. The fundamental basis for this periodicity is the structure of the atom which we developed in this chapter. In order to interpret atomic emission spectra, we need to understand the properties of light, and this means both its wave and photon (quantum) properties which led to a quantum model of the atom. The discovery of the wave nature of electrons led to a wave description of atoms that accounts for both the kinetic and potential energy of an electron wave interacting with a nucleus. We characterized the structure of these atoms in terms of electron shells. The electron shell model is useful for understanding and predicting a great many periodic patterns in atomic and elemental properties, but it is too simple to explain more subtle periodicities and the spins of elemental atoms. For these we need the more complete description of electron waves (orbitals) based on wave equations that describe the substructure of the shells.

Check your understanding of the ideas in this chapter by reviewing these expected outcomes of your study. You should be able to:

- identify and describe periodic patterns in atomic and elemental properties and by extrapolation or interpolation predict values of these properties for elements that have not been measured [Sections 4.1 and 4.10].
- explain the relationship between atomic emission and absorption of light [Section 4.2].
- calculate the wavelength, frequency, or speed of propagation of a wave, given two of the three variables or the information necessary to derive them [Section 4.3].
- describe in pictures and/or words how superimposition of waves produces diffraction patterns when waves pass through a grating [Section 4.3].
- identify, given the characteristics of a source or detector of radiation, where in the electromagnetic spectrum the radiation will be found [Section 4.3].
- calculate the wavelength, frequency, or energy of a photon, given two of the three variables or the information necessary to derive them [Section 4.4].
- show how the results of photoelectric-effect experiments can be explained by Planck's quantum hypothesis and how the wave model fails [Section 4.4].
- identify and explain whether a phenomenon is a result of the wave or photon (quantum) properties of light [Section 4.4].
- explain how the line emission and absorption of light by atoms requires that the energies of atoms be quantized [Section 4.5].

- use an energy level diagram for an atom and emissions from some known energy level changes to predict the emissions from other energy level changes [Section 4.5].
- calculate the de Broglie wavelength of any particle of known mass and velocity and identify where in the electromagnetic spectrum it will be found [Section 4.6].
- characterize a standing wave in terms of its amplitude, wavelength, and nodal properties [Sections 4.7 and 4.11].
- explain why the uncertainty principle leads to a probability picture of an electron wave in an atom instead of more easily visualized orbits of an electron [Section 4.7].
- describe in pictures and/or words how vibrating objects like strings and guitar bodies are related to the probability model of electron waves (orbitals) in atoms [Sections 4.7 and 4.11].
- calculate or predict the direction of change for the kinetic energy, if the mass or velocity of the object changes [Section 4.8].
- calculate or predict the direction of change for the potential energy of a system if the variables that describe the potential energy (charge and distance) are changed [Section 4.8].
- use pictures, words, and/or equations to explain how the potential and kinetic energies of electron waves prevent atoms from collapsing and determine their size [Section 4.8].
- explain how the balance of potential and kinetic energies of electron waves together with the exclusion principle lead to a shell structure for atoms [Sections 4.9, 4.10, and 4.11].
- show how the periodic properties of atoms and elements provide evidence for (or against) an electron shell and subshell model of atomic structure [Sections 4.10 and 4.11].
- show how the electron shell model for atoms explains and predicts periodic properties of atoms and elements, including, but not limited to, atomic size, ionization energies, and electronegativities [Section 4.10].
- describe the atomic orbitals derived from the Schrödinger wave equation for principal quantum numbers 1 and 2 [Section 4.11].
- write the electron configurations for atoms of the first 20 elements, using the energy levels for multielectron atoms and their degeneracies and accounting for the exclusion principle and electron-electron repulsion energies [Section 4.11].
- predict, from its electron configuration, the number of unpaired electron spins a ground-state atom has [Section 4.11].
- describe how the electron shell model and the orbitals and energies derived from solutions to the wave equation are complementary and together provide a way to explain all the periodic properties of atoms and elements discussed in the chapter [Section 4.1, 4.9, 4.10, and 4.11].

Structure of Atoms   Chapter 4

Section 4.13. EXTENSION — Energies of a Spherical Electron Wave

4.63. Investigate This *How is rotation affected by size?*

For this investigation, use a small, sturdy turntable and two equal masses (about 1 kg) that are easy to hold, one in each hand. Stand on the turntable and have a partner get you spinning around at a moderate speed while you are holding the masses at arm's length out to your sides. Bring the masses closer to your body and observe the effect this has on your speed of rotation. Straighten your arms again and observe your speed once more.

4.64. Consider This *Why is rotation affected by size?*

(a) To what does "size" refer to in this title and the title of Investigate This 4.63? Explain your answer clearly.

(b) What is the relationship between size and speed of rotation? In Investigate This 4.63, is it harder to hold the masses close to or far from your body while you are spinning? Why do you think it is harder one way than another? What other experiences have you had that were similar to this investigation?

(c) Can you see a relationship between the results in Investigate This 4.63 and our discussion of the energies of an electron wave? Explain your response.

Kinetic energy of an electron wave

From Section 4.8 we have:

$$KE = \frac{m \cdot u^2}{2} \tag{4.14}$$

The larger the mass, m, and/or the faster the motion, u^2 (velocity squared), the larger the kinetic energy. The dependence of kinetic energy on mass and velocity is similar to the dependence of momentum, $m \cdot u$, on mass and velocity. Indeed, we can write the equation for the kinetic energy of an object in terms of its momentum:

$$KE = \frac{(m \cdot u)^2}{2m} \tag{4.26}$$

To get the kinetic energy of an electron wave, we substitute the momentum for an electron wave, from the de Broglie wavelength relationship, into equation (4.26):

$$m \cdot u = \frac{h}{\lambda} \quad \text{(de Broglie wave)} \tag{4.27}$$

$$KE \text{ (electron wave)} = \frac{(h/\lambda)^2}{2m} = \frac{h^2}{2m \cdot \lambda^2} \qquad (4.28)$$

The kinetic energy of the electron wave is inversely proportional to the square of its wavelength. For a spherical electron wave, we can substitute the radius of the sphere for the wavelength in equation (4.28):

$$KE \text{ (spherical electron wave)} \propto \frac{1}{R^2} \qquad (4.29)$$

The smaller the radius of the electron wave (smaller volume of space occupied), the higher its kinetic energy. The observations you made in Investigate This 4.63 are analogous to this result for the kinetic energy of an electron wave. As you drew the masses closer to your body, you spun faster. As an electron is constrained to be closer to the nucleus — that is, as the size of the spherical wave decreases — the electron acts as though it moves faster; its kinetic energy increases.

4.65. Check This *Kinetic energy and the size of an electron wave*

Is the relationship in equation (4.29) consistent with the curve shown for the kinetic energy in Figure 4.27. Explain your response.

4.66. Worked Example *Kinetic energy of an electron wave*

What is the kinetic energy of an electron wave that has a wavelength, λ, of 100. pm?

Necessary Information: We need the wavelength of the electron wave from the problem statement as well as the electron mass, 9.11×10^{-31} kg, from Table 3.1 and Planck's constant, 6.63×10^{-34} J·s. To be sure the units work out, we need to know that the units of joules are kg·m²·s⁻².

Strategy: We substitute our known values of mass, wavelength, and Planck's constant into equation (4.28) to find the kinetic energy.

Implementation:

$$KE \text{ (electron wave)} = \frac{h^2}{2m \cdot \lambda^2} = \frac{(6.63 \times 10^{-34} \text{ J·s})^2}{2(9.11 \times 10^{-31} \text{ kg}) \cdot (100 \times 10^{-12} \text{ m})^2} = 2.41 \times 10^{-17} \text{ J}$$

Does the answer make sense? We can compare the kinetic energy for this electron wave, with the kinetic energy we got for the electron in Worked Example 4.44, 4.02×10^{-19} J, which we know from Worked Example 4.40 has a wavelength of 780 pm. The shorter wavelength electron

Structure of Atoms Chapter 4

in the present example has the higher kinetic energy. Note that, for electrons moving freely in space, the wave and particle descriptions for the kinetic energy give the same result.

4.67. Check This *Kinetic energy of matter waves*

In Check This 4.41(c), you found the wavelength of a thermal neutron. Equation (4.28) is applicable to any matter wave. What is the kinetic energy of a thermal neutron matter wave? of a mole of neutron matter waves? How does your kinetic energy per mole compare to what you got in Check This 4.45(b)? Is this result you expect? Why or why not?

Potential energy of an electron wave

From Section 4.8 we have:

$$PE \propto \frac{(+1)(-1)}{r} = -\frac{1}{r} \tag{4.16}$$

Although we cannot locate the electron and measure its distance, r, from the nucleus, the radius of the spherical electron wave, R, is proportional to the average electron-nucleus distance. Thus, we can also express the potential energy in terms of R:

$$PE \propto -\frac{1}{R} \tag{4.30}$$

Total energy of an atom

We know that the total energy, E, is the sum of the kinetic and potential energies:

$$E = KE + PE \tag{4.17}$$

The proportionalities in equations (4.29) and (4.30) are such that we get the result used in Section 4.8; the total energy for a nucleus-electron orbital system can be written as:

$$E = KE + PE \propto \frac{1}{R^2} - \frac{1}{R} \tag{4.18}$$

In order to do quantitative calculations of energies in the atom, we would need to know the proportionality in equation (4.18). Our purpose, however, has been to understand why atoms do not collapse and equation (4.18) provides a qualitative answer to that question. R, the radius of the electron wave, is a positive number. The total energy, E, is made up of two terms, the kinetic energy ($KE \propto \frac{1}{R^2}$) which is always positive, and the potential energy ($PE \propto -\frac{1}{R}$) which is always negative. The total energy depends on R and is the net balance between the positive kinetic energy and negative potential energy, as shown in Figure 4.27.

Index of Terms

ψ, electron wave function, 4-65
ν, wave frequency, 4-15
λ, wavelength, 4-14
absorption spectrum, 4-12
alkali metal, 4-6
alkaline earth metal, 4-6
amplitude, 4-14
Bohr quantum model, 4-34
cancellation, 4-16
de Broglie wavelength, 4-38
degenerate energy levels, 4-66
diffraction pattern, 4-16
electromagnetic radiation, 4-20
electromagnetic spectrum, 4-21
electromagnetic wave theory, 4-19
electromagnetic waves, 4-20
electron configuration, 4-67
electron spin, 4-53
electron wave function, 4-65
electronegativity, 4-58
excited state, 4-35
exclusion principle, 4-53
frequency, 4-15
ground state, 4-35
Heisenberg uncertainty principle, 4-43
hertz, 4-23
homonuclear bond, 4-57
ionization energy, 4-9
kinetic energy, 4-45, 4-74
momentum, 4-38
n, principal quantum number, 4-66

noble gases, 4-8
node, 4-14
orbital, 4-44
Pauli exclusion principle, 4-53
periodic table, 4-6
periodicity, 4-6
photoelectric effect, 4-28
photon, 4-30
Planck's constant, 4-26
potential energy, 4-46, 4-76
principal quantum number, 4-66
quantized, 4-26
quantum, 4-26
quantum model, 4-34
reinforcement, 4-16
Schrödinger wave equation, 4-65
shell model, 4-56
speed of light, 4-22
spin, 4-53
spin-paired electrons, 4-54
standing waves, 4-41
superimposition, 4-16
total energy, 4-47, 4-76
traveling waves, 4-41
u, wave velocity, 4-15
uncertainty principle, 4-43
wave equation, 4-65
wave mechanics, 4-42
wave velocity, 4-15
wavelength, 4-14

Chapter 4 Problems

Section 4.1. Periodicity and the Periodic Table

4.1. The modern Periodic Table is based on the structure of atoms. This was not the evidence that Mendeleev used when he proposed the periodic arrangement of elements in 1869.

(a) Why didn't Mendeleev base his table on structure of atoms?

(b) What was the basis for Mendeleev's periodic table?

(c) Why were there missing elements in Mendeleev's periodic table?

4.2. Consider these data for the atomic radii of the elements in the first three periods.

Atomic Radii (pm)

Element	Radius	Element	Radius	Element	Radius
H	37	N	75	Al	118
He	32	O	73	Si	111
Li	134	F	71	P	105
Be	90	Ne	69	S	102
B	82	Na	154	Cl	99
C	77	Mg	130	Ar	97

(a) Use these data to find a periodic pattern in the atomic radii of the first thee periods of main group elements.

(b) Display the data graphically.

(c) Which display, the tabulated values or your graph, best facilitates understanding of the relationships among the data? Explain your reasoning.

(d) What do you predict that the atomic radius will be for potassium? for calcium? What is the basis for your prediction?

(e) Use a print or web-based reference to check the predictions made in part (d). Were your predictions accurate? Why or why not?

4.3. Figure 4.4 shows the values for the *first* ionization energies for elements with atomic numbers 1 through 20.

(a) How do you predict the value for the *second* ionization energy for helium (loss of an electron from $He^+(g)$) would compare with its first ionization energy? Write equations representing the first and second ionizations to help explain your reasoning. *Hint:* Use equation (4.1) as a model.

(b) How do you predict that the values of the *first*, *second*, and *third* ionization energies for beryllium would compare? Write equations representing the first, second and third ionizations to help explain your reasoning.

4.4. Figure 4.4 shows the trend in first ionization energies for the first 20 elements. Here is an alternate representation of first ionization energy data.

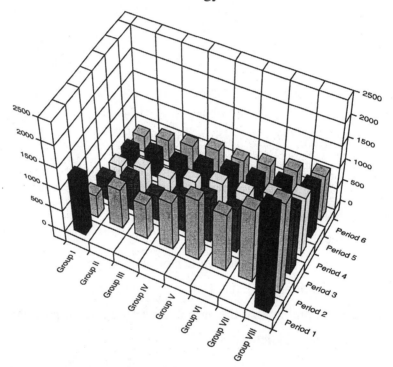

Compare and discuss the two different representations, paying particular attention to:
(i) how well the patterns across periods are represented.
(ii) how well the patterns within a family (group) are represented.
(iii) how easy it is to discern a shell structure for atoms.

4.5. Values for the second ionization energies of the first twenty elements are repeated here from Table 4.2. These are the energies for reactions such as $Na^+(g) \rightarrow Na^{2+}(g) + e^-(g)$.

Second Ionization Energy (IE_2), kJ·mol^{-1}

Element	IE_2	Element	IE_2	Element	IE_2
H		O	3390	P	1904
He	5248	F	3370	S	2250
Li	7300	Ne	3950	Cl	2297
Be	1757	Na	4560	Ar	2666
B	2430	Mg	1450	K	3052
C	2350	Al	1820	Ca	1145
N	2860	Si	1580		

(a) Why is no value given for the second ionization energy of the hydrogen atom?

(b) Plot these second ionization energies as a function of atomic number, as was done for first ionization energies in Figure 4.4. How does your plot compare to Figure 4.4? How do you explain the similarities? the differences?

4.6. WEB Chap 4. Click on the circled "P" in the left-hand menu to access the interactive periodic table. On the tabs at the top of the window, click on "Plot Data."

(a) On the plot of molar mass (relative atomic mass) as a function of atomic number, can you find pairs of relative atomic masses that are "out of order," that is, decrease as the atomic number increases? Check these out on the periodic table on the inside front cover of the text to confirm that the relative atomic masses of these elemental pairs are out of order.

(b) Could the elemental pairs you found in part (a) have caused problems for Mendeleev when he constructed his periodic table? Explain the problems he might have encountered. Is there any evidence in his table (shown in the chapter opening illustration) to suggest that he had such problems?

Section 4.2. Atomic Emission and Absorption Spectra

4.7. Why is it necessary to heat the sodium vapor sample represented in Figure 4.6(b)? *Hint:* Use a reference handbook to find the properties of sodium.

4.8. Explain why atoms at the surface of the sun emit light, whereas atoms in a layer further away absorb light. Relate your explanation to the experiments represented in Figure 4.6.

Structure of Atoms   Chapter 4

Section 4.3. Light as a Wave

4.9. Sketch a wave and label or define:

(a) wavelength

(b) amplitude

(c) node

(d) cycle

4.10. (a) At its closest approach to Earth, Mars is 56 million km from Earth. How long does it take a radio message from a space probe on Mars to reach the Earth when the two planets are this distance apart?

(b) Does your answer in part (a) suggest problems controllers on Earth might have in maneuvering a remote-controlled vehicle, Figure 1.39, on the Martian surface? Explain.

4.11. Calculate the wavelength and identify the type of radiation that has a frequency of:

(a) 101 MHz

(b) 25×10^{15} Hz

(c) 200 GHz [1 gigahertz (GHz) = 10^9 Hz]

4.12. Calculate the frequency and identify the type of radiation that has a wavelength of:

(a) 400 nm

(b) 25 km

(c) 4.5×10^{-4} m

4.13. A wave has a wavelength of 0.34 m and a frequency of 0.75 s^{-1}.

(a) What is the speed of the wave?

(b) Is the wave electromagnetic radiation? Explain.

4.14. What is the wavelength of microwave radiation with a frequency of 1.145×10^{10} s^{-1}? *Note:* Microwave radiation has frequencies in the range 10^9 to 10^{12} s^{-1}.

4.15. Ultraviolet-visible (UV-Vis) spectroscopy is typically used to determine concentrations of protein samples. If a protein absorbs light at 280 nm, what frequency does this correspond to?

4.16. Radio and television antennas are designed so that the length of a crossbar is approximately equal to the wavelength of the signal received. If you know that an antenna receives a signal with frequency of 3×10^2 MHz, calculate the length of a crossbar in the antenna.

4.17. WEB Chap 4, Sect 4.2.4.

(a) How is the correct set of oscillating arrows related to the representation of the electromagnetic wave? Be as specific as possible in your description.

(b) How is the correct set of oscillating arrows related to the description of electromagnetic radiation given in this chapter? Explain clearly.

Section 4.4.   Light as a Particle: The Photoelectric Effect

4.18. Explain how the predictions of the wave theory of light differed from the observed distribution of wavelengths of emission from glowing bodies.

4.19. (a) Explain how the photoelectric effect was evidence for the particulate nature of light.

(b) Explain why the wave theory failed to explain the photoelectric effect.

4.20. Explain the role Einstein's explanation of the photoelectric effect played in the development of our understanding of the dual nature of light.

4.21. Classical physics assumed that atoms and molecules could emit or absorb any arbitrary amount of radiant energy. However, Planck stated that atoms and molecules could emit or absorb energy only in quanta. Give a few examples from everyday life that illustrate the concept of quantization.

4.22. What is the energy of a photon with a frequency of 1.255×10^6 s^{-1}?

4.23. The energy of one photon associated with a typical microwave oven is approximately 1.64×10^{-24} J.

(a) What is the frequency of the radiation associated with one photon of this microwave radiation?

(b) How does this frequency compare with the frequency of green light with a wavelength of 515 nm?

(c) Which has more energy — a photon of green light or a photon of microwave energy?

4.24. A 500-MHz nuclear magnetic resonance (NMR) spectrometer is routinely used to help determine the structures of biomolecules.

(a) Calculate the wavelength (in meters) of this instrument.

(b) Using Figure 4.13, determine the region of the electromagnetic spectrum in which this instrument operates.

(c) To how much energy, in kJ·mol^{-1}, does 500 MHz correspond?

4.25. Calculate the energy of one photon and one mole of photons with:
  (a) a frequency of 101 MHz.
  (b) a wavelength of 400 nm.

4.26. A student calculated that the smallest increment of energy (quantum of energy) that can be emitted from a yellow light with wavelength of 589 nm is 4.2×10^{-19} J. Verify the student's answer.

4.27. In Investigate This 4.29, you explored the effect of light on silver chloride. How can this same effect be used to produce the lenses for eyeglasses that change with the light level? *Hint*: Think about reaction (4.7) in Consider This 4.30. You may want to research your prediction in print or web-based references to confirm or refute your suggestion.

4.28. Based on the results of Investigate This 4.29, why does a photographic darkroom use a red safelight?

4.29. In a *Science* article, scientists reported variations in the concentrations of thorium and potassium on the moon's surface. They used gamma ray spectrometry that recorded the gamma ray emission from potassium at approximately 1.4 MeV and from thorium at approximately 2.6 MeV. Calculate the frequencies of these two gamma ray emissions. Calculate the wavelengths of these two gamma ray emissions. *Note:* 1 MeV = 1×10^6 eV (electron volt) and 1 eV = 1.602×10^{-19} J.

Section 4.5. The Quantum Model of Atoms

4.30. (a) Explain what happens to electrons when an atomic emission spectrum is produced.
  (b) Explain what happens to electrons when an atomic absorption spectrum is produced.
  (c) How are these two types of spectra related?

4.31. Choose the best phrase to complete this sentence.
The intensity of a spectral line in an atomic emission spectrum can be directly related to:
  (i) the number of energy levels involved in the transition that gives rise to the line.
  (ii) the speed with which an electron undergoes a transition from one energy level to another.
  (iii) the number of electrons undergoing the transition that gives rise to the line.

4.32. Choose the best phrase to complete this sentence.

When electrons are excited from a ground state to an excited state,

(i) light is emitted.

(ii) heat is released.

(iii) energy is absorbed.

(iv) an emission spectrum results.

4.33. Neon lights glow because an electrical discharge is passed through a low pressure of neon in the glass tubing. In an experiment, emission from a neon light was dispersed by a prism to form a spectrum. The spectrum that was formed was not continuous but consisted of several sharp lines. Explain why a line spectrum was produced.

4.34. When energy is added to an electron it may become excited and move farther from the nucleus. That energy is released when the electron returns to its original energy. What happens when you add so much energy that the electron completely leaves the atom?

4.35. To distinguish between a solution of sodium ions and a solution of potassium ions, a simple flame emission test can be used. A drop of the solution to be tested is held in burner flame and the color of the flame is observed. (See Problem 3.4 at the end of Chapter 3.) A solution of sodium ions gives a bright yellow flame, while a solution of potassium ions gives a violet flame.

(a) To what approximate wavelengths in the electromagnetic spectrum do these emissions correspond?

(b) How, if at all, are these data related to Figure 4.6? Explain.

(c) Which ion's emission has the higher energy? Explain your answer.

4.36. The emission spectrum of mercury contains six wavelengths in the visible range: 405, 408, 436, 546, 577, and 579 nm.

(a) To what color does each of these wavelengths correspond?

(b) Which wavelength corresponds to the largest difference in energy between the atomic states responsible for the emission?

(c) Which wavelength corresponds to the smallest difference?

4.37. How do the line spectra of hydrogen, helium, mercury, and neon support the idea that the energy of electrons is quantized?

4.38. For the following electron transitions (between the energy levels on this energy level diagram) say whether the transition would result in absorption or emission of a photon, and rank the relative wavelengths of the photons absorbed or emitted from shortest to longest.

(a) $E_1 \rightarrow E_2$
(b) $E_2 \rightarrow E_1$
(c) $E_4 \rightarrow E_3$
(d) $E_4 \rightarrow E_5$
(e) $E_6 \rightarrow E_2$
(f) $E_6 \rightarrow E_1$

Section 4.6. If a Wave Can Be a Particle, Can a Particle Be a Wave?

4.39. Since de Broglie's hypothesis applies to all matter, does any object of a given mass and velocity give rise to a wave? Is it possible to measure the wavelengths of any moving object?

4.40. Calculate the wavelength of a baseball of mass 0.5 kg traveling at 30 m·s^{-1}. Why does the wavelength of macroscopic objects not affect the behavior of the object?

4.41. Calculate the wavelength of an electron of mass 9.1×10^{-31} kg traveling at 1.5×10^6 m·s^{-1}. Is this wavelength significant relative to the size of an atom?

4.42. Explain why the distinction between a wave and a particle is meaningful in the macroscopic world and why the distinction becomes blurred at the atomic level. *Hint:* Consider the wavelengths of macroscopic and atomic level particles.

4.43. (a) What is the de Broglie wavelength of an alpha particle that is traveling at a velocity of 1.5×10^7 m·s^{-1}? The mass of an alpha particle is 6.6×10^{-24} g.
(b) In what region of the electromagnetic spectrum does the wavelength of this alpha particle fall?

Chapter 4

Structure of Atoms

Section 4.7. The Wave Model of Electrons in Atoms

4.44. The sounds you hear when musical instruments are played are created by standing waves in the instruments. The air molecules in contact with the instrument are set in motion by the standing waves and form traveling waves. The traveling waves reach us, set our eardrums vibrating, and our brains interpret the motion as music. We have seen that standing waves on a guitar string are responsible for the sound produced by the guitar. What standing waves are responsible for the sound produced by these other musical instruments? Explain your answers.

(a) drums

(b) flutes

(c) tuning forks

4.45. What contributions did these scientists make to our understanding of atomic and electronic structure?

(a) Bohr

(b) Heisenberg

(c) Einstein

(d) de Broglie

4.46. What does the Heisenberg uncertainty principle state about what we can and cannot know about an electron's behavior?

4.47. Why did the de Broglie wave model and Heisenberg's uncertainty principle give rise to a new approach in which it was not appropriate to imagine electrons moving in well-defined orbits about the nucleus?

4.48. Why is it that we can calculate exactly the position and the momentum of a rolling ball but we cannot calculate the same for an electron that exhibits wave properties?

Section 4.8. Energies of Electrons in Atoms: Why Atoms Don't Collapse

4.49. Discuss the factors that affect the size of a hydrogen atom with regard to the kinetic and potential energy of the electron wave and the nucleus.

4.50. If particle **A** has a mass of 1.56×10^{-25} kg and particle **B** has a mass of 4.25×10^{-24} kg, which particle has the greater kinetic energy if they are both traveling at a velocity of 3.15×10^5 m s^{-1}? Explain your reasoning.

4.51. How is the potential energy of an electron wave related to its distance from the nucleus?

Structure of Atoms   Chapter 4

4.52. In Figure 4.27, why is the potential energy negative and the kinetic energy positive?

4.53. WEB Chap 4, Sect 4.7.1. What is the significance of the red circle on this interactive page? Explain how it is related to the energies that are represented on the graph.

Section 4.9. Multi-electron Atoms: Electron Spin

4.54. Explain which property of an electron is responsible for its interaction with a magnetic field.

4.55. Examine Figure 4.28. Explain why the magnet does not seem to affect helium atoms but does affect lithium atoms.

4.56. What experiment might you perform to tell whether atoms of an element have odd numbers of electrons?

4.57. Experiments on nitrogen atoms show that the ground state atom has three electrons with unpaired spins.
(a) How many of the nitrogen atom's electrons are spin paired? Explain how you get your answer.
(b) What does the electron spin data tell you about the orbital descriptions of the three electrons with unpaired spins? Explain.

Section 4.10. Periodicity and Electron Shells

4.58. Define the Pauli exclusion principle and how it can be used to explain the very low ionization energy of the lithium atom.

4.59. Why is the second ionization energy always greater than the first ionization energy for an atom?

4.60. (a) What is the same for each of these ions and atom: Na^+, Ne, F^-, and O^{2-}?
(b) Which of the species in part (a) has the highest ionization energy? Explain your answer.

4.61. Discuss the following properties and their trends in relationship to the periodic table.
(a) ionization energy
(b) atomic radius
(c) electronegativity

4.62. What is the trend in ionization energy as you go from left to right in a row on the periodic table? How is the shell model for atoms related to this trend?

4.63. What is the trend in ionization energy as you go from top to bottom in a group on the periodic table? How is the shell model for atoms related to this trend?

4.64. How many valence electrons does each of these elements have?

(a) oxygen

(b) sodium

(c) chlorine

(d) argon

4.65. The empirical formulas of the simplest hydrides (binary compounds with hydrogen) of the second period elements are given in this table:

element	Li	Be	B	C	N	O	F	Ne
hydride	LiH	BeH_2	BH_3	CH_4	NH_3	OH_2	FH	none

(a) Describe any pattern you see in these data.

(b) Does the hydride of hydrogen, H_2, fit the pattern(s) you described in part (a).

(c) What do you predict for the empirical formulas of the hydrides of aluminum, silicon, and phosphorus? Explain the basis of your prediction and, if possible, relate it to periodicity and the shell model of the atom.

4.66. WEB Chap 4. Click on the circled "P" in the left-hand menu to access the interactive periodic table. On the tabs at the top of the window, click on "Plot Data."

(a) Click on the "Properties" button and then on "1st Ionization Energy" to select this property to be plotted. Compare the plot with Figures 4.4 and 4.30 to be sure you identify the same trends in all of them. What are some of these?

(b) Click on the "Properties" button and then on "2nd Ionization Energy" to select this property to be plotted. Compare the plot with the one you made in Problem 4.5. Are the conclusions you drew from the first 20 elements in Problem 4.5 valid for the rest of the elements as well? Explain why or why not.

(c) Select the "3rd Ionization Energy" to be plotted. How does this plot compare to those for the first and second ionization energies? How do you explain the similarities? the differences?

4.67. WEB Chap 4. Click on the circled "P" in the left-hand menu to access the interactive periodic table. On the tabs at the top of the window, click on "Plot Data." Write a problem about periodicity that is based on analysis of one or more of the data plots accessible from this window. Have a classmate solve your problem.

4.68. One kind of periodic table is a three-dimensional model of eight stacked, round wooden discs with the symbols for the elements on the discs. Shown here are the symbols on the second tier of discs. How does this periodic table represent the electron shell model? Do you see any problems with this representation compared to Figure 4.31? Explain your responses.

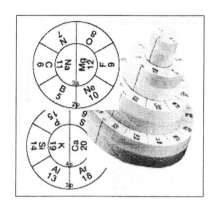

4.69. Two views of a three-dimensional version of the periodic table are shown here (and with other views in the chapter opening illustration).

(a) What parts of this table correspond to a conventional table, as on the inside front cover of this book? Explain the correspondences.

(b) What parts of this table correspond to the spiral table, Figure 4.34? Explain the correspondences.

Section 4.11. Wave Equations and Atomic Orbitals

4.70. Answer each of the following with TRUE or FALSE. If a statement is false correct it.

(a) The Schrödinger wave equation relates the wave properties of the electron to its kinetic energy.

(b) The wave function (ψ) is a mathematical description of an allowed energy state (orbital) for an electron.

(c) The wave function (ψ) is proportional to the probability of finding the electron in a given location near the nucleus.

(d) Orbitals with the same energy are called degenerate.

(e) Orbitals can be thought of as a boxes into which electrons can be placed.

(f) Each calculated energy level corresponds to a spatial probability distribution for the electron.

4.71. Verify the ground-state electron configurations for the elements listed below. For the configurations which are incorrect, explain what mistakes have been made and write the correct electron configurations.

(a) Al $1s^2\ 2s^2\ 2p^4 3s^2 3p^3$

(b) P $1s^2 2s^2 2p^6 3s^2 3p^2$

(c) B $1s^2 2s^0 2p^3$

4.72. How many valence electrons are located in an s orbital in each of the following elements?

(a) potassium

(b) fluorine

(c) magnesium

(d) boron

4.73. How many valence electrons are located in a p orbital in each of the following elements?

(a) phosphorus

(b) aluminum

(c) lithium

(d) bromine

(e) sulfur

4.74. Atoms **A**, **B**, and **C** have the electron configurations given below. Which atom has the *largest third ionization energy*? Explain your answer.

A = $1s^2 2s^2 2p^6 3s^2$ **B** = $1s^2 2s^2 2p^6 3s^2 3p^4$ **C** = $1s^2 2s^2 2p^6 3s^2 3p^6$

4.75. Using Figure 4.37 as a template and Worked Example 4.60 as a guide, draw electron energy diagrams for the sulfur atom and the sulfide ion.

4.76. Using Figure 4.37 as a template and Worked Example 4.60 as a guide, draw electron energy diagrams for Li and Na atoms.

4.77. (a) What does the series of atoms from H through Ar have in common with the series of monocations He^+ through K^+ and the series of dications Li^{2+} through Ca^{2+}? Explain.

(b) How does your answer in part (a) relate to your results and answers to Problems 4.5 and 4.66? Explain.

4.78. WEB Chap 4. Click on the circled "P" in the left-hand menu to access the interactive periodic table. On the tabs at the top of the window, click on "Electron Configuration."

(a) How does the graphic in the electron configuration window compare to Figure 4.37? What does the vertical direction on the graphic represent? Explain.

(b) Move the cursor over the blank periodic table until the pointer is on lithium. Click to select lithium and describe what happens. How does the result compare to the graphics in Worked Example 4.60? Explain.

(c) Use the electron configuration window to compare with your results from Consider This 4.62(a). How do they compare?

4.79. WEB Chap 4. Click on the circled "P" in the left-hand menu to access the interactive periodic table. On the tabs at the top of the window, click on "Electron Configuration." The energy levels represented in the electron configuration window (as rows of boxes) are enough to describe the ground state electron configurations of all the known elements.

(a) Examine the electron configurations of the first series of transition metals, Sc through Zn. What conclusion(s) can you draw about the relative energies of $4s$ and $3d$ electrons? Explain how you reach your conclusion(s).

(b) How are the electron configurations and the spins of the electrons for the first series of transition metals related to the rule you found in Consider This 4.62? Explain your reasoning and, if necessary, suggest how to modify your rule to make it more general, now that you have more data.

(c) Do elements in the second transition metal series, Y through Cd, follow the same electron configuration and electron spin pattern as the first series? If so, is this what you expected? Explain. If not, explain what factor(s) might cause differences.

Section 4.13. EXTENSION — Energies of a Spherical Electron Wave

4.80. (a) At what value of R does the total energy curve in Figure 4.27 reach a minimum?

(b) At the minimum in the total energy, what is the relationship of the kinetic energy to the potential energy of the proton-electron wave system? Show your work. *Hint:* Use the value of R you got in part (a) in equations (4.29) and (4.30). Assume that the proportionality is the same in both cases and let the proportionality constant be H.

(c) Express the value of the total energy at its minimum for the proton-electron wave system as a function of H from part (b). Use the ionization energy for the hydrogen atom to determine the value of H in kJ·mol^{-1}.

4.81. (a) For one-electron ions such as He$^+$ and Li^{2+}, you can use the same analysis of the energies as we used for the hydrogen atom, with the exception that you must account for the higher nuclear charge. The potential energy, equation (4.16), depends on the nuclear charge. Write equation (4.16) for a one-electron ion which has a nuclear charge +Z.

(b) What is the potential energy of a spherical electron wave attracted to the nucleus in part (a)? Write this potential energy as a function of R, the radius of the wave, H, the proportionality constant from Problem 4.80(b), and Z.

(c) What is the total energy of a one-electron ion with a nuclear charge +Z? Write the energy as a function of R, H, and Z. *Hint:* The kinetic energy of the electron wave does not depend on the nuclear charge.

(d) Find the value of R that makes the total energy in part (c) a minimum. *Hint:* There are at least three ways to do this problem. (i) If you are familiar with calculus, you can take the derivative of E with respect to R, dE/dR, set the derivative equal to zero, and solve for R. (ii) Use a graphing calculator or computer graphing program to graph the total energy function and find the minimum by tracing the curve. (iii) Assume (correctly) that the relationship between the kinetic and potential energies at the energy minimum, which you found in Problem 4.80, is the same for all one-electron atoms and ions. Use this relationship with R as an unknown and solve for R.

(e) Using your result from part (b), write the minimum for the total energy of a one-electron ion as a function of H and Z. Use your equation to predict the ionization energies for He$^+$, Li^{2+}, and B^{4+}. How do your predictions compare with the experimental values in Table 4.2? Can you predict other values in the table? Explain.

General Problems

4.82. This image of a white blood cell was made by a scanning electron microscope (SEM). Consult print or web-based resources to find out how an electron microscope differs from the traditional light-based microscope you have probably used in a biology class.

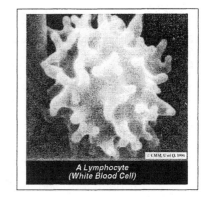

A Lymphocyte
(White Blood Cell)

Structure of Atoms   Chapter 4

4.83. Einstein's relativity theory shows that the effective mass of a moving particle, m_{moving}, is related to its rest mass, m_{rest}, its velocity, u, and the speed of light, c, by this equation:

$$m_{moving} = \frac{m_{rest}}{\sqrt{1-(u/c)^2}}$$

(a) What is the effective mass of a 145 g baseball thrown at 44 m·s^{-1}? Will a batter have to be concerned about relativistic effects in the pitched ball? Explain.

(b) What is the effective mass of the moving alpha particle in Problem 4.43(a)? How does this result affect your answer to Problem 4.43(b)? Explain. *Hint:* Use the effective mass in the de Broglie equation.

(c) What are the effective mass and the wavelength of an electron moving at 50% the speed of light.

4.84. Particle accelerators can accelerate protons, electrons, and other charged particles to extremely high energies. These high energy particles are sometimes fused with lighter elements, producing synthetic heavier elements.

(a) Calculate the wavelength (in meters) of a proton that has been accelerated to 50% the speed of light. *Hint:* See Problem 4.83.

(b) A mole of electrons is accelerated to 90% the speed of light. What kinetic energy (in kJ·mol^{-1}) do these electrons have? What is the kinetic energy of a single accelerated electron? *Hint:* See Problem 4.83.

4.85. WEB Chap 4. Click on the circled "P" in the left-hand menu to access the interactive periodic table. On the tabs at the top of the window, click on "Electron Configuration." Examine the representation of the periodic table at the bottom of the electron configuration window.

(a) Explain the relationship of this periodic table to the one on the inside front cover of this book.

(b) Explain the relationship of this periodic table to the spiral table shown in Figure 4.34. How are corresponding parts of the atomic shell structure represented in the two tables?

(c) Explain the relationship of this periodic table to the three-dimensional table shown in Problem 4.69 and the chapter opening illustration. How are corresponding parts of the atomic shell structure represented in the two tables?

Chapter 5. Structure of Molecules

Section 5.1. Isomers 5-6

Section 5.2. Lewis Structures and Molecular Models of Isomers 5-9

    Atomic connections in isomers 5-10

Reflection and projection   *5-15*

Section 5.3. Sigma Molecular Orbitals 5-15

    Molecular orbitals 5-15

    Localized, one-electron σ (sigma) molecular orbitals 5-18

    Sigma bonding orbitals 5-18

    Sigma nonbonding molecular orbitals 5-20

Section 5.4. Sigma Molecular Orbitals and Molecular Geometry 5-21

    The sigma molecular framework 5-21

    Molecular shapes 5-22

    Bond angles 5-23

    Molecules containing third period elements 5-24

    Why are there only four sigma orbitals on second period atoms? 5-26

Reflection and projection   *5-27*

Section 5.5. Multiple Bonds 5-28

    Double bonds and molecular properties 5-29

    Triple bonds 5-31

    Multiple bonds in higher period atoms 5-33

Section 5.6. Pi Molecular Orbitals 5-33

    Pi orbital standing wave 5-34

    Sigma-pi molecular geometry with one pi orbital 5-34

    Triple bonds: sigma-pi geometry with two pi orbitals 5-37

Reflection and projection   *5-38*

Section 5.7. Delocalized Orbitals 5-39

    Delocalized π molecular orbitals 5-41

    Bond order 5-42

    Orbital energies 5-42

    Metallic properties 5-44

    Metallic bonding: Delocalized molecular orbitals 5-46

Structure of Molecules — Chapter 5

Section 5.8. Representations of Molecular Geometry 5-47
- Tetrahedral representation 5-48
- Trigonal planar representation 5-49
- Lewis structures and molecular shape 5-50
- Condensed structure for 1-butanol 5-51
- Skeletal structure for 1-butanol 5-53

Reflection and projection *5-55*

Section 5.9. Stereoisomerism 5-56
- Cis- and trans-isomers 5-57
- Stereoisomers 5-58
- Polarized light and isomerism 5-58
- Tetrahedral arrangement around carbon 5-62

Reflection and projection *5-64*

Section 5.10. Functional Groups — Making Life Interesting 5-65
- Alkanes 5-65
- Functional groups 5-66
- Alkenes 5-66
- Functional groups containing oxygen and/or nitrogen 5-66
- Alcohols and ethers 5-68
- Carbonyl compounds 5-69
- Carboxyl compounds 5-71
- Multiple functional groups 5-72

Section 5.11. Molecular Recognition 5-72
- Noncovalent interactions 5-73
- Molecular recognition 5-73

Reflection and projection *5-74*

Section 5.12. Outcomes Review 5-74

Section 5.13. EXTENSION — Antibonding Orbitals: The Oxygen Story 5-76
- Paramagnetism of oxygen 5-77
- Antibonding pi molecular orbitals 5-77
- Molecular orbital model for oxygen 5-78
- Ground-state oxygen reacts slowly 5-79

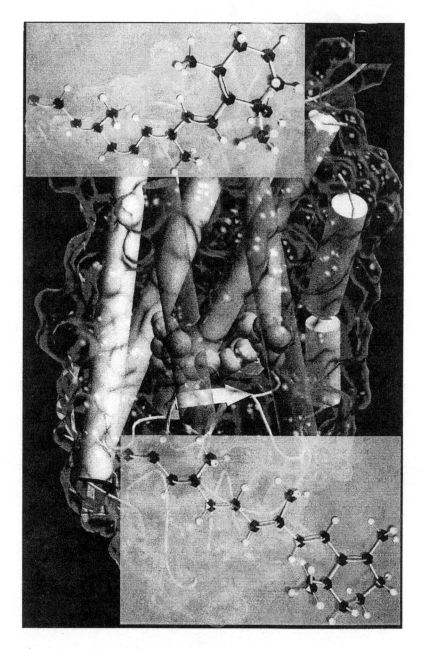

The central representation here is rhodopsin, the molecule required for vision. The gray cylinders represent α-helices that are part of the structure of the protein opsin, a chain of about 350 amino acids. The green structure nested among the helices is a molecule of retinal which is bonded at its oxygen end to one of the helices to form rhodopsin. The shape of the retinal molecule is also represented by the ball-and-stick model at the upper left. When the retinal absorbs a quantum of visible light its shape changes to that shown on the lower right.

Chapter 5. Structure of Molecules

> To every Form of being is assigned ...
> An active Principle.
>
> *William Wordsworth (1770-1850)*, The Excursion

When you read, you probably do not think much about the chemistry taking place in your eyes. You can read these words because the light that reaches your retina causes molecules of retinal to change shape, as shown by the ball-and-stick models on the facing page. This change in shape of a relatively small molecule forces the helices of the rhodopsin molecule to move, thus changing the shape of the whole protein molecule. A very large number of these rhodopsin molecules are present in membranes in your retinas. The changes triggered by the change in shape of the retinal are responsible for the signal that is sent to your brain when light strikes your retina. The chemistry of vision is just one example of what is captured poetically in Wordsworth's words: the functions and reactivity of molecules depend on their form, their structure and bonding.

In order to understand the kinds of electron and atomic core rearrangements that have to occur to change the shape of the retinal molecule, you have to know what holds the molecule together in the first place. The goal of this chapter is to help you develop an understanding of the bonding that holds atoms together to form molecules, the structures (shapes) of molecules, and the basis for the reactions between molecules.

At the heart of molecular structure is the role of electrons in holding molecules together. ***The fundamental principle of molecular structure and bonding is that the negative electrons and positive nuclei attract one another.*** In Chapter 4, you saw that electrons in atoms behave more like vibrating guitar strings than they do like marbles or baseballs. In this chapter, we will build on those wave mechanical models for atoms to understand how valence electrons bind atomic cores together in molecules. You have already seen examples of the results of wave-mechanical calculations in computer-generated representations such as that of the water molecule we introduced in Chapter 1. Some of these examples described the consequences of unequal charge distribution in chemical bonds, the polarity of molecules. In this chapter, we will continue to emphasize the polarity of molecules, since polarity is the basis for understanding the reactions of molecules that we will examine in Chapter 6.

Molecular wave mechanics can help us understand molecular structure and reactivity. To set the stage for this exploration, we'll first revisit some properties of compounds we have seen previously and introduce some new compounds, many of which are isomers. And we will extend the writing of Lewis structures to more complicated molecules that we will come to understand using the wave mechanical model of bonding.

Section 5.1. Isomers

5.1. Investigate This *How do different alcohols react in a Breathalyzer® test?*

Do this as a class investigation and work in small groups to discuss and analyze the results. Add about 1.0 mL of acetone to each of six small test tubes and then add *1 drop* of the Breathalyzer® reagent, a solution of the orange dichromate ion, $Cr_2O_7^{2-}$, in sulfuric acid, to each of the test tubes. CAUTION: Sulfuric acid solutions are very corrosive to many materials, including skin. Dichromate is a suspect carcinogen. Be careful not to spill any reagent and wear protective plastic gloves when handling the solution. **Leave one of the filled test tubes as a control** and add *1 drop* or a *tiny* crystal of the following five compounds to separate test tubes containing the reagent: ethanol, *n*-butyl alcohol, *sec*-butyl alcohol, *iso*-butyl alcohol, and *tert*-butyl alcohol. Observe the solutions and record your observations.

5.2. Consider This *Are all alcohols detected by the Breathalyzer® test?*

(a) In the Breathalyzer® test for blood alcohol level, ethanol in the breath of the person being tested reacts with dichromate ion. The orange dichromate, $Cr_2O_7^{2-}$, is changed to the green chromic ion, Cr^{3+}, as shown in the photos. Do you find evidence for this reaction in Investigate This 5.1? Explain.

(b) Do any other alcohols react with the Breathalyzer® reagent? What is your evidence? Which alcohols?

(c) Do any alcohols *not* react with the Breathalyzer® reagent? What is your evidence? Which alcohols?

(d) Can you suggest an explanation for differences, if any, among the alcohols you tested?

In Chapter 1 you found that the boiling points of compounds varied with the polarity of their molecules. We attributed the higher boiling points of more polar compounds to the stronger polar attractions, especially hydrogen bonds, among the molecules in their liquid phase, which meant that more energy was required for the molecules to move into the gas phase. For nonpolar compounds, boiling points generally increase with molar mass, a rough indicator of the number of electrons in the molecules. The more electrons that are available, the greater the dispersion forces between molecules and the more energy required to get them into the gas phase. The boiling points

> The prefixes for the butyl alcohols are short for:
> n = normal   sec = secondary
> iso = isomeric   tert = tertiary
> We have used these names to draw attention to the fact that these compounds are isomers, but will rarely use the names again.
> The structural names we will usually use are based on the names of the simplest hydrocarbons:
> methane CH_4
> ethane CH_3CH_3
> propane $CH_3CH_2CH_3$
> butane $CH_3CH_2CH_2CH_3$
> pentane $CH_3CH_2CH_2CH_2CH_3$

and other properties of the alcohols you used in Investigate This 5.1 are given in Table 5.1. The table shows that all four of the butyl alcohols have the same molecular formula. Compounds with the same molecular formula but different properties are called **isomers**.

Table 5.1. Selected properties of a few alcohols.

Alcohol	Structural name	Molecular formula	Melting point, °C	Boiling point, °C	Solubility, g in 100 mL H_2O
ethyl	ethanol	C_2H_6O	–114.5	78.5	miscible
n-butyl	1-butanol	$C_4H_{10}O$	–89.8	118.0	9
sec-butyl	2-butanol	$C_4H_{10}O$	–114.7	99.5	12.5
iso-butyl	2-methyl-1-propanol	$C_4H_{10}O$	–108	108.1	10
tert-butyl	2-methyl-2-propanol	$C_4H_{10}O$	25.6	82.6	miscible

> **5.3. Consider This** *How do the properties of the butyl alcohol isomers differ?*
>
> **(a)** Are the properties of some of the isomers of butyl alcohol similar to one another? If so, which ones? Which one(s) are quite different? Explain the rationale for your choices.
>
> **(b)** How do the properties of the isomeric butyl alcohols compare to the properties of ethanol (ethyl alcohol)? Are any of the differences or similarities surprising? Explain why or why not.
>
> **(c)** Imagine adding a column to Table 5.1 that gives the result (positive or negative) of the Breathalyzer® test from Investigate This 5.1 for each alcohol. Would this property fit the patterns you have found in parts (a) and (b)? Explain why or why not.

You have found that *tert*-butyl alcohol (2-methyl-2-propanol) is strikingly different from the other $C_4H_{10}O$ isomeric alcohols. Its melting point is more than 100 °C higher than any of the others; it is a solid at room temperature, whereas the others are liquids. Its boiling point is lower than any of the others. In fact, its boiling point is almost the same as that of ethanol, even though it has a higher molar mass than ethanol. Unlike the other $C_4H_{10}O$ isomers and ethanol, *tert*-butyl alcohol does not react with the Breathalyzer® reagent.

We have seen similar great differences between isomers before. Look back at Table 1.3 in Chapter 1, Section 1.11, to review the properties of ethanol, CH_3CH_2OH, and dimethyl ether, CH_3OCH_3. Ethanol is a liquid at room temperature and dimethyl ether is a gas. Ethanol is miscible with water, but only about 7 g of dimethyl ether dissolves in 100 mL of water. These compounds have the same molecular formula, C_2H_6O, and different properties, so they are isomers.

Consider one more example of isomers, three compounds with the molecular formula, C_5H_{12}. Table 5.2 gives the melting and boiling points of these isomers.

Table 5.2. Melting and boiling points of the C_5H_{12} isomers.

Structural name	Melting point, °C	Boiling point, °C
pentane	–129.7	36.1
2-methylbutane	–159.9	27.9
2,2-dimethylpropane	–16.6	9.5

5.4. Check This *Comparison of the C_5H_{12} and $C_4H_{10}O$ isomers*

(a) Do you see any similar patterns among the C_5H_{12} and $C_4H_{10}O$ isomers? If so, what are they? Explain the rationale for your response.

(b) There is some similarity in the structural names of the two sets of isomers. Do these similarities reflect the patterns you found in part (a)? Explain why or why not.

(c) The molar masses of these two sets of isomers are almost the same. (Prove this for yourself.) What factor do you think is mainly responsible for the differences in boiling points between the two sets of isomers? Explain your reasoning.

Based on what you learned in previous chapters, you have probably concluded that the large differences in properties within a set of isomeric compounds, such as the butyl alcohols, must reflect differences in their molecular structures. Some of our tasks in the rest of the chapter will be to discover what these molecular structures are, to develop the bonding model that is responsible for them, and to try to relate observed properties to the structures. We will begin by reviewing and extending our simplest molecular bonding model, the Lewis structure, and showing how you can use your molecular model kits to translate them into three dimensions.

Section 5.2. Lewis Structures and Molecular Models of Isomers

We have seen that molecules can be thought of as collections of **atomic cores** that are held together by **valence electrons**. The valence electrons in molecules include *all* those contributed from the valence shells of the atoms that form the molecule. For example, a molecule of methane, CH_4, has eight valence electrons. The

> Recall that the term "atomic core" is not in general use. We are using it here to remind you that every atom except hydrogen has deeply buried electrons that do take part in chemical reactions. Later, we will use the conventional term "atom"; you will have to interpret from the context, if atomic core is what is really meant.

carbon atom contributes four electrons, and each of the four hydrogen atoms contributes one electron.

One of the earliest bonding models of the 20th century, which we have already been using, is the Lewis model. Lewis used the known formulas for a large number of compounds, the nuclear model of the atom, and the periodic table as the basis for his model. As you saw, Lewis structures

> **Personal Tutor**
> We introduced Lewis structures in Chapter 1, Section 1.4, and asked you to write several of them in Chapters 1 and 2. If necessary, you should review these and the Personal Tutor before studying this section.

for molecules containing the second-period atoms, C, N, O, and F, have four pairs of valence electrons (an octet of electrons) around these atomic cores. The tetrahedral arrangement of bonds (sticks) around the atom centers in your molecular model kit is consistent with these Lewis structures. (In Section 5.9 we will discuss the logic used by van't Hoff and Le Bel, who proposed the tetrahedral arrangement around carbon more than forty years before Lewis developed his model.) For all of the Lewis structures you have written so far, you were given a structural formula or molecular model to show the connectivity of the atomic cores in the molecule.

5.5. **Check This** *Lewis structures of the C_2H_7N isomers*
(a) Write Lewis structures for ethyl amine, $CH_3CH_2NH_2$, and dimethyl amine, CH_3NHCH_3.
(b) Use your molecular model kit to build models of these two molecules. How would you describe the shape of each molecule? Are the shapes about the same or different. Explain.

Structure of Molecules Chapter 5

Atomic connections in isomers

Sometimes you will be faced with the problem of writing a Lewis structure (or structures) without knowing the connectivity of the atomic centers in the molecule(s). In these cases, you will have to choose the connectivity based on what you know about how atomic cores (centers) are connected in other molecules. We will use isomers as examples to show how to approach such problems.

5.6. Worked Example *Lewis structures and molecular models for C_3H_8O isomers*

Write Lewis structures and build the corresponding molecular models for as many C_3H_8O isomers as you can.

Necessary information: We need to know that C, H, and O atoms have, respectively, 4, 1, and 6 valence electrons to be used in the Lewis structures. We also need to recall that a carbon atomic center usually forms two-electron bonds to *four* other atomic centers and oxygen usually forms two-electron bonds to *two* other atomic centers.

Strategy: Calculate the number of valence electrons in the molecules. Connect the four second period atomic centers in as many ways as possible with two-electron bonds. Add all the hydrogens with two-electron bonds to give four atomic centers on each carbon and two on the oxygen. Use any remaining valence electrons to complete octets on each second-period element.

Implementation: The atoms, three C, eight H, and one O, have a total of 26 valence electrons.

One way to proceed systematically to find different connectivities is first to connect the carbons to one another in as many ways as possible and then connect the oxygen to each carbon in turn (In the following partial Lewis structures, we use lines to represent two-electron bonds.):

(a) C–C–C–O   (b) C–C(–O)–C   (c) O–C–C–C

Structures (a) and (c) are identical, since they can be rotated and superimposed on one another. Thus, we have only two different structures here, (a) and (b). Next, we need to consider connecting two carbons to the oxygen. There is one such structure:

(d) C–C–O–C

Add hydrogens and two-electron bonds to structures (a), (b), and (d):

```
    H H H                     H O-H                  H H   H
    | | |                     | | |                  | |   |
(a) H-C-C-C-O-H         (b) H-C-C-C-H           (d) H-C-C-O-C-H
    | | |                     | | |                  | |   |
    H H H                     H H H                  H H   H
```

Each structure has 22 electrons in two-electron bonds. The carbons have octets of electrons, but the oxygens are missing four electrons and we have four valence electrons still to distribute:

(a′)
```
    H H H
    | | | ..
H−C−C−C−O−H
    | | | ..
    H H H
```

(b′)
```
       :Ö−H
    H  |  H
    |  |  |
H−C−C−C−H
    |  |  |
    H  H  H
```

(d′)
```
    H H    H
    | |  .. |
H−C−C−O−C−H
    | |  .. |
    H H    H
```

Molecular models of these three isomeric compounds are shown in Figure 5.1.

Does the answer make sense? All the Lewis structures have 26 electrons, the usual number of atomic centers bonded to carbon and oxygen, and an octet of electrons around each second-period atomic center. Since the C_2H_6O isomers are an alcohol and an ether, we would expect to find alcohol and ether molecules when a carbon and two hydrogens (CH_2) are added to the C_2H_6O structure and this is what we find. There are two isomeric alcohols because there are two different places to bond the oxygen to the carbon chain.

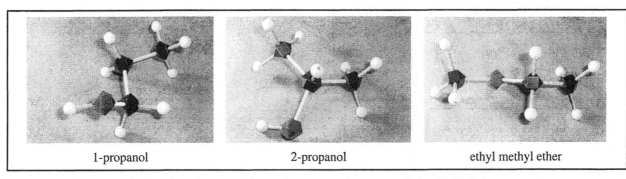

1-propanol   2-propanol   ethyl methyl ether

Figure 5.1. Molecular models and structural names of the C_3H_8O isomers.

5.7. **Check This** *Lewis structures and molecular models for C_3H_9N isomers*
Write Lewis structures and build the corresponding molecular models for as many C_3H_9N isomers as you can. Explain the procedure you use to find all the different connectivities for the Lewis structures. *Hint:* There are four isomers, three of which are analogous to the structures we found in Worked Example 5.6.

Worked Example 5.6 and Check This 5.7 demonstrate the importance of using systematic procedures for finding all the possible connectivities among the second (and higher) period elements when writing Lewis structures. In the carbon-containing molecules we have been using as examples, the number of possibilities grows rapidly as the number of carbons increases, because the carbons do not have to be connected one after another, but can branch. One system for finding all the ways the carbons in a set of isomers can be connected is illustrated in Worked Example 5.8.

5.8. Worked Example *Lewis structures and molecular models for C_5H_{12} isomers*

Write Lewis structures and build the corresponding molecular models for as many C_5H_{12} isomers as you can.

Necessary information: We need to know that C and H atoms have, respectively, 4 and 1 valence electrons to be used in the Lewis structures and that a carbon atomic center forms two-electron bonds to four other atomic centers.

Strategy: The approach to this problem is the same as in Worked Example 5.6.

Implementation: The five C and twelve H atoms have a total of 32 valence electrons.

We begin by connecting the carbons to form the longest possible chain of carbon atoms:

(a) C–C–C–C–C

Then we remove one of the carbons and add it to interior positions on the chain:

```
            C                      C
            |                      |
(b) C–C–C–C         (c) C–C–C–C
```

The longest carbon chain is highlighted in red. Structures (b) and (c) are identical; they can be interconverted by simply flipping one of them over. Thus, (b) and (c) represent a single structure: a four-carbon chain with another carbon bonded to one of the two identical interior positions.

Next, we remove two carbons from the longest chain and then add them first together and then separately to the interior positions on the chain:

```
      C                             C
      |                             |
      C                     C–C–C
      |                             |
(d) C–C–C              (e)     C
```

Structure (d) is identical to structures (b) and (c): a four-carbon chain (shown in red) with another carbon bonded to one of the two identical interior positions. Structure (e) is different from any of the others. Its longest carbon chain, three carbons, is shown in red.

> The longest continuous chain of carbons can be "bent" as shown in structure (d). There may be more than one equivalent longest chain as in:
> ```
>       C              C              C
>       |              |              |
>   C–C–C          C–C–C          C–C–C
>       |              |              |
>       C              C              C
> ```
> All three of these chains (and others you can find) are equivalent. Use models to prove this.

No further steps are possible to form different connections of the carbon atoms. Each structure uses all the carbons and eight of the valence electrons. Use the remaining 24 valence electrons to add the 12 hydrogen atomic cores with two-electron bonds:

(a′)
H H H H H
H–C–C–C–C–C–H
H H H H H

(b′)
```
      H
   H  |  H
    \ C /
H   H | H   H
 \  | | |  /
H–C–C–C–C–H
 /  | | |  \
H   H H H   H
```

(e′)
```
    H H H
    H\|/H
     \C/
  H   |   H
H–C–C–C–C–H
  H   |   H
     /C\
    H/|\H
    H H H
```

Molecular models of these three isomeric compounds are shown in Figure 5.2.

Does the answer make sense? All atom centers and valence electrons are accounted for in each Lewis structure. See Check This 5.9 for further consideration of these isomers.

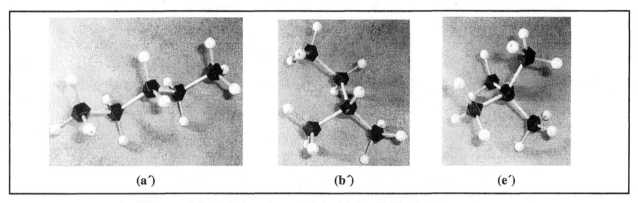

Figure 5.2. Molecular models of the C_5H_{12} isomers.
Letter designations for the models are from Worked Example 5.8.

5.9. Check This *More on the C_5H_{12} isomers*

(a) Make three molecular models of structure (a), the five-carbon chain, in Worked Example 5.8. Detach a carbon atomic center and one bond stick from one end of one of the models and reattach it to make structure (b). Repeat with a second model to make structure (c). Detach two carbon atomic centers and bonds sticks from the remaining model of structure (a) and reattach them to make structure (d). Show that the three structures you have made are identical to one another, that is, that they can be superimposed on one another. This will prove that structures (b), (c), and (d) all represent the same isomer.

(b) The three Lewis structures written at the end of Worked Example 5.8 and the models in Figure 5.2 represent molecules of the compounds in Table 5.2. Which structure corresponds to which compound? Are the names a clue? Explain.

Structure of Molecules Chapter 5

> **5.10. Consider This** *What are the Lewis structures and molecular models for the $C_4H_{10}O$ isomeric alcohols?*
>
> **(a)** Write the Lewis structures and make molecular models for the isomeric alcohols with the molecular formula, $C_4H_{10}O$.
>
> **(b)** Assign the correct structural name in Table 5.1 to each of the molecular models you made in part (a). Explain how you make each assignment.
>
> **(c)** Is one of the structures rather different from the others? If so, how is it different? Is the difference reflected in the properties of the compound? Explain.
>
> **(d)** The molecular structure of vitamin A (retinol) is shown below. Your body converts it into the retinal molecule shown in the chapter opening illustration. Describe the similarity between the two molecules. Explain how you could use your results for the $C_4H_{10}O$ isomers to predict what would be observed if vitamin A is mixed with an acidic solution of dichromate ion, $Cr_2O_7^{2-}$, as in Investigate This 5.1.
>
> retinol [structural formula of retinol]

In both sets of isomers we looked at, the $C_4H_{10}O$ and C_5H_{12} compounds, one of the isomers was quite different from the others. For example, one isomer in each set melts at a much higher temperature and boils at a lower temperature than the others. And, in the case of the alcohols, this isomer did not undergo the same reaction as the others. The molecular models you made for the isomeric C_4H_8O alcohols and those shown in Figure 5.2 for the C_5H_{12} isomers show that one of the isomers in each set is quite compact, more-or-less spherical, compared to the others. On the basis of their structural names, you probably assigned these compact structures to the "odd" isomer in each set, 2-methyl-2-propanol and 2,2-dimethyl propane.

> **5.11. Check This** *Boiling points of the C_4H_8O alcohols*
>
> Recall from Chapter 1, that alcohol molecules form hydrogen bonds to one another which must be broken for them to vaporize. Also, we recall that there are attractive dispersion forces between molecules that depend on less directed contacts with one another. The boiling point data in Table 5.1 show that 2-methyl-2-propanol vaporizes more easily than the other isomers. Do the structures of the molecules suggest an explanation for the lower boiling point of this isomer? Or, conversely, for the higher boiling points of the others? Why or why not?

Chapter 5 — Structure of Molecules

Reflection and projection

You have found that isomeric compounds have the same molecular formula but different chemical and physical properties. You have also found that there are several different ways to connect atomic cores as you write Lewis structures or make the molecular models for the molecules of isomeric compounds. The different molecular structures can give isomers very different observable properties and you can often use the molecular structures to rationalize these differences.

So far, our arguments about structure have all rested on the molecular models we build from our model kit. How do we know these structures are valid? What is the basis for these models? In Chapter 1, Section 1.5, we suggested that four pairs of valence electrons stack in a tetrahedral array, like four balls, around a central positive atomic center, getting the negative electrons as close as possible to the positive nuclear charge, while remaining equidistant from one another. This arrangement is the basis for the tetrahedral array of holes drilled in the atomic centers that represent the second period elements in your model kit. Why should a model based on stacking of balls around a central point be applicable to bonding in molecules? Applying the ideas from wave mechanics developed in the last chapter can help us answer this question. But keep in mind that the Lewis model, with no reference to electron waves, is a simple way to describe connectivity that you should continue to use.

Section 5.3.   Sigma Molecular Orbitals

Molecular orbitals

Descriptions of electron waves in molecules are governed by the same wave mechanical principles we discussed for atoms in Chapter 4. But molecular electron waves are necessarily more complicated. An atom, for example, has only *one* anchor point, its nucleus, for the electron waves. On the other hand, methane, a very simple molecule, has *five* nuclear centers to attract electrons (and to use as anchor points for standing waves). The standing electron waves that define regions of space where there is a high probability of finding electrons in molecules are referred to as **molecular orbitals**, paralleling the way atomic orbitals describe the location of an atom's electrons.

The increased complexity of molecules forces scientists to use approximations to calculate and describe simple pictures of molecular orbitals. For example, in the calculations, an electron wave can be limited to interaction with a pair of adjacent atomic cores, instead of with all the atomic cores in the molecule. Figure 5.3 compares the electrostatic interactions in the simplest atom, an H atom, and the simplest molecule (molecular ion), an H_2^+ ion. For an atom with one electron and one proton (atomic core), Figure 5.3(a) shows the single electrostatic interaction, *PE*, the potential energy of attraction between the proton and electron. Figure 5.3(b) shows the interactions among an electron and two protons (atomic cores). There are two attractive potential energies, PE_A and PE_B, one for each atomic core with the electron. *The negative potential energies of attraction of the two atomic cores and the electron are what hold the atomic cores together in the molecule.* This is the same electrostatic attractive potential energy that is responsible for the stability of the atom, but now there are two attractions, instead of one.

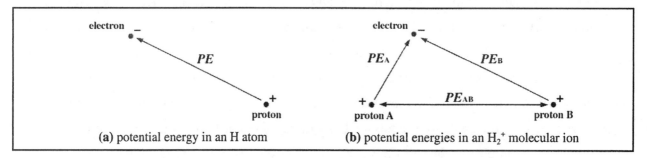

Figure 5.3. Potential energies among atomic cores and an electron.
Attractions between the positive cores and the electron are represented by red single-headed arrows. The repulsion between the positive cores in (b) is represented by the double-headed blue arrow.

The total energy of the atom or the molecule is sum of these negative potential energies of attraction plus other positive contributions. There is a positive (destabilizing) contribution from the kinetic energy, *KE*, of the electron orbital; *KE* gets larger as the orbital gets smaller. The total energy for the atom is:

$$E \text{ (H atom)} = PE + KE \tag{5.1}$$

The negative (stabilizing) contributions to the total energy are shaded in yellow and the positive (destabilizing) contributions are shaded in blue. In the molecule, there is also a positive contribution to the total energy from the potential energy of repulsion, PE_{AB}, between the positively-charged atomic cores. The total energy for the molecule is:

WEB Chap 5, Sect 5.3
Study animations representing the energy changes involved in molecular orbital formation.

$$E \text{ (}H_2^+ \text{ molecular ion)} = PE_A + PE_B + PE_{AB} + KE \tag{5.2}$$

5.12. Consider This   *How do the PEs and KE vary with the proton-proton distance?*
   WEB   Chap 5, Sect 5.3.1. Drag proton B towards proton A, as directed. Explain the affect this change in proton-proton distance has on:
   (a) the potential energy of the interaction between proton B and the electron.
   (b) the potential energy of the interaction between proton A and the electron.

(a) the potential energy of the interaction between proton B and proton A.

(a) the kinetic energy of the electron cloud.

The potential, kinetic, and total energies for an atomic system were plotted in Chapter 4, Figure 4.27, as functions of the size of the atom. The corresponding plots for a molecule are shown in Figure 5.4 as functions of the distance between the atomic cores, the **bond length**. The shorter the bond length, the greater the attractive potential energies, PE_A and PE_B. The nearer the electron is to the atomic cores, the more favorable (more negative) are the attractive potential energies. But, also, the shorter the bond length, the higher the atomic core repulsion, PE_{AB}. Figure 5.4 shows that the attractive potential predominates for this orbital. A shorter bond length increases the kinetic energy, KE, because the electron is being constrained in a smaller volume between the atomic cores. The minimum in the total energy, E, corresponds to bond formation at a bond length that gives the lowest total energy.

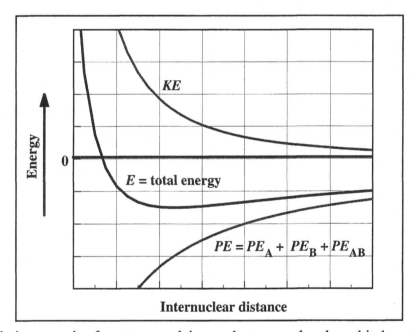

Figure 5.4. Relative energies for a two-nuclei-one-electron molecular orbital system.
As the nuclei get closer, the electron orbital is constrained to a smaller volume and its kinetic energy increases. Only relative energies and distances are shown; the plots are similar for all bonding molecular orbitals.

5.13. **Check This** *Potential and kinetic energy combinations*

Choose three internuclear distances on the plots in Figure 5.4 and check whether equation (5.2) is obeyed in each case. What is the result?

Structure of Molecules Chapter 5

Localized, one-electron σ (sigma) molecular orbitals

Solutions to the wave equation that account for the interactions illustrated in Figure 5.3(b) and by equation (5.2), are a set of **localized, one-electron molecular orbitals**. The lowest energy molecular orbital for the H_2^+ molecular ion is represented in Figure 5.5, together with the lowest energy orbital of the H atom for comparison. Recall from Chapter 4 that atomic *s* orbitals are spherically symmetric about the nucleus, as shown for the 1*s* orbital in Figure 5.5(a). The lowest energy molecular orbital shown in Figure 5.5(b) is cylindrically symmetric about the **bond axis**, the imaginary line joining the nuclei. That is, if you rotate the molecule about the bond axis, the molecular orbital shown in the figure looks the same, no matter how far the molecule is rotated. This cylindrical symmetry about the bond axis is analogous to the spherical symmetry of atomic *s* orbitals. Thus, σ, Greek lower case sigma, corresponding to the Roman "*s*", has been chosen to name these **σ (sigma) molecular orbitals**.

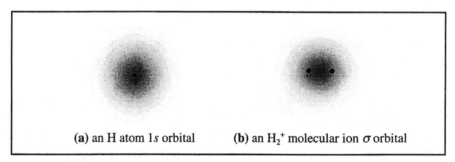

(a) an H atom 1*s* orbital   (b) an H_2^+ molecular ion σ orbital

Figure 5.5. Representation of a 1*s* atomic orbital and a σ molecular orbital.
The size of the protons, dark dots, is greatly exaggerated.

5.14. **Consider This**   *What are the relative sizes of atomic and molecular orbitals?*

(a) In Figure 5.5, the lowest energy σ molecular orbital for H_2^+ is shown as somewhat smaller than the H atom 1*s* orbital. Which orbital electron has the higher kinetic energy? Give your reasoning.

(b) How is it possible for the molecular orbital to be smaller than the atomic orbital? Clearly explain how the interactions among the protons and electron can lead to this result.

Sigma bonding orbitals

Bonding orbitals concentrate electron density *between* atomic cores. They are called bonding orbitals because these electrons *always* contribute to lowering the energy of the molecule, thus making it more stable. Bonding molecular orbitals have lower energies than atomic orbitals in the separated atoms. The lower energy is a result of attractions between the electron and *two* atomic cores *which makes the molecule more stable than its atoms*. Figure 5.6 shows this energy relationship between atoms and molecules. In σ **(sigma) bonding orbitals**, the electron density is concentrated directly between the atomic cores, as you see in Figure 5.5(b). If you imagine looking from one core (proton) toward the other, the electron orbital would block your view. This is the most favorable arrangement for coulombic attraction between the electron orbital and the nuclei. Therefore, *σ bonding orbitals form the strongest bonds and all molecules have σ bonding orbitals*.

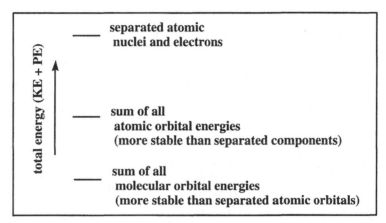

Figure 5.6. Relative total energies for a molecule, its atoms, and their components.

In Figure 5.5, note that, like the proton in the H atom, the two protons in the H_2^+ molecular ion are "inside" the σ electron orbital. Though much of the electron density is between the protons, serving to bond them together, some of it surrounds the protons as well. The same is true for the second-period elements: the major portion of the sigma bonding orbital electron density is between the atomic cores. In Figure 5.7, to emphasize this point, very little of the electron density is shown "outside" the second period atomic cores. The figure shows pictorial representations of sigma bonding molecular orbitals for second period elements bonded to hydrogen (a proton), Figure 5.7(a), or another second period element, Figure 5.7(b).

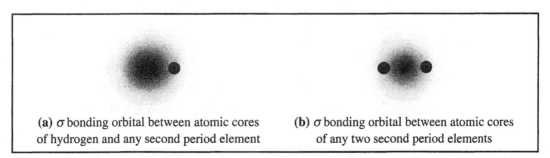

Figure 5.7. Sigma bonding molecular orbitals involving second period elements.
Second period atomic cores are represented by a nucleus surrounded by its dense core electron orbital.

Each of the molecular orbitals pictured in Figures 5.5 and 5.7 can accommodate two electrons with opposite spins. These are the electron-pair bonds that we have been showing as a line between the elemental symbols in Lewis structures. Now we see that the line represents a roughly spherical region of electron density in the space between the atomic cores and that the attractions of the cores for the electrons in this space are what bonds the cores together. If there are two or more σ bonding orbitals attracted to a single atomic core, the exclusion principle acts, in effect, to keep the orbitals from occupying the same space. This is the basis for the simple model of connected balloons or stacked balls that we used in Chapter 1, Section 1.5, to explain molecular geometry. Using the electron wave model, we interpret the balls as sigma orbitals, Figure 5.8, and provide a justification for what was, previously, just a physical analogy.

> Recall the warning from Chapter 4 about the way chemists refer to electrons and orbitals. Electrons are said to "occupy" orbitals or to be "in" orbitals. We will use this shorthand at times. But keep in mind that *an orbital is a description of an electron wave and electron probability density*; it has no existence apart from the electron.

Figure 5.8. Tetrahedral arrangement of four sigma orbitals around an atomic core.

Sigma nonbonding molecular orbitals

Sigma bonding orbitals explain the lines we use for bonds in Lewis structures, but they do not explain the electrons that we show as pairs of dots associated with only one elemental symbol, for example, oxygen and nitrogen in Section 5.2. These electrons are in orbitals that are much like atomic orbitals, since they are mainly attracted by a single atomic core. Since these electrons are not between atomic cores they do not contribute to holding atoms together, but they don't weaken the bonds either, so they are called **nonbonding electrons**. However, these electrons are in molecules and are, in principle, affected by all the atomic cores, so their orbitals are called **sigma nonbonding molecular orbitals**, and are symbolized as σ_n. In Figure 5.9, representations of a σ and a σ_n orbital are shown for comparison on a second-period atomic core. In the next section we will begin to see how sigma orbitals determine the shape of molecules.

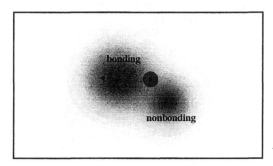

Figure 5.9. Sigma bonding and nonbonding molecular orbitals on the same atomic core.

5.15. **Consider This**   *What is the symmetry of an σ_n orbital?*

Sigma molecular orbitals have cylindrical symmetry. How would you define the axis of symmetry for a σ_n orbital? Explain your reasoning.

Section 5.4.   Sigma Molecular Orbitals and Molecular Geometry

5.16. **Investigate This**   *What are the geometries of second-period hydrides?*

(a) Write Lewis structures for the hydrides of carbon (CH_4), nitrogen (NH_3), and oxygen (H_2O).

(b) Make molecular models of the hydrides in part (a). Use the paddles in your model kit to represent any nonbonding electrons in the Lewis structures. How would you describe the geometry (shape) of the electron pairs around the second-period atom center in each of your models? Explain your responses.

(c) How would you describe the geometry of the atom centers with respect to one another in the model of each molecule? Explain your responses.

(d) For a given hydride, are your answers to parts (b) and (c) the same? Explain why or why not.

The sigma molecular framework

The distribution of electrons in molecules is governed by the same interactions we discussed for atoms. Electrons will occupy as many as possible of the lowest energy molecular orbitals; these provide the strongest attractions between the atomic cores and electrons that hold the molecule together. This means that as many σ bonding orbitals as possible are formed, each containing two spin-paired electrons. As a consequence, in the vast majority of molecules, *there is a spin-paired, two-electron, σ bonding orbital between every pair of bonded atomic cores.* This is usually referred to as a molecule's **σ framework** or **σ bonding framework**. Although other orbitals are often present as well, the σ framework is responsible for a large fraction of the stability of a molecule, which is represented by the molecular energy level in Figure 5.6. Our task now is to learn how these orbitals are related to the geometry of the molecule.

There are several possible combinations of σ and σ_n molecular orbitals around the second-period elemental atoms in molecules that interest us, but they all have one characteristic: *the number of bonding orbitals plus the number of nonbonding orbitals always adds up to four for each second-period element (carbon through fluorine) in the molecule.* Each of these is a two-electron orbital. Once again, you can see the connection to Lewis structures in which you write structures with four electron pairs around each second-period element. The tetrahedral geometry shown in Figure 5.8 is best for getting four electron orbitals grouped as close as possible to the positive nucleus.

Molecular shapes

The shape or geometry of a molecule is defined by the spatial relationship of the atomic nuclei to one another. The distinction between *molecular shape* and *arrangement of orbitals* is important. The shape of a molecule is often different from the geometry of the orbitals. The examples in Figure 5.10 are a reminder of this important point. Here we have three different molecular shapes, tetrahedral, **trigonal pyramidal** (a flattened triangular pyramid), and bent, all arising from the same tetrahedral arrangement of orbitals. Study the figure carefully to learn the relationships that lead to these results.

> WEB Chap 5, Sect 5.4
> Study animations of different molecular shapes arising from the same orbital arrangement.

5.17. Check This *A fourth combination of sigma orbitals*

(a) A fourth possible combination of sigma molecular orbitals is missing from Figure 5.10. An example of a molecule with this combination of orbitals is hydrogen fluoride, HF. Write the Lewis structure for HF. What is the combination of σ and σ_n orbitals for the fluorine? What is the shape of the molecule? Draw orbital structures for HF like those in Figure 5.10. Could you have predicted the shape without any reference to orbitals? Give your reasoning.

(b) WEB Chap 5, Sect 5.4.2. Look at the movies. Explain clearly why HF, H_2O, and NH_3 have different molecular shapes, even though they all have four sigma molecular orbitals.

Chapter 5   Structure of Molecules

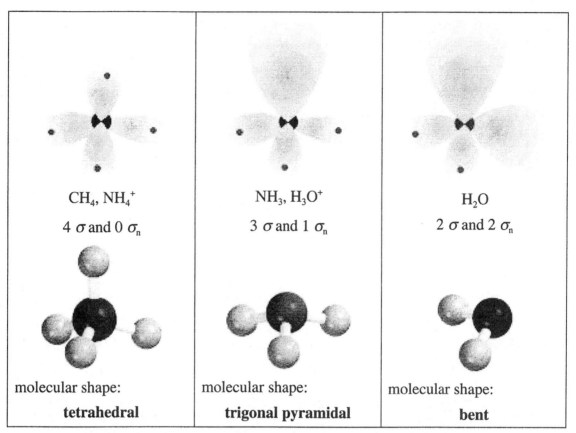

Figure 5.10. Molecular shapes of molecules with four sigma molecular orbitals.
The top representation in each panel shows the orbitals and nuclei in perspective. The dots in the orbitals are the H nuclei. The bottom representation uses ball-and-stick models to show the molecular shape. The combination of σ and σ_n orbitals for each structure is given in the middle of each panel.

Bond angles

The molecular shapes in Figure 5.10 can also be characterized by their **bond angles**, the angles between imaginary lines connecting the nuclei. You can think of the sticks connecting the atomic centers in a ball-and-stick molecular model as representing these imaginary lines. If the orbitals retained their tetrahedral geometry in all three molecules, you would expect the H–C–H, H–N–H, and H–O–H bond angles all to be 109.5°. Experimentally, the bond angles in CH_4, NH_3, and H_2O are found to be 109.5°, 107°, and 104.5°, respectively. The agreement of the experiment with the expectation is quite good. The decreasing angle, from CH_4 to H_2O, can be rationalized as an effect of increasing number of σ_n nonbonding orbitals. Close to the second-period element, the σ_n nonbonding orbitals take up larger volumes of space than σ bonding orbitals. Bonding electrons are attracted by two positive centers and nonbonding electrons by only one. The attraction of the bonding electrons to two centers results in a more elongated orbital less concentrated at the second-period element. Thus, the regular tetrahedral arrangement of equal-sized σ bonding orbitals, will be distorted when both σ bonding and σ_n nonbonding orbitals are

Structure of Molecules   Chapter 5

present. The effect of the distortion is to make the angle(s) between atoms bonded to the central atom more acute, that is, smaller than 109.5°, as observed.

Molecules containing third period elements

Recall from Figure 3.27 in Chapter 3, that the most abundant elements in living organisms are hydrogen, oxygen, carbon, and nitrogen. The geometry of many molecules in organisms is based on the tetrahedral arrangement of sigma molecular orbitals around second period elements that we have just discussed. Figure 3.27 also shows, however, that several other elements, including third period elements, are present in living systems. And, if we consider the vast amount of nonliving matter on Earth, silicon, a third period element, is the second most abundant element, after oxygen, in the Earth's crust. Silicon is an important part of the structure of many minerals and manufactured products, including all the glass we see around us. Compounds containing, phosphorus, sulfur, and chlorine are also found in minerals, in the sea, and in all organisms. Here we will examine a few simple compounds of phosphorus and sulfur to see how they are the same and/or different from the compounds of elements in the second period.

5.18. Worked Example *The structures of PF_3 and PF_5*

Write Lewis structures and construct molecular models for a molecule of PF_3 and of PF_5.

Necessary information: We need the number of valence electrons in P and F, 5 and 7, respectively. We also need to know that third (and higher) period elements are not limited to an octet of electrons around the atomic core.

Strategy: Our approach is the same as in Worked Examples 5.6 and 5.8, except that we are not limited to only four pairs of electrons around the phosphorus atomic core.

Implementation: PF_3 has 26 valence electrons and PF_5 has 40 valence electrons.

First, we connect the F atomic cores to the P atomic core with two-electron bonds:

$$\begin{array}{c} F-P-F \\ | \\ F \end{array} \qquad \begin{array}{c} F \quad F \\ \diagdown \diagup \\ F - P - F \\ | \\ F \end{array}$$

Next, we give each of the second period elements, the Fs, an octet of electrons:

The PF_5 structure has a total of 40 electrons in σ and σ_n orbitals, so the structure is complete as shown. The PF_3 structure has a total of only 24 electrons in σ and σ_n orbitals; the remaining two electrons can complete an octet in a σ_n orbital on the P:

:F—P—F:
  |
 :F:

Molecular models of the two structures are shown in Figure 5.11. The PF$_3$ structure is analogous to the NH$_3$ structure shown in Figure 5.10; the molecular geometry is trigonal pyramidal. The PF$_5$ structure is **trigonal bipyramidal.** Imaginary lines connecting the Fs form two three-sided pyramids joined at their bases.

Does the answer make sense? All valence electrons are accounted for in both structures. In the case of PF$_5$, this gives phosphorus five electron-pair σ bonds. The next paragraph and Figure 5.12 examine further the geometry of these five σ bonds.

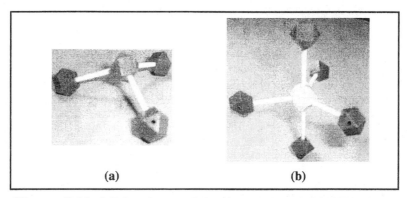

Figure 5.11. Molecular models of (a) PF$_3$ and (b) PF$_5$.

5.19. Check This *The structure of SF$_6$*

Sulfur hexafluoride, SF$_6$, is quite an unreactive gas. Write a Lewis structure for SF$_6$ and construct a molecular model that distributes the Fs as symmetrically as possible around the S.

Stacking four balls around a central point in an arrangement that gets them all as close as possible to the point, results in the tetrahedral array shown in Figure 5.8, and an explanation for the geometry around atom cores surrounded by a total of four σ plus σ_n orbitals. If we examine the arrangements of five and six balls around a central point, such that all the balls are as close as possible to the point, we find the arrays shown in Figure 5.12. For five balls, three are in a triangular array with the remaining two above and below the plane of the triangle and nestling in the center of the triangle. Imaginary lines joining the centers of the five balls form a trigonal bipyramid, shown by the red lines in Figure 5.12(a). This trigonal bipyramidal array represents the five σ bonding orbitals in PF$_5$, Figure 5.11(b). The distances from the central point to the centers of each ball in the trigonal bipyramid are not the same; the two balls above and below the triangular plane are further from the central point.

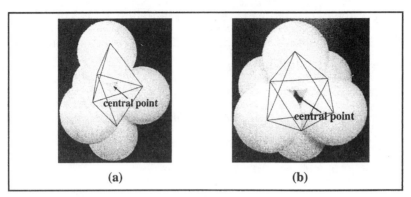

Figure 5.12. (a) Trigonal bipyramidal and (b) octahedral arrays of balls around a central point.
The red structures represent lines connecting the centers of the balls.

When we stack six balls, four are in a square arrangement with the remaining two above and below the plane of the square and nestling in the center of the square. This is an **octahedral** array. Imaginary lines joining the centers of the six balls form a regular octahedron. The distance from the central point to the center of each ball in an octahedral array is the same; all six positions (balls) around the central point are equivalent.

> **5.20. Consider This** *How are all six octahedral positions equivalent?*
>
> **(a)** Use six Styrofoam® balls of the same size to construct an octahedral array of balls. Use toothpicks to hold four of the balls together in a square and then attach the other two to opposite faces of the square. Explain to other members of your group how all six balls are in equivalent positions in this arrangement. *Hint:* Close your eyes. Have a group member rotate the model around one of its axes. Open our eyes. Can you whether or not the model has been rotated?
>
> **(b)** Is your octahedral array of balls the same as the arrangement of σ bonding orbitals in the SF_6 model you made for Check This 5.19? Should it be? Explain why or why not.

Why are there only four sigma orbitals on second period atoms?
More than four electron-pair sigma orbitals can be accommodated around third (and higher) period elements. Why are there no molecules with more than four sigma orbitals around second period elements? The answer involves the energies of the interactions among the atomic cores and the sigma orbital electrons. Second period atoms have tiny cores — the nucleus and only two core electrons. When more than four two-electron sigma orbitals are stacked around this tiny core, their size prevents them all from getting as close as the four orbitals in a tetrahedral array.

Thus, the potential energy of attraction between the core and each orbital is reduced. In order to get closer, the orbitals would have to become smaller, but this would increase their kinetic energy, which would further destabilize the molecule. No stable molecule is formed with more than four pairs of electrons around a second period element.

Third period atoms have a core that is a good deal larger — the nucleus and ten core electrons. Five or six two-electron sigma orbitals can get essentially as close to this larger core as can four orbitals, so there is little or no loss of attraction for the larger number of electrons. The electrons are, of course, all repelling one another, so there is a point beyond which no more can be packed around a core of a given size and charge.

5.21. **Check This** *Sigma orbitals around third and higher period atoms in molecules*
Fluorine forms compounds with almost every other element. The other halogens react with fluorine to form compounds containing various numbers of fluorine atoms. The most highly fluorinated compounds of each of the halogens are ClF_3, BrF_5, and IF_7. Write Lewis structures for these molecules. How many electron pairs surround each of the central atoms? Explain why Cl and Br do not react to form compounds with more F atoms.

Reflection and projection

A summary of the main points of Sections 5.3 and 5.4 can help put the ideas in perspective.

• *A molecule is held together by the attractions between atomic cores and valence electrons.* Although repulsion between nuclei and the kinetic energy of the electron waves have to be accounted for, the attractive contributions to the potential energy make the total energy of the molecule lower than the energy of its separated atoms.

• *All molecules have sigma bonding orbitals.* There is a two-electron, σ bonding orbital between every pair of bonded atomic cores in a molecule.

• *Many molecules also have two-electron, sigma nonbonding orbitals.* Nonbonding electrons do not contribute to holding the molecule together, but σ_n *orbitals, along with the σ orbitals, determine the three-dimensional geometry of molecules.*

• Although the arrangement of sigma orbitals determines the geometry or shape of a molecule, *the descriptions of molecular geometry are based on the positions of the nuclei with respect to one another, not the arrangements of electron orbitals.*

These ideas, especially those that connect the stability and geometry of molecules with the simplest molecular orbitals, the sigma orbitals, are powerful. They enable you to make predictions about geometries and to understand the basis for the structures you have made with your molecular models. However, just as *s* orbitals are not the only atomic orbitals, sigma orbitals, important as they are, are not the whole story of molecular bonding. It should not be surprising that other, more complicated bonding arrangements and electron waves are also found in molecules. These are the topics of the next sections.

Section 5.5. Multiple Bonds

5.22. Investigate This *Do carbon-hydrogen compounds react with permanganate?*

Do this as a class investigation and work in small groups to discuss and analyze the results. Add *one drop* of each of the following three liquids to 2 mL of 95% ethanol in separate small test tubes: hexane (C_6H_{14}), hexene (C_6H_{12}), and turpentine [mostly pinene ($C_{10}H_{16}$)]. To each test tube add *one drop* of 0.1 M potassium permanganate, $KMnO_4$, solution. Observe and record the appearance of the mixtures. Discuss your results to be sure your group agrees on the observations.

5.23. Consider This *What compounds react with permanganate?*

(a) The permanganate ion, MnO_4^-, is reddish purple; manganese dioxide, MnO_2, is a dark brown/black solid. Was there any evidence that permanganate reacted with the samples in Investigate This 5.22? If so, what was the evidence?

(b) Did all the samples react? If not, which ones reacted and which did not? Are there any similarities among those that reacted? Explain your response.

5.24. Worked Example *The Lewis structure and molecular model for ethene*

Write the Lewis structure and make a molecular model of ethene, C_2H_4 or CH_2CH_2.

Necessary information: We need the number of valence electrons in C and H, 4 and 1, respectively and the rule that second period elements have an octet, four pairs, of electrons in Lewis structures.

Strategy: The approach is the same as in Worked Examples 5.6 and 5.8.

Implementation: Ethene, C_2H_4, has 12 valence electrons.

First we will bond the carbons with a two-electron bond and then bond the four hydrogens, two to each carbon (as indicated by the structural formula in the problem statement) to get:

Chapter 5 — Structure of Molecules

```
H—C—C—H
  |   |
  H   H
```

Ten of the valence electrons are used to form this σ bonding framework. The question we face is how to give each of the Cs an octet of electrons, when there are only two electrons left and there are two Cs. The answer to this dilemma is to write the remaining two electrons between the two Cs, so that each has an octet of electrons (circled in red):

```
H—C=C—H
  |   |
  H   H
```

This is a correct Lewis structure. The octet rule does not forbid sharing of more than one pair of electrons between atomic cores.

Figure 5.13 shows how to use the bent bonds in your molecular model set to represent the structure of ethene.

Does the answer make sense? All the valence electrons are accounted for and all the second period elements have an octet of electrons. We need to investigate the bonding model further to find out if this Lewis structure and the molecular model make sense.

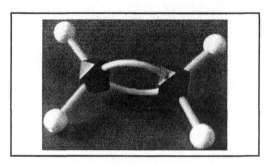

Figure 5.13. Molecular model of the double bond in ethene using "bent" bonds.

5.25. Consider This *What are the Lewis structure and molecular model for methanal?*

(a) Write the Lewis structure for methanal, CH_2CO. How is your Lewis structure different from that for ethene in Worked Example 5.24? How is it the same?

(b) Make a molecular model of methanal. If necessary, pattern your model after the one in Figure 5.13. Explain the reasoning you use to build your model.

Double bonds and molecular properties

Both the Lewis structure for ethene in Worked Example 5.24 and the one you wrote for methanal in Consider This 5.25 have two electron pairs, — two lines — between the second

Structure of Molecules   Chapter 5

period atomic cores. Because there are two pairs of bonding electrons between the atoms, this is called a **double bond**. For both compounds, you used the curved connectors in your molecular model set to represent the double bond as a pair of **bent bonds**, which are required to make two connections between tetrahedral atom centers. This is the simplest way to construct a double bond with your molecular models. However, before we use this model further we need to know if it represents the observed properties of double bonds.

5.26. Consider This   *What are the structures and reactivity of C_6H_{14} and C_6H_{12}?*

(a) Write Lewis structures for C_6H_{14} and C_6H_{12} molecules that have the carbons bonded in a six-carbon chain. Is there more than one way to write a Lewis structure for either of these molecules? If so, what are the possible structures? Do your structures have any common feature(s)? What is(are) it(they)?

(b) Make molecular models corresponding to your Lewis structures in part (a). What is(are) the difference(s) among your structures? Can you correlate the molecular difference(s) with the reactivity of the C_6H_{14} and C_6H_{12} compounds in Investigate This 5.22? Explain why or why not.

(c) From your results in the investigation, what might you conclude about the structure of pinene? Explain the reasoning for your response.

(d) The structure of vitamin A (retinol) is shown in Consider This 5.10. Predict what you would observe if vitamin A is mixed with permanganate ion solution, as in Investigate This 5.21. Explain the reasoning for your prediction.

In Investigate This 5.22, you found that compounds with double bonds react with the clear, purple permanganate solution to produce a muddy brownish mixture, whereas those without double bonds (including ethanol) did not react. The representation of the double bond in Figure 5.13 (and the ones you made in Consider This 5.25 and 5.26) suggests that the electron density in a double bond is not all concentrated directly between the atomic cores, but lies outside the bond axis. This implies that the electrons in the double bond are on the "outside" of the molecule and not as tightly held as sigma electrons, so they should be better able to interact with other molecules. We predict that compounds with double bonds will be more reactive than similar compounds without double bonds, as you observed in Investigate This 5.22.

5.27. Consider This   *What is the geometry of double bonded molecules?*

(a) Examine molecular models of ethene and methanal. How would you describe the geometry of the atomic centers with respect to one another? Explain your reasoning.

(b) In your models, what are the H-C-H bond angles?

Experiments show that the two carbons and four hydrogen atoms bonded to them in ethene all lie in the same plane, that is, the doubly-bonded atoms and the atoms bonded to them have **planar geometry**. You can see this planar geometry in Figure 5.13, it is the geometry you found in Consider This 5.27 for both ethene and methanal, and is represented again in Figure 5.14. The experimentally observed H-C-H bond angles in ethene and methanal are 117° and 116°, respectively. The models you examined in Consider This 5.27 were made with tetrahedral carbons, so they have H-C-H angles of 109.5°. The agreement between the observed angles and our model is not perfect, but the model is not far off.

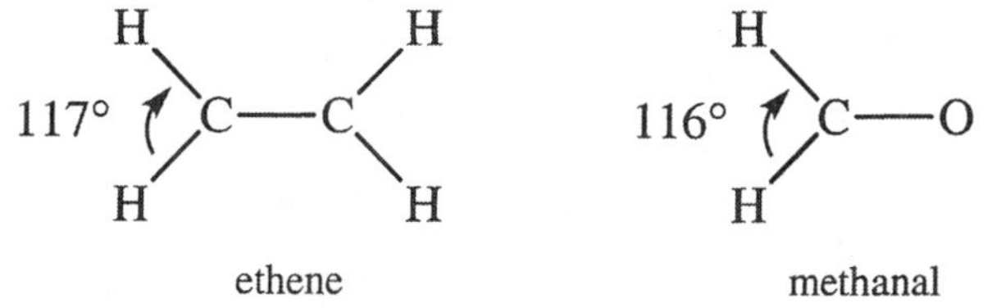

Figure 5.14. All the atom centers in ethene and methanal lie in a plane.
The plane of the paper is the molecular plane in these structures. Only the sigma bonds are represented.

5.28. **Check This**   *Other bond angles in ethene and methanal*
   (a) What is the experimental H-C-C bond angle in ethene? the experimental H-C-O bond angle in methanal? Explain how you arrive at your answers.
   (b) What is the H-C-C bond angle in your ethene model? the H-C-O bond angle in your methanal model? How do these angle compare with the experimental angles from part (a)? Explain how you arrive at your answers.

Triple bonds

You have seen that second-period atoms can share two pairs of electrons to form double bonds. Can they share a third pair as well?

5.29. **Worked Example**   *Lewis structure for the hydrogen cyanide molecule*
   Write the Lewis structure for the hydrogen cyanide molecule, HCN. Make a molecular model representing this structure and predict the geometry of the molecule.
   Necessary information: We need to know that H, C, and N have 1, 4, and 5 valence electrons, respectively, and that second period atoms have an octet of electrons.

Structure of Molecules — Chapter 5

Strategy: The approach is the same as in Worked Examples 5.6 and 5.8.

Implementation: The HCN molecule has 10 valence electrons.

First we connect the atomic cores with two-electron bonds: H–C–N

There are six electrons left to distribute to give each second-period atom an octet and six are needed by the N alone, which would leave the C with only four electrons. If the C and N share four of the six electrons (two more bonding pairs) and the leftover pair are assigned to the N, then both C and N will have an octet: H–C≡N:

Figure 5.15 shows how to represent the structure of HCN using the bent bonds in your molecular models. The model, with three bent bonds, forces the three atoms to lie on a line, so we predict that the molecule is linear.

Does the answer make sense? The second period atoms in our structure have octets of electrons with three pairs of bonding electrons shared between the C and N. The N also has one unshared (nonbonding) pair of electrons, which is typical for N in the molecules we have seen before. Experimentally, the HCN molecule is found to be linear; our shape prediction is correct.

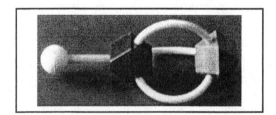

Figure 5.15. Molecular model of the triple bond in hydrogen cyanide using "bent" bonds.

5.30. Check This *Lewis structure for the ethyne (acetylene) molecule*

Write the Lewis structure for the ethyne (acetylene), HCCH, molecule. Make a molecular model representing this structure and predict the geometry of the molecule.

The Lewis structures and molecular models of hydrogen cyanide and ethyne have three bonding electron pairs between the second-period atoms. In Figure 5.15, we have interpreted these as three bent bonds, a **triple bond**. You can see that the electron density in the triple bond is not concentrated directly between the atomic cores, so it should be available to interact with other molecules, just as in double-bonded molecules. Experimentally, we find that compounds whose molecules contain triple bonds between carbons are even more reactive than corresponding doubly-bonded compounds.

Chapter 5 — Structure of Molecules

Multiple bonds in higher period atoms

Multiple bonds between third and higher period atoms are less common than between second period atoms. This is probably because the molecular orbitals in molecules with higher period atoms are larger and, in multiple bonds, more available to react than those in analogous second period molecules. Reactions to form products with more σ bonding orbitals are favored and these are the compounds usually observed.

Section 5.6. Pi Molecular Orbitals

The bent bonds shown in Figure 5.13 and 5.15 are consistent with the reactivity and geometry of double and triple bonds and are a convenience that enables us to use tetrahedral atom centers to build models of molecules with multiple bonds. However, when we consider the wave mechanical description of the double (or triple) bond we get a somewhat different picture. One way to visualize the wave mechanical formation of the second bond between two second period elemental atoms is to combine electron waves from the two atoms. Recall the wave interference (diffraction) patterns you saw in Chapter 4, Section 4.3, and the explanation in Figure 4.9. When two wave crests coincide, the waves reinforce one another. An application of this idea to electron waves is shown schematically in Figure 5.16 for two second period atomic cores.

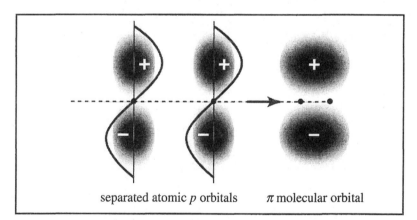

separated atomic *p* orbitals π molecular orbital

Figure 5.16. Electron wave interference (reinforcement) to form a pi bonding orbital.
Reinforcement of electron wave amplitude between the atomic cores (black dots) for waves with matching + and – amplitudes The dashed lines represent the nodal planes of the orbitals.

The second period atomic cores in Figure 5.16 are already bonded by a sigma orbital that is not shown in the figure. A second bonding orbital, a **π (pi) bonding molecular orbital**, is formed when the crests and valleys of electron waves in *p* orbitals on each atom reinforce one another between the atomic nuclei. (The Greek lower case pi, corresponds to the Roman "*p*.")

Note that the electron density in the π orbital is not concentrated directly between the nuclei, but is parallel to the bond in two sausage-shaped lobes that are somewhat outside the bond axis. The bond axis lies in the **nodal plane** of the π orbital, that is, the plane where the electron density of the orbital goes to zero.

Pi orbital standing wave

A possible confusion about a pi orbital arises because it seems to be in two separate regions of space. The two regions are parts of a whole orbital, that is, a single standing wave describing one or two electrons (depending on how many electrons are in the orbital). As an analogy, look at one of the standing waves on the acoustic body of a guitar, Figure 5.17, which repeats Figure 4.35(b). When one side of the body face is moving out toward you, the other side is moving away from you. The movements are connected. One side can't move out unless the other is moving in. This is a single wave, even though there is a node down the center where the face doesn't move at all. Similarly, the π-electron wave is a single wave. The signs shown in Figure 5.16 are analogous to the out and in movements of the guitar face.

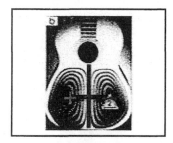

Figure 5.17. A standing wave on the acoustic body of a guitar.

Sigma-pi molecular geometry with one pi orbital

Look again at the first structure written in Worked Example 5.24: H–C–C–H with H H below. Each carbon is surrounded by three σ bonding electron pairs. Figure 5.18 shows how we would represent these three electron-pair orbitals as a trigonal planar arrangement of closely packed balls around a central point (an atomic core). Imaginary lines joining the centers of the three balls form an equilateral triangle. Figure 5.19(a) shows both of the carbons in ethene and the associated five σ orbitals with the carbon cores sharing one of them. This represents the sigma framework for a molecule of ethene.

> **WEB Chap 5, Sect 5.6**
> Try animations and exercises to help get a three-dimensional feel for sigma-pi geometries.

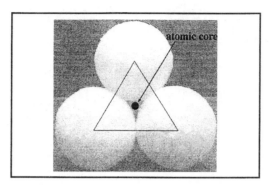

Figure 5.18. Trigonal planar arrangement of three sigma orbitals around an atomic core.

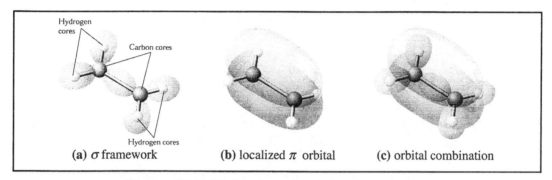

Figure 5.19. Molecular orbitals in ethene.
The six atomic cores define a molecular plane (see Figure 5.14). The σ framework is cut by this plane. The localized π orbital lies above and below the molecular plane and is cut by a plane perpendicular to the molecular plane.

The second bond between the carbons in ethene, the pi molecular orbital, is a little more difficult to represent. Figure 5.19 shows how the sigma bonding orbital framework and the pi bonding orbital combine to give the orbital arrangement for ethene: three σ orbitals in a trigonal plane surrounding each carbon and a π orbital in a plane perpendicular to the molecular plane and parallel to the C–C bond axis. The planar arrangement of the atoms results from the planar arrangement of the sigma orbitals. The reason for this planar arrangement of σ orbitals is that this geometry gives the strongest pi bonding, that is, the best reinforcement of the electron waves that form the pi molecular orbital and therefore the lowest energy for the molecule. The orbital represented in Figure 5.19(b) is called a **localized π molecular orbital**, because it is associated with (localized between) two sigma-bonded atoms.

5.31. **Check This** *Representations of the bonding in ethene*
WEB Chap 5, Sect 5.6.1. Load the movie and exercise "Molecular representations of Ethene." Place the cursor in the middle of the image in the movie. Move the cursor up and down. Describe what you see. What information can you gather from each of the different representations? Which representation do you find most useful? Explain why.

The structure of ethene represented in Figure 5.19(c) is consistent with the properties of compounds containing double bonds that we discussed above. The pi bonding electrons are not concentrated directly between the carbon atomic cores and are, therefore, more accessible for reaction, thus making these compounds more reactive, as you have found. For sigma orbitals in the trigonal planar arrangement shown in Figure 5.18, the predicted H-C-H bond angle is 120°, which is close to the experimental value of 117°.

5.32. Check This *The sigma-pi representation for methanal*

Explain how you would modify the illustrations in Figure 5.19 so that they represent the sigma and pi molecular orbitals in methanal. Make a sketch of your modifications.

Another way to model the ethene structure with your molecular models is shown in Figure 5.20. For this model, you use the dark gray, trigonal atom centers for the carbons, straight connectors for the five sigma bonding electron pairs, and four paddles, two on each carbon center, that altogether represent the pi bonding electron pair. Colored tape connects the paddles to highlight the two lobes of the pi orbital. This model is closer to the conventional representation of the sigma-pi bonding in ethene, as in Figure 5.19(c), than the model with bent bonds in Figure 5.13. However, in most cases, we will use bent bonds for models of double bonds, because they are easier to make and give much the same information about the molecular properties.

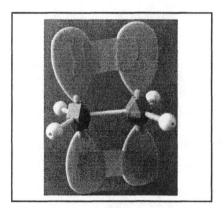

Figure 5.20. Molecular model of ethene showing the sigma framework and pi orbital.

5.33. Check This *The sigma framework and pi molecular orbital in methanal*

(a) Use your molecular models to make a model of methanal patterned after the one for ethene in Figure 5.20. How is your model similar to that for ethene? how is it different? Explain.

(b) Did you represent the sigma nonbonding electron pairs in your methanal model? If so, how? If not, can you think of a way to do so? Compare your solution to the ones others choose.

Triple bonds: sigma-pi geometry with two pi orbitals

To complete the comparison of triple bonds with double bonds, we also need to consider an interpretation in terms of sigma and pi bonding. The incomplete Lewis structure in Worked Example 5.9, H–C–N, shows that the central atom, the carbon, has two sigma bonding orbitals. Figure 5.21 shows how we would represent these two electron-pair orbitals as a linear arrangement of closely packed balls on opposite sides of a central point (an atomic core).

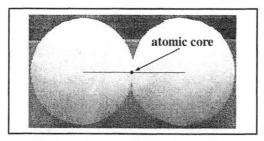

Figure 5.21. Linear arrangement of two sigma orbitals and an atomic core.

Once again, we can consider reinforcing combinations of electron waves from the atoms creating π-bonding orbitals. The central atom has two sigma orbitals and can form two pi bonding orbitals. In the sigma-pi representation, a **triple bond** between two atoms consists of a σ bonding orbital and two π bonding orbitals. Figure 5.22 shows the sigma-pi bonding in HCN. Sigma bonds are represented by the sticks in the ball-and-stick structure that shows how the atomic cores lie on a line. The two π orbitals are localized between the C and N atomic cores and lie in planes that are perpendicular to each other. Figure 5.22 displays one π orbital as transparent and the other π orbital as mesh to make it easier to see them both.

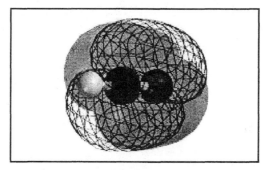

Figure 5.22. Sigma-pi representation of the triple bond in hydrogen cyanide.
The π molecular orbitals on HCN are localized between the C and N.

Structure of Molecules Chapter 5

5.34. Check This *The sigma-pi representation for ethyne*

Explain how you would modify the illustration in Figure 5.22 so that it represents the sigma and pi molecular orbitals in ethyne, HCCH. Make a sketch of your modifications.

Reflection and projection

In molecules, atoms of second-period elements can share one, two or three pairs of bonding electrons to form single, double, and triple bonds. We have shown two ways to represent multiple bonds: as bent bonds between the atoms or as a σ bonding orbital plus one or two π bonding orbitals. In both representations, multiple bonds are characterized as having a good deal of electron density that is away from the bond axis and not directly between the bonded atomic cores. These electrons are readily available for interactions with other molecules, so compounds with multiple bonds are more reactive than similar compounds without multiple bonds.

Both representations of multiple bonds also explain and predict the correct geometry of molecules with multiple bonds. The generalization we made earlier in Section 5.4 is still valid for molecules with multiple bonds: *the shape of a molecule is determined by its framework of σ and σ_n bonding orbitals*. Table 5.3 summarizes the geometries we have discussed so far for molecules composed of second period elements (and hydrogen). Often, we would like to draw these structures in a way that represents their shapes as accurately as possible. In order to do so, we need to introduce the methods chemists use to show three-dimensional structures on a two-dimensional sheet of paper. Before doing that, however, we will look at another important characteristic of molecular orbitals: the fact that they are not always localized between two atomic cores. In the next section, we will extend our discussion of pi orbitals to include molecules that cannot be described by a single Lewis structure, but require introduction of molecular orbitals that extend over three or more atoms. We will finish the section with a brief discussion of metallic bonding, in which electron orbitals extend through the entire crystal.

Chapter 5   Structure of Molecules

Table 5.3. Combinations of localized molecular orbitals around second-period elements.
The σ orbital geometry refers to the bold, blue elemental atom in the examples.

number of σ and σ_n orbitals	σ orbital geometry	number of π orbitals	π orbital arrangement	examples
4	tetrahedral 109.5° angles	0	—	CH_4, NH_4^+, NH_3, H_2O, H_3O^+ See Figure 5.10
3	trigonal planar 120° angles	1	perpendicular to σ framework; parallel to σ bond axis	H_2CCH_2, H_2CO
2	linear 180° angles	2	perpendicular to each other; parallel to σ bond axis	HCN, HCCH

Section 5.7.   Delocalized Orbitals

5.35.   Worked Example   *Lewis structure for the carbonate ion*

Write the Lewis structure for the carbonate ion, CO_3^{2-}. All the oxygens are equivalent.

Necessary information: Carbon and oxygen atoms have 4 and 6 valence electrons, respectively, and the ion has two more electrons than are provided by these valence electrons. The equivalent oxygens must all be bonded in the same way to the carbon.

Strategy: The approach is the same as in Worked Examples 5.6 and 5.8.

Structure of Molecules — Chapter 5

Implementation: The carbonate ion has 24 valence electrons, 22 from the atoms and 2 more that account for the –2 charge on the ion.

First we will bond the oxygens to the carbon with two-electron bonds and then, since the oxygens are equivalent, give each oxygen an octet of electrons:

(a) $\left[\ddot{\mathrm{O}}-\mathrm{C}(-\ddot{\mathrm{O}})-\ddot{\mathrm{O}} \right]^{2-}$

This Lewis structure uses all 24 valence electrons and has all three O atoms equivalent. However, the C atom has only six electrons (three pairs), so we need to find a way to write a structure with an octet of electrons on the C as well as on the O atoms. We can do this by using a double bond between the C atom and one of the O atoms:

(b) $\left[\ddot{\mathrm{O}}-\mathrm{C}(=\ddot{\mathrm{O}})-\ddot{\mathrm{O}} \right]^{2-}$

This structure gives all four second period atoms an octet of electrons, but implies that one oxygen is different from the others. Since we know that the oxygens are equivalent, this structure does not fit the experimental facts. Also, there is no reason to choose one oxygen over another to make the double bond. Any of these three structures gives all the atoms an octet of electrons:

(b) (c) (d)

None of the four structures, (a) through (d), satisfies all the criteria for Lewis structures as well as the experimental observation that the oxygens are equivalent.

Does the answer make sense? We do not yet have a satisfactory Lewis structure for the carbonate ion.

5.36. Check This *Lewis structure for the nitrate ion*

Write the Lewis structure for the nitrate ion, NO_3^-. All the oxygens are equivalent. Does your structure have the same problems we found for carbonate in Worked Example 5.35? Why or why not?

5.37. Consider This *Do Lewis structures give correct geometries for CO_3^{2-} and NO_3^-?*

Both carbonate and nitrate ions are planar. All four atoms lie in the same plane. Do the Lewis structures written in Worked Example 5.35 and Check This 5.36 predict this geometry? Show why or why not for both ions.

Worked Example 5.35 and Check This 5.36 show that, for some ions (and molecules), a single Lewis structure is not adequate to satisfy both the rules we use to write the structures and all of the observed properties of the ion or molecule. For carbonate and nitrate ions, the geometry of the ions is correctly predicted by the sigma framework we can derive from the Lewis structures we write, but the equivalence of the oxygen atoms is not explained. Each of the Lewis structures (b), (c), and (d) we wrote in Worked Example 5.35 has two single bonds and one double bond from C to O. Since exactly the same bonding is involved, the energies of the three structures are identical. Whenever we can write Lewis structures that are equivalent in energy, but have double bonds between different pairs of atoms in the molecule, none of the individual structures correctly represents the molecule.

Delocalized π molecular orbitals

To account for the properties of molecules for which no single Lewis structure is correct, we need to consider pi orbitals that extend over more than two atomic cores. Figure 5.23(a) shows the **delocalized π molecular orbital**, describing two electrons, for the nitrate ion. The delocalized π orbital is spread across the nitrogen core and *all three* oxygen cores in the nitrate ion. For comparison, Figure 5.23(b) shows the two-center, localized π molecular orbital in methanal.

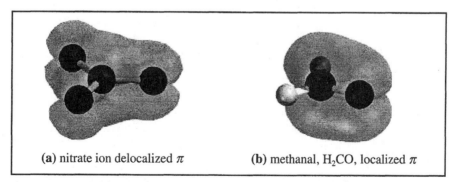

(a) nitrate ion delocalized π   (b) methanal, H_2CO, localized π

Figure 5.23. Delocalized and localized π molecular orbitals.
Both molecules have planar σ frameworks. Because the delocalized π orbital in the nitrate ion is "shared" among three N–O pairs, each is assigned one-third of its bonding.

Structure of Molecules   **Chapter 5**

Bond order

It's awkward always to refer to a σ or π bonding orbital between two (or more) atomic cores, so we almost always use the shorthand versions, "σ bond" and "π bond," with the understanding that these refer to the orbitals. In methanal, the carbon and oxygen are joined by a σ bond and a π bond, totaling two bonds. In the nitrate ion, there is a σ bond between the nitrogen and each oxygen. The delocalized π orbital is spread over several atomic cores. There are three N–O bonds among which to divide the π bonding. Each bond is assigned one-third of the π bond. Thus, each oxygen is joined to the nitrogen by a σ bond and one-third of a π bond, totaling $1\,^1/_3$ bonds.

The sum of the number of two-electron, bonding molecular orbitals that are shared between two atomic cores in a molecule is called the **bond order**. In molecules with only sigma orbitals, there is one, two-electron, σ orbital shared by each pair of bonded atomic cores; the *bond order is one* and this is called a **single bond**. In molecules like ethene and ethanal, a pair of atomic cores shares both a σ orbital and a π orbital; the *bond order is two* and this is called a **double bond**. This is the nomenclature we have been using to describe the bonding in Lewis structures. For molecules (or ions) with delocalized π orbitals, you have to divide up the π bonding interactions among the pairs of atomic cores. In the nitrate ion, NO_3^-, for example, each N–O bond is a $1\,^1/_3$ bond; the bond order is one-and-one-third. There is no simple way to express non-integral bond orders nor is there any way to write a single Lewis structure for a molecule or ion like nitrate. A Lewis structure can't represent non-integral numbers of electrons in bonds.

5.38. Consider This   *What are the bonding and structure of the CO_2 molecule?*

(a) Write the Lewis structure for CO_2. Both oxygen atoms are bonded to the carbon.

(b) How many σ orbitals does the carbon atom have in CO_2? What geometry do you predict for the CO_2 molecule? Explain how you make your prediction.

(c) If possible, make a molecular model of the CO_2 molecule. If you can make a model, does it confirm the geometry you predicted in part (b)?

Orbital energies

You can write a Lewis structure for CO_2 with two bonding pairs of electrons between the carbon and each of the oxygens. You can also make a bent-bond model of the molecule that shows its linear structure, the three atom cores lying along a line. However, these models do not adequately describe all the properties of carbon dioxide. In particular, carbon dioxide is more stable, that is, has a lower total energy, than is predicted for a molecule with two localized π bonds between carbon and oxygen. In Chapter 7, we will discuss bond formation energies quantitatively; here we will look only at why carbon dioxide has such a low total energy.

The sigma-pi bonding in carbon dioxide involves two sigma and two pi bonds on the carbon, which, as Table 5.3 reminds you, gives the molecule the linear structure you found in Consider This 5.38. Instead of two localized π orbitals, the wave mechanical solution for bonding in CO_2 gives two *delocalized* π molecular orbitals that extend across all three atoms, as shown in Figure 5.24. The two delocalized π orbitals lie in planes that are perpendicular to each other, just like the two localized π orbitals in hydrogen cyanide, Figure 5.22. Again, one π orbital is shown as transparent and the other as mesh to make it easier to see them both.

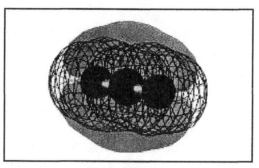

Figure 5.24. Delocalized π molecular orbitals in carbon dioxide.
One of the π orbitals is shown as a mesh to distinguish if from the other.

5.39. Check This   *Bond order in carbon dioxide*

When the delocalized π orbitals in Figure 5.24 are taken into account, what is the bond order for the bond between carbon and one of the oxygen atoms in carbon dioxide? Explain clearly how you arrive at your answer. Does your Lewis structure from Consider This 5.38 show this same bond order? Why or why not?

A molecule (or ion) with delocalized π orbitals has a lower total energy than a molecule with an equal number of localized π orbitals. The energy is lower for two reasons. First, the electrons are interacting with more than two positive atomic cores, so the potential energy of attraction is more favorable (more negative) for the delocalized electrons. Second, the delocalized orbital is larger, so the kinetic energy of the electron wave is lower (less positive). As always, when an electron is less constrained, its kinetic energy is lower and its total energy is lower (more favorable). The delocalized π orbitals in carbon dioxide, the carbonate ion, the nitrate ion, and many other molecules and ions make these molecules and ions more stable than they would be with the localized orbitals we write in their Lewis structures. Figure 5.25 is a qualitative energy level diagram that shows the relative energies of molecular orbitals.

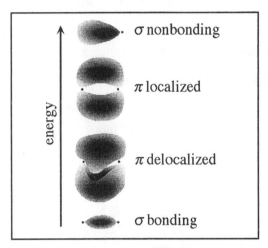

Figure 5.25. Relative energies of molecular orbitals.
Sigma nonbonding orbitals have the same energy as the atomic orbitals on the atoms. Pi delocalized orbitals are spread over three or more atomic cores. The atomic cores are represented by the dots.

5.40. Check This *Structure of the cyanate ion and the ozone molecule*

Write a Lewis structure for the ozone molecule, O_3. What geometry does your structure predict for ozone? The three atoms are connected in a chain and the end atoms are equivalent. Is your structure consistent with these experimental facts? Explain why or why not. Would you expect ozone to have delocalized π electrons? What is the bond order between the oxygens in ozone? Explain the reasoning for your responses.

Metallic properties

5.41. Investigate This *What happens when substances are cut?*

Do this as a class investigation and work in small groups to discuss and analyze the results. Place a small piece of sodium metal in a shallow dish and try cutting it in two with a sharp knife. Record your observations, including the ease of cutting, and then return the pieces of sodium to their storage container. CAUTION: Sodium metal can cause severe burns by reacting with the moisture on your skin. Handle the metal with tongs.

Place a crystal of rock salt, sodium chloride, in a shallow dish and try cutting it in two with a sharp knife. Record your observations, including the ease of cutting.

Chapter 5 — Structure of Molecules

> **5.42. Consider This** *What substances can be cut in two?*
>
> **(a)** In Investigate This 5.41, did sodium metal and sodium chloride behave the same or differently when you tried to cut the solids in two? Explain how they were the same or different.
>
> **(b)** Try to think of a molecular level explanation for the similarities and/or differences in the behavior of sodium metal and sodium chloride.

Recall from Chapter 2, Sections 2.3 and 2.4, especially Figures 2.9 and 2.13, that ionic solids (crystals) are three-dimensional arrays of positive and negative ions held together by coulombic attraction. These attractions are relatively strong for nearest-neighbor ions and hold the ions tightly together. When you attempt to cut an ionic crystal in two, these attractions strongly resist being broken. It takes considerable effort to cut into the crystal, which often shatters into several pieces as the crystal structure is disrupted by the applied force.

On the other hand, many metals are relatively easy to cut with a strong scissors and, as you found in Investigate This 5.41, some metals, such as sodium, are soft enough to be cut easily with a knife. This difference in behavior between metals and ionic crystals, must be due to a difference in the forces that hold the atoms together in a metal compared to those that hold the ions together in an ionic crystal. An obvious difference between an elemental metal, such as sodium, copper, aluminum, or iron, and an ionic crystal is that the metal contains only one kind of atom. Since most metals are relatively dense solids, the elemental spherical atoms must be packed closely together, as represented for two-dimensional packing in Figure 5.26.

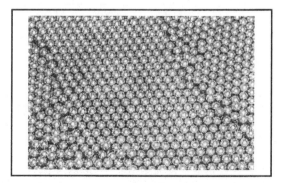

Figure 5.26. Two-dimensional close packing of spherical objects.

> **5.43. Check This** *Three-dimensional packing of spheres*
>
> How do you have to pack the plastic balls that represent hydrogen atom centers to get them to fit in the space allotted for them in your model kit? Use drawings and words to describe the packing and to relate it to the two-dimensional case in Figure 5.26.

Structure of Molecules   Chapter 5

Metallic bonding: Delocalized molecular orbitals

It is unlikely that oppositely charged ions of the same element, such as Na^+ and Na^-, are responsible for the binding in a metal. We know from our previous discussions of compounds that metals always are present as positively-charged ions in ionic compounds. This is because metals give up one or more of their valence electrons relatively easily, as you can recall from their low ionization energies shown in Chapter 4, Figures 4.4 and 4.30. Consider what can happen when metal atoms are stacked together, as represented in Figure 5.26 and your sketches for Check This 5.43. Assume that the bright spot in the center of each sphere in Figure 5.26 represents the positively-charged atomic core of the metal atom (nucleus plus inner-shell electrons) and that the rest of the copper-colored sphere represents its valence electrons.

The atomic cores of the metal atoms are about as close together in this array as are the atomic cores in the molecules we have been discussing so far in this chapter. We have seen that, when the cores are this close together, the valence electrons are attracted by and interact with more than one positive center. In a uniform stack of metal atoms (a crystal of the metal), the valence electrons from one atom can interact with *all* the other atomic cores in the crystal. The electron waves, orbitals, that describe these electrons are spread out through the entire crystal. You can think of the metal crystal as a single molecule with enormously delocalized electron orbitals. This **metallic bonding model**, represented in Figure 5.27, is usually described as a "sea of electrons" surrounding the positive metal atomic cores. In a sense, all the positive centers attract all the valence electrons, which holds the structure together.

Figure 5.27. A representation of metallic bonding.
Yellow positive cores are immersed in a sea of electrons shown as a diffuse white cloud.

The metallic bonding model explains why it is easy to cut many metals without shattering them. Since the delocalized electron orbitals are spread out through the crystal, there is little directionality to the bonding. Adjacent atomic cores resist being pushed apart by a blade, but these pair-wise interactions are relatively weak. When enough force is applied they are pushed apart and the electron orbitals easily readjust to the new configuration of atomic cores.

5.44. Check This   *More properties of metals*

Metals are **ductile**, which means they can be pulled into long wires without breaking, like the copper wire shown in this photograph. Metals are also **malleable**, which means they can be pounded into sheets without shattering and pressed into useful shapes such as spoons and automobile fenders. Ionic crystals like the copper sulfate pentahydrate, $CuSO_4 \cdot 5H_2O$, in the photograph, do not have these properties. How does the metallic bonding model explain these properties of metals? Why do the crystals shatter?

Other familiar properties of metals, for example, their luster or shininess, their high heat conductivity, and their high electrical conductivity, can also be interpreted with this metallic bonding model based on delocalized molecular orbitals. We will include discussions of metallic properties in Problems and later chapters, as appropriate, but for now will go on to the discussion of drawing molecular structures.

Section 5.8.   Representations of Molecular Geometry

Lewis structures and ball-and-stick models simplify the molecular orbital model by using a line or a stick to represent the sausage-shaped σ bond between pairs of atoms. You can think of the pairs of dots on individual atoms in Lewis structures as two-electron σ_n orbitals. With your ball-and-stick models you can use the paddles to represent nonbonding electrons. In Figure 5.28, the orbital model, Lewis structure, and ball-and-stick models for the ammonia molecule are compared. Of the four representations shown, the Lewis structure is obviously the easiest to write, since

WEB Chap 5, Sect 5.8
Manipulate several interactive representations of molecular geometry.

it is written with letters, lines, and dots. However, Lewis structures are used only to show connectivity of atoms; the Lewis structure doesn't show the three-dimensional shape of the molecule, the trigonal pyramid formed by the nitrogen and three hydrogens.

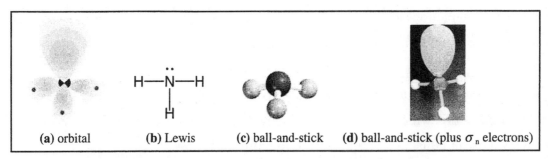

Figure 5.28. Comparison of models for the ammonia molecule.

Tetrahedral representation

You can retain *some* of the simplicity of a Lewis structure, written much as it is in Figure 5.28(b), *and* show the three-dimensional shape of molecules. To do so, you start by analyzing the Lewis structure to see how many σ orbitals there are around the atom of interest. In ammonia, there are four σ orbitals around the nitrogen atomic core: three σ and one σ_n. You know from Section 5.4 that four σ orbitals will be arranged tetrahedrally about the nitrogen atom. To represent the three-dimensional tetrahedral structure in two dimensions requires some way to show perspective. You can do this by using different kinds of lines to represent σ bonds. Figure 5.29 shows how solid lines, dashed lines, and wedge lines are used to represent perspective and give a three-dimensional character to a drawing of the ammonia molecule. This form of molecular representation is called a **3-d structure**.

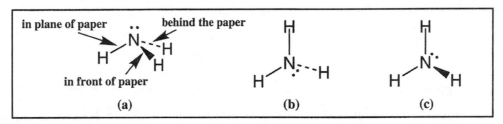

Figure 5.29. 3-d structures of the ammonia molecule.

The kind of line you use in a 3-d structure depends upon whether the bond lies in the plane of the paper or is in front of or behind the plane, as you look at the molecule. Solid lines are bonds in the plane of the paper. Dashed lines represent bonds that extend away from you behind the paper. Wedges represent bonds that extend toward you in front of the paper. Usually, you choose to orient your structure so that as many of the bonds as possible lie in the plane of the paper in these 3-d structures. This often, but not always, makes it easier for you to interpret the structure. Note that little attempt is made to show the nonbonding electron pair in perspective. 3-d structures are designed to show molecular shape, the orientation of the atoms with respect to one another, not orbital geometry. The shape is, however, determined by σ orbital geometry.

5.45. Check This *Interpreting 3-d structures*

(a) Which, if any, of the 3-d structures in Figure 5.29 is easiest for you to interpret as the trigonal pyramidal structure for ammonia? Why is it easiest for you?

(b) WEB Chap 5, Sect 5.8.1. Load the movie "Molecular representations of Methane." Place the cursor in the middle of the image in the movie. Move the cursor up and down. Describe what you see. What information can you gather from each of the different representations? Which representation do you find most useful? Explain why.

(c) Draw 3-d structures for the chloromethane molecule, CH₃Cl, that are comparable to each of the 3-d structures in Figure 5.29. Make a ball-and-stick model of the chloromethane molecule. Practice your understanding of the structure of molecules by writing a description of how the drawings correlate with the three-dimensional model.

Trigonal planar representation

The orbital model, Lewis structure, ball-and-stick model, and 3-d structure for the methanal molecule, H₂CO, are compared in Figure 5.30. To generate the 3-d structure, start with the Lewis structure and figure out how many σ orbitals there are around the carbon atomic core. In this case, there are three σ orbitals: each H–C single bond is a σ bonding orbital and the double bond is composed of a σ bonding and a π bonding orbital. You know that three σ orbitals will be in a trigonal planar arrangement around the carbon atom. It's easy to draw the planar structure, Figure 5.30(d), with all four atoms in the plane and bond angles of 120°. (Or you can draw the structure with the experimental bond angles of 116° and 122°, respectively, for the H–C–H and H–C–O angles.)

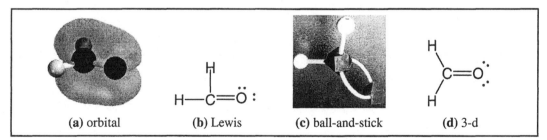

(a) orbital   (b) Lewis   (c) ball-and-stick   (d) 3-d

Figure 5.30. Comparison of models for the methanal, H₂CO, molecule.

The Lewis and 3-d structures in Figure 5.30 look quite similar. Be cautious; the similarity is misleading because the Lewis structure shows connectivity, not geometry. The 3-d structures show connectivity, and, in addition, molecular shape — note the planar structure and correct bond angles. In the case of planar and linear molecular structures, the Lewis and 3-d structures

Structure of Molecules Chapter 5

can be easily confused since, none of the wedged and dashed bonds are necessary for the 3-d structures.

5.46. Check This   *3-d structures*

(a) Write a Lewis structure for a molecule of methanol, CH_3OH. How many sigma molecular orbitals are there around each second-period atom in the molecule? How many of the other atoms *must* lie in a plane that contains the carbon atom? How many of the other atoms *can* lie in this same plane? Draw a 3-d structure that has as many atoms in the plane as possible. If necessary, use wedge and dashed bonds to show where the other atoms are relative to the plane. You might find that making a ball-and-stick model will help you.

(b) Write a Lewis structure for a molecule of formic (methanoic) acid, HC(O)OH. How many sigma molecular orbitals are there around each second-period atom in the molecule? How many of the

> Formic acid was first extracted from ants (*formica* = ant). It causes the sting of an ant bite.

other atoms *must* lie in a plane that contains the carbon atom? How many of the other atoms *can* lie in this same plane? Draw a 3-d structure that has as many atoms in the plane as possible. If necessary, use wedge and dashed bonds to show where the other atoms are relative to the plane. You might find that making a ball-and-stick model will help you.

(c) Draw the 3-d structure for ethyne, HCCH.

Lewis structures and molecular shape

We set out in this section to learn to translate Lewis structures into 3-d structures that show molecular shape. Molecular shape is determined by the σ orbitals and Lewis structures show you how many σ orbitals each atom has. The spatial orientation of π bonds is not shown in 3-d structures. You have to remember that a second line between atoms is a π bond that lies parallel to the bond axis and in a plane perpendicular to the molecular plane (the plane of the paper). If there are three lines between atoms, the second and third are π bonds that lie parallel to the bond axis and in two perpendicular planes that contain the bond axis; see Figures 5.22 and 5.24.

5.47. Consider This   *How do you draw 3-d structures for larger molecules?*

(a) Write the Lewis structure and draw the 3-d structure for propene, CH_3CHCH_2. Propene is like ethene, Figures 5.19 and 5.20, with a H_3C- in place of one of the hydrogens. Can all the carbons in the molecule lie in a plane? If so, that should make them easier to draw. If they can lie in a plane, is there more than one way for them to lie in a plane? You might find it helpful to

Chapter 5 — Structure of Molecules

make ball-and-stick models of these molecules to help visualize in three dimensions what you're drawing on a plane.

(b) Write the Lewis structure and draw the 3-d structure for 1-butene, $CH_3CH_2CHCH_2$. Answer the same questions for 1-butene as for propene in part (a).

Consider This 5.47 probably convinced you that drawing 3-d structures for molecules with even as few as four carbons is pretty tedious. Also, most of us have a hard time actually visualizing the three-dimensional structure when there are so many dashes and wedges and elemental symbols to keep track of. Biologically important molecules are even more complex and often contain extended structures of carbon atoms, with nitrogen, oxygen, and other atoms at strategic sites. To get around these representational problems, chemists use a progressive system of shorthand to represent complex structures, with each step showing less detail. To understand structures that are represented using abbreviated symbolism, you need to be able to recognize and understand the detail that is omitted along the way. That's why you should also continue to practice writing Lewis structures, visualizing the structures they represent, making molecular models, and drawing the 3-d structures; it's the 3-d structures that are being abbreviated. We will exemplify the successive steps in this system by considering the structure of 1-butanol, one of the compounds whose properties you examined in Sections 5.1 and 5.2.

Condensed structure for 1-butanol

Figure 5.31 shows a ball-and-stick model, a 3-d structure, and a condensed structure for 1-butanol. In 3-d structures, we often do not show the nonbonding electrons, since we rarely try to show their geometry. Also, we don't need to show the lines representing bonds to hydrogen, since there is always one σ bond to hydrogen. Thus, in the **condensed structure**, we eliminate the lines to H. The H symbols are still shown associated with the atom cores to which they are bonded and placed where they will not hide essential features of the skeleton of the molecule, which is shown in red in Figure 5.31(b) and (c).

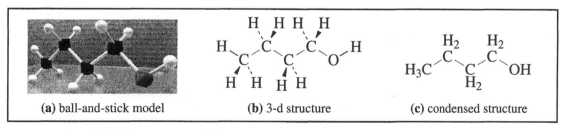

Figure 5.31. Structural representations of the 1-butanol molecule.
The red bonds in structures (b) and (c) denote the bonds that will form the skeletal structure, Figure 5.32.

5.48. Worked Example — *3-d and condensed structures for 2-propanol*

The Lewis structure for 2-propanol is given in Worked Example 5.6 and a ball-and-stick model is shown in Figure 5.1(b). Draw 3-d and condensed structures for 2-propanol.

Necessary information: All the connectivity information we need is in Worked Example 5.6.

Strategy: For the 3-d structure, choose the longest carbon chain and orient these carbons so they lie in a plane. Attach other non-hydrogen atoms to the appropriate carbon(s) in the chain, using appropriate dashed or wedge bonds. Add the hydrogen atoms, using solid, dashed, or wedge bonds to give the appropriate geometry around each second (or higher) row atom.

Implementation: The 2-propanol molecule has only sigma electrons, so the geometry around each carbon is tetrahedral. The longest carbon chain is three carbons, which we show lying in the plane of the paper and then add the oxygen bonded at the middle carbon using either a wedge or dashed bond:

$$\begin{array}{c} O \\ | \\ C-C-C \end{array}$$

We complete the 3-d structure by adding seven hydrogen atoms to give tetrahedral bonds around the carbons and bond the eighth hydrogen atom to oxygen:

For the condensed structure, we eliminate the bond lines between hydrogen atoms and other atoms and write the Hs next to the atoms to which they are bonded:

$$H_3C-HC(OH)-CH_3$$

Does the answer make sense? Compare the 3-d structure to the ball-and-stick model in Figure 5.1(b) to see whether the drawing is a good representation of the geometry of the model. Note that we could have chosen to show the oxygen atom bonded with a dashed bond (in back of the paper). This would still be the same structure shown here, because they can be converted into one another by simply rotating the molecule in space. Use molecular models to prove this.

5.49. Check This — *3-d and condensed structures for the C_5H_{12} isomers*

Lewis structures and ball-and-stick models for the C_5H_{12} isomers are shown in Worked Example 5.8 and Figure 5.2. Draw 3-d and condensed structures for each isomer. You might find that making ball-and-stick models will help you.

5.50. Consider This *How many hydrogen atoms are bonded to a carbon?*

Do any of the carbon-containing molecular structures we have shown in this or the preceding chapters have σ nonbonding orbitals on carbon? In structures such as those in Worked Examples 5.6 and 5.8, Consider This 5.10, and Figure 5.31, can you tell how many hydrogen atoms will be σ bonded to a particular carbon atom without counting the number of hydrogens? Explain how.

Skeletal structure for 1-butanol

The final step in this process of abbreviation requires you to account for the observations and analysis you did in Consider This 5.50. *Carbon atoms in molecules do not have σ_n electrons. If a carbon atom has hydrogens bonded to it, we know that the carbon atom is σ bonded to enough hydrogen atoms to give the carbon four bonding orbitals.* Therefore, we can condense the notation

> Carbon monoxide, CO, is the single notable exception to the statement that carbon never has σ_n electrons. There are also reactive fragments of molecules (ions and free radicals), in which carbon has nonbonding electrons.

even more by omitting the hydrogen symbols on carbons. Further, we understand that the end of a line or the intersection of two or more lines represents a carbon atom and any attendant hydrogen atoms, so we can also omit the letter symbol for carbon. We must keep the letter symbol for oxygen (or any other element) to distinguish the atom at that position from carbon. The result is shown in Figure 5.32. To the practiced eye, this **skeletal structure** gives a reasonable idea of the three dimensional structure of 1-butanol.

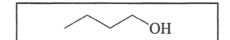

Figure 5.32. Skeletal structure of 1-butanol.

5.51. Worked Example *The skeletal structure for 2-propanol*

Draw the skeletal structure for 2-propanol.

Necessary information: The 3-d and condensed structures are in Worked Example 5.48.

Strategy: For the skeletal structure, start with either the 3-d or condensed structure. Eliminate all hydrogen atoms and their bonds on carbons. Eliminate all the carbon Cs and extend the remaining bonds to meet where the Cs were. Retain wedge and dashed bonds, if any.

Implementation: Begin with the condensed structure from Worked Example 5.48 and carry out the necessary changes:

Structure of Molecules — Chapter 5

$$CH_3-CH(OH)-CH_3 \rightarrow \text{(skeletal structure of 2-propanol with OH)}$$

Does the answer make sense? Make a ball-and-stick model of 2-propanol, remove all the hydrogen atoms and their bonds attached to carbon, and compare the result with this skeletal structure to see that it represents the 2-propanol skeleton.

5.52. Check This *Skeletal structures for the C_5H_{12} isomers*

(a) What does the line at the left end of the 1-butanol skeletal structure, Figure 5.28, represent? Explain clearly.

(b) Draw skeletal structures for the C_5H_{12} condensed structures you drew in Check This 5.48. *Note:* If you need to represent a –CH_3 group attached to a carbon chain, but not in the plane of the chain, use a dashed line or a wedge instead of just a solid line.

5.53. Consider This *Can different skeletal structures represent the same molecule?*

(a) Parts of a molecule can rotate with respect to one another without breaking any sigma bonds. This sort of motion is going on all the time. The rotation does not create a new molecule, but it does create a new shape and therefore a new skeletal structure. The skeletal structure for a possible different shape of 1-butanol is: (skeletal structure with OH) Make a model of 1-butanol and test whether this new shape and the one in Figure 5.32 can be interconverted without breaking any bonds.

(b) In Check This 5.50, you drew skeletal structures for the C_5H_{12} isomers. For each isomer, draw another different skeletal structure that can be obtained from the ones you already have by rotating one part of the molecule with respect to another. Is it possible to do this for all three isomers? Why or why not? You might find it helpful to use your models of the molecules.

5.54. Check This *Interpreting skeletal representations*

(a) The skeletal structures of 2-methylpropane and 2-methylpropene are (skeletal) and (skeletal with double bond), respectively. Draw condensed structures for each molecule. What are the molecular formulas for the molecules? Explain the reasoning for your responses.

(b) The condensed structure for retinol is shown in Consider This 5.10(d). Draw the skeletal structure for retinol. Is your skeletal structure related to the molecular models of retinal shown in the chapter opening illustration? Explain your response.

Reflection and projection

Delocalized bonding is the last bonding concept you need to interpret the bonding and behavior of the molecules you will meet in this book. When an orbital (or orbitals) spreads out and interacts with several atomic cores, the energy of the electrons is more negative (more favorable) because they are attracted by more positive charges and their kinetic energy is lower. For molecules (not metals), however, the major contributors to the favorable energy relative to their atoms are their σ bonds, which also determine their geometries. Metals lack a directional sigma bonding framework. They are bonded by non-directional delocalized molecular orbitals which give them their characteristic properties, including cutting without shattering, ductility, and malleability.

One of the most important skills for any scientist working with molecular materials is the ability to visualize their structures in three dimensions. From the very beginning of this text, we have emphasized the critical role played by molecular structure in determining the properties of substances. Now that we have developed a model that explains and predicts molecular structure, it is important to be able to represent the structures in a way that aids three-dimensional visualization. Lewis structures combined with the structural rules based on the σ molecular framework are used to draw 3-d structures that are built around individual atom centers linked together to form more complex structures.

As the molecules of interest become ever larger, the focus on geometry about individual atoms becomes so complex that it can obscure, rather than clarify, structural visualization. We introduced a series of abbreviated representations for carbon-containing molecules that range from condensed to 3-d to skeletal structures. In the final step, atomic symbols for all carbons and hydrogens bonded to carbon disappear and you are left with intersecting lines representing the carbon skeleton of the molecule. As you use these abbreviations, always keep in mind what is missing. Continue to practice going back and forth from complete structural representations and models to the skeletal structures.

Molecular structures are, in and of themselves, interesting and often quite beautiful. In the next section, we will extend our discussion of structure to a different form of isomerism.

Structure of Molecules

Chapter 5

Section 5.9. Stereoisomerism

5.55. Investigate This *How many C_4H_8 molecular structures are possible?*

(a) Work in small groups on this investigation. Use your molecular models to make as many models of isomers of C_4H_8 as you can. If necessary, use bent bonds to represent multiple bonds.

(b) Draw skeletal structures of all the molecules you found in part (a). Share your results with the entire class and come to an agreement on the number of possible isomers. *Hint*: Carbon atoms can be connected in rings of three, four, or more carbons. (You can even think of the double bond as a two-membered ring.)

5.56. Consider This *What are the condensed formulas for the C_4H_8 isomers?*

(a) Recall that condensed or line formulas are written to try to show the connectivity among the atoms in a molecule in a compact notation on a single line. For example, one of your C_4H_8 isomers has the condensed formula: $CH_2C(CH_3)CH_3$. What are the condensed formulas of the other molecules whose models you made in Investigate This 5.55?

(b) Do all your isomers have different condensed formulas? If two (or more) of the models you made have the same condensed formula, why are they isomers? What is the criterion you use to call them isomers?

All of the isomers you have encountered so far in this and preceding chapters have been **structural isomers**, isomers with different connections among the atoms. These isomers all have different condensed formulas. Now you have found two different molecular models with the same atomic composition, C_4H_8, and the *same* condensed formula, $CH_3CHCHCH_3$. Compare your molecular models with the sigma-pi representations in Figure 5.33. Carbon chains are numbered beginning with the carbon at one end. The "2" in the name 2-butene indicates that the double bond connects the *second* and third carbons in the four-carbon chain. The lobes of the π bonding orbital lie above and below the plane of the four carbon atoms in Figure 5.33.

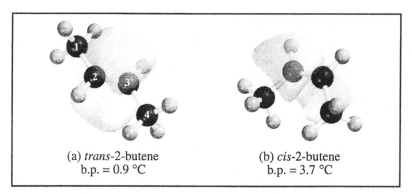

(a) *trans*-2-butene
b.p. = 0.9 °C

(b) *cis*-2-butene
b.p. = 3.7 °C

Figure 5.33. Sigma-pi bonding models of *trans*- and *cis*-2-butene.
The π bonding orbital is superimposed on the σ framework to show its location between the second and third carbons with its lobes above and below the plane of the molecule.

Cis- and trans-isomers

The double bond in 2-butene imparts important structural properties as well as the chemical reactivity associated with the multiple bond. In Figure 5.34, we see that the two -CH$_3$ groups in *trans*-2-butene are on opposite sides of the double bond (*trans* = across from; one on the left and one on the right). In *cis*-2-butene, the -CH$_3$ groups are on the same side of the double bond (*cis* = next to; both on the right). *Trans*-2-butene cannot be changed to *cis*-2-butene (and *vice versa*) without breaking the π bond between the second and third carbon atoms. The same is true for your bent-bond models. The only way to convert one model into the other is to break and reform bonds. Experiments show that the conversion requires about 270 kJ·mol^{-1}, which is far more energy than the molecules have available at room temperature. If the two isomers could rotate easily around the double bond, 2-butene would have a single boiling point. The observation that *cis*-2-butene and *trans*-2-butene have different boiling points, Figure 5.33, is experimental evidence that **cis-** and **trans-isomers** are different compounds that do not interconvert at room temperature.

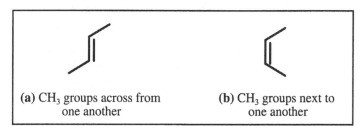

(a) CH$_3$ groups across from one another

(b) CH$_3$ groups next to one another

Figure 5.34. Geometry of (a) *trans*-2-butene and (b) *cis*-2-butene.
All four carbon atoms in each skeletal structure lie in the plane of the paper.

Structure of Molecules   Chapter 5

5.57. Check This   *Cis-trans isomers*

(a) Use your molecular models to make as many models of isomers of $C_2H_2Cl_2$ as you can. Are any of your structures cis-trans isomers? Draw skeletal structures of all the isomers and identify the cis-trans isomers, if any. Explain the reasoning for your answers.

(b) The structure of retinol, vitamin A, in Consider This 5.10 is sometimes called "all-*trans*-retinol." Draw a skeletal structure and explain why this is an appropriate name for this isomer.

(c) The structural change in retinal illustrated in the chapter Frontispiece is a *cis-trans* conversion caused by the absorption of energy from a photon of light. Draw skeletal structures of the two isomers that show where the *cis-trans* change occurs.

Stereoisomers

Cis-trans isomers are an example of **stereoisomers**, isomers whose atoms are connected in the same sequence, but whose shape in space is different and not interconvertible. The names of

> "Stereo-" is from Greek *stereos* = solid, three-dimensional. You are probably familiar with the word stereophonic referring to the perception of sound as coming from many directions.

cis-trans stereoisomers are the same (2-butene, for example), which reflects their connectivity; the prefix *cis-* or *trans-* reflects their different shapes. If a pair of structures has the same elemental composition and the same connectivity, one way to test whether they are stereoisomers is to try to superimpose models of the molecules. If the models cannot be made to superimpose on one another without breaking and remaking bonds, they are stereoisomers.

5.58. Check This   *Identifying isomers*

Five skeletal structures are shown below. What is the elemental composition of each? Which structures could be isomers of one another? Within a set of possible isomers, which *pairs* are structural isomers? which are stereoisomers? which are identical structures? Give the reasoning for your answers.

(i)   (ii)   (iii)   (iv)   (v)

Polarized light and isomerism

In Chapter 4, Section 4.1, we discussed how the atomic masses proposed by Cannizzaro in 1858 led within a few years to the development of the periodic table. These atomic masses, combined with the many known mass percent compositions of compounds, enabled scientists to determine the molecular formulas for these compounds. Thus, by 1860, scientists knew the formulas for many compounds and they also knew that isomers existed; the same molecular formula could represent more than one compound. Since the isomers had different properties, they assumed that the structures had different fixed geometric shapes. They did not know what held the atoms together in these different shapes; they simply assumed that something did.

One of the properties that was different for some isomers was their interaction with polarized light. Recall from Chapter 4, Section 4.3, that a light wave consists of oscillating electric and magnetic fields perpendicular to one another. A **polarizer**, like the lenses in polarized sun glasses, allows light to pass only if the electric field is aligned in a particular direction, as in Figure 5.35; the transmitted light is said to be **polarized**. If a second polarizer is aligned in the same direction as the light, the **polarized light** will pass through. If the second polarizer is turned 90° from this alignment, none of the polarized light can get through; the polarizers are then "crossed." Figure 5.35 illustrates the principle of the **polarimeter**, an instrument invented early in the 19th century that is used to study the effect of substances on a beam of polarized light. In Figure 5.35(a) the pair of polarizers is aligned so the maximum amount of light is transmitted. Investigate This 5.59 gives you an opportunity to use a setup like Figure 5.35 to investigate the effect of water and a sugar solution on polarized light.

> WEB, Chep 5, Sect 5.9.1-4
> Learn more about polarized light and the polarimeter using these animations.

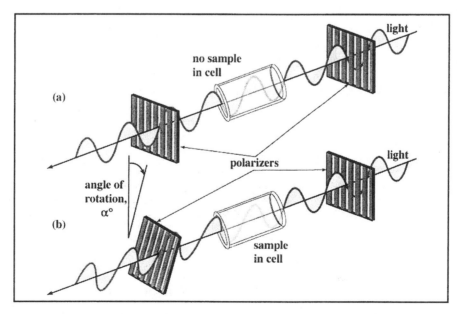

Figure 5.35. Principle of operation of a polarimeter.
(a) The two polarizers are aligned; light passes through to the observer. (b) A sample rotates the plane of polarization by $\alpha°$. The observer has to rotate the second polarizer by $\alpha°$ to align it with light leaving the sample.

5.59. Investigate This   *Do solutions affect polarized light?*

Do this as a class investigation and work in small groups to discuss and analyze the results. You need two, 10-cm square sheets of polarizing filter, two glass Petri dishes about 6-cm in diameter, and a sheet of light cardboard with an 8-cm diameter hole in the center. Place the cardboard on an overhead projector and put one of the squares of polarizing filter over the hole. Make a mark on the filter near the edge of the projected image. Place the second polarizing filter on top of the first and rotate the second filter until the image on the screen is as dark as possible. Make a mark on the second filter directly over the mark on the first filter.

Fill one of the Petri dishes about half full of water and sandwich the dish between the two polarizing filters. Make sure the mark on the top filter is directly over the mark on the bottom filter and check to see whether the image on the screen is still as dark as possible in the center where the dish of water is between the filters. If the image isn't as dark as possible, rotate the top filter until the image is as dark as possible and make another mark on the top filter to indicate the new alignment of the filters. Repeat the procedure with the second Petri dish about half full of light corn syrup (a concentrated solution of sugars) from the grocery store. Record your observations.

5.60. Consider This   *What solutions affect polarized light?*

(a) In Investigate This 5.59 you use the polarizers in their crossed condition because it is easier to tell when the image on the screen is darkest rather than lightest. If a sample between crossed polarizers changes the direction of polarization of the polarized light from the first filter, then the second filter has to be rotated to bring it into the crossed orientation. Does water change the direction of polarization of the light beam? What is the evidence for your answer? How would you describe the effect of water on polarized light?

(b) Does the solution of sugars change the direction of polarization of the light beam? What is the evidence for your answer? How would you describe the effect of a sugar solution on polarized light? Can you think of any further experiments you could do to find out more about the effect of water and sugar solutions on polarized light? If so, try them.

Some substances, when placed in a beam of polarized light, rotate the plane of polarization of the light. If such a sample is placed between aligned polarizers, the light passing through the sample will no longer be aligned with the second polarizer; the polarizer will have to be *rotated* to bring it back into alignment, as in Figure 5.35(b). The effect of the sample substance on polarized light is called **optical rotation**. Substances that rotate the plane of polarized light are said to be **optically active**. Different compounds rotate polarized light by different amounts. The amount of the rotation also depends

> The Latin words for right (*dexter*) and left (*laevus*) are often used to describe optical rotation: right, dextrorotatory, and left, levorotatory. Solutions of glucose and fructose are dextrorotatory and levorotatory, respectively. Their optical activity is what led to the names "dextrose" and "levulose" for these sugars.

on the number of molecules of the optically active compound that are in the polarized light beam. The rotation can be either clockwise (right) or counterclockwise (left), with respect to the unrotated light.

5.61. Check This *Direction of optical rotation*

(a) Figure 5.35(b) shows the angle of rotation of the sample as $\alpha°$. Is the rotation dextrorotatory or levorotatory? Explain your response.

(b) WEB Chap 5, Sect 5.9.2. The angle of rotation of the sample in the illustration is $\alpha°$. Is the rotation dextrorotatory or levorotatory? Explain your response.

(c) In Investigate This 5.59, was the rotation of the sugar solution dextrorotatory or levorotatory? Explain.

By the middle of the 19th century, many optically active compounds had been discovered. Most of these came from living matter, like the sugars in corn syrup that you used in Investigate This 5.59. While investigating the crystallization of such compounds, Louis Pasteur (French chemist, 1822-1895) observed that some solutions that were not optically active produced two different crystal forms, like those shown in Figure 5.36. Imagine that the dotted line between the two crystals is a mirror and you can see that *each is the mirror image of the other*. Pasteur separated the crystals, made solutions of both forms and found that the solutions were optically active. The startling result was that one solution rotated polarized light clockwise and the other rotated it counterclockwise by exactly the same amount. A mixture of equal amounts of the two solutions did not rotate polarized light.

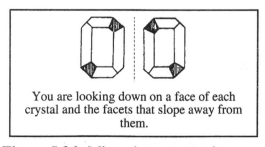

You are looking down on a face of each crystal and the facets that slope away from them.

Figure 5.36. Mirror image crystals.

Structure of Molecules   Chapter 5

Pasteur's observations and subsequent experiments by others convinced 19th century scientists that optical isomers always came in pairs. One of the isomers rotated polarized light one direction and the other in the opposite direction. As far as chemists could tell, all the chemical and other physical properties of optical isomers were identical; the isomers differed only in their optical rotation. The identical chemistry suggested that the isomers had basically the same arrangement of their constituent atoms, that is, they are stereoisomers. The mirror-image crystals formed by a pair of optical isomers, like those Pasteur had found, somehow seemed to reflect a fundamental mirror-image relationship between the molecular structures of the isomers.

5.62. Investigate This   *Are there stereoisomers of lactic acid?*

Work in small groups and compare the structures each group member makes. The Lewis structure for lactic acid (present in sour milk and your muscles after hard exercise) is given below. Make a molecular model of lactic acid. If you rotate parts of your model appropriately (without breaking any bonds), can you superimpose it on the models other students built? If not, how many different molecular structures are there for lactic acid? How are they related to one another? Explain using 3-d structural drawings to show the relationship(s).

$$\begin{array}{c} H \;\; :\!O\!: \\ | \quad \| \\ H-\!\ddot{\underset{..}{O}}\!-\!C\!-\!C\!-\!\ddot{\underset{..}{O}}\!-\!H \\ | \\ H-C-H \\ | \\ H \end{array}$$

Tetrahedral arrangement around carbon

In 1874, Jacobus van't Hoff (Dutch chemist, 1852-1911) and Jules Le Bel (French chemist, 1847-1930) independently proposed that a tetrahedral arrangement of four atoms or groups of atoms bonded about a central carbon atom explained the observations on optical isomers. The simplest cases are represented by compounds like lactic acid, in which there is *a central carbon atom with four different atoms or groups attached*. In Investigate This 5.62, you constructed models with tetrahedral arrangements around carbon and found that there are two structures for lactic acid that are not superimposable on one another. These are **optical isomers**, stereoisomers that are mirror images of each other and differ only in their effect on polarized light.

> **WEB Chap 5, Sect 5.9.5-7**
> Study animations of optical isomers and the effects of their structure.

Chapter 5 — Structure of Molecules

5.63. Consider This *Can you identify optical isomers?*

These problems model the kind of logic that van't Hoff and Le Bel went through to develop the tetrahedral-carbon model.

(a) Hold your model of lactic acid by the hydrogen connected to the central carbon and look down the H–C axis toward the C. Rotate the model so that the acid group, –C(O)OH, is at the 12-o'clock position. Which group, the –CH₃ or the –OH is at the 4-o'clock position? Make a drawing with the C in the middle, the acid at the 12-o'clock position and the other two groups at the correct 4-o'clock and 8-o'clock positions. Next to your first drawing, draw its mirror image. Find another student with a model that is not superimposable on yours. Check to see whether your mirror-image drawing describes the other student's model.

(b) Another possible arrangement of four atoms or groups around the central carbon is a square planar arrangement. You can treat the Lewis structure in Investigate This 5.62 as a representation of a square planar lactic acid structure. Is the mirror image of this structure different from the original? Why or why not? If you have trouble deciding, construct models of the original and its mirror image. For the central atom, you will have to select one of the atom centers that has holes in a square planar arrangement. How does your answer help to decide whether a square planar structure is a possibility for lactic acid?

Although the electrical nature of matter that we discussed in Chapter 1 was well known to the scientists of the later 19th century, the substructure of the atom had yet to be discovered. In spite of this and without the many instruments available today, these scientists were able to deduce the three-dimensional structure of many carbon-containing molecules like lactic acid. Further, they knew many other compounds that had more than one carbon with four different groups attached, notably sugars, and they were able to predict the number of mirror-image isomers these compounds have. Essentially all the arrangements of groups attached to second-period atoms were known before the bonding models we have discussed in this chapter had been developed. Indeed, this structural knowledge helped 20th century scientists develop these bonding models.

> Late in the 19th century, there were lively controversies about the structures of metal-containing complex ions. Some of these problems were resolved by arguments based on optical rotation of the complexes. We will discuss metal-ion complexes in Chapter 6.

5.64. Check This

(a) WEB Chap 5, Sect 5.9.5. View the animation of the stereoisomers of phenylalanine. Using the Lewis structure shown on the page to model the groups on the central carbon, draw 3-d

Structure of Molecules   Chapter 5

structures corresponding to the representations of the isomers at the end of the animation. Show clearly how and why these are stereoisomers and why they are optical isomers.

(b) WEB Chap 5, Sect 5.9.6. Which 3-d structure for phenylalanine from part (a) corresponds to the isomer shown in this animation of its effect on plane polarized light? Describe how the animation would be different if you substituted the other isomer in the path of the polarized light.

(c) The sugars in the corn syrup in Investigate This 5.59 have more complicated structures than lactic acid and phenylalanine. The skeletal structure for one form of glucose, an ingredient in corn syrup, is:

Here is a simple way to test whether this structure represents an optical isomer. Cut two identical regular hexagons out of paper. On one hexagon, put a red dot at the corner which is an oxygen in the glucose six-membered ring. Put a black dot at each of the other corners where an –OH group is out of the plane of the paper toward you in the glucose structure. Use the second hexagon to make a mirror image of the first. Are the two hexagons superimposable so that all dots of the same color match? What does your result tell you about the stereoisomers of glucose? Is this consistent with your results in Investigate This 5.59? Explain.

Reflection and projection

Isomers are compounds with the same elemental composition, but different properties. The molecules from a pair of isomeric compounds can always be distinguished from one another because the isomeric molecules, or models of the molecules, cannot be superimposed on one another. Structural isomers have different connectivities among the atoms, as well as different shapes. Stereoisomers have the same connectivities, but different shapes. There are two kinds of stereoisomers, *cis-trans* isomers and optical (mirror-image) isomers. You can often distinguish *cis-trans* isomers by structural drawings, because the structures are usually planar or close to planar and telling one side from the other is easy. These double-bonded structures are relatively rigid and it is this rigidity of the chain of double bonds in retinal that causes its change in shape, shown in the chapter opening illustration, to have such a strong effect on the shape of the much larger protein molecule to which it is bonded. Optical isomers are trickier to distinguish because the mirror-image relationship of the isomers is often not so easy to see without using models or some other physical representation of the shapes. In many carbon-containing compounds, you can tell whether optical isomerism is possible by searching the structure for carbon atoms to which four different atoms or groups are bonded.

Understanding molecular structure is important, because the reactions and interconversions of structures are what most interest chemists and biologists. In order for a molecule to react, there must be sites within it that can interact strongly with other molecules. Almost invariably, the interactions are between positive and negative centers in the reacting molecules. Such interactions are quite analogous to the reactions of cations and anions discussed in Chapter 2. In the next section, we'll examine several kinds of atomic groupings in molecules that provide sites for reaction.

Section 5.10. Functional Groups — Making Life Interesting

Alkanes

As you have seen in this and previous chapters, the structure of a molecule determines its chemical and as well as physical properties. In this chapter, we have emphasized the structure and bonding in carbon-containing compounds. A large number of these compounds also contain hydrogen. In most carbon-containing molecules, hydrogen is the element that occupies all the σ orbitals that the carbon atoms are not using to bond to other elements. (This fact is what enables us to leave the hydrogens off skeletal structures.)

Compounds that contain *only* hydrogen and carbon are called **hydrocarbons** to emphasize the fact that only those two elements are present. Hydrocarbons in which *every* carbon atom has tetrahedral geometry are called **alkanes**. The C_5H_{12} isomers we discussed in Sections 5.1 and 5.2 and the hexane in Investigate This 5.22 are examples of alkanes. The names of all alkanes are distinguished by the "-ane" ending. You should be able to recognize a structure as an alkane, but we will not present the rules for naming them or any other hydrocarbons.

In general, the chemistry of the alkanes is not very eventful because both carbon–carbon and carbon–hydrogen σ bonding orbitals are quite stable. Alkanes do burn in air, however, reacting with oxygen to produce carbon dioxide, water, heat energy, and light energy. Hydrocarbon fuels are of immense worldwide economic importance. These fuels, often referred to as "fossil fuels," originated from once-living organisms that decayed anaerobically ("without air") over millions of years. Natural gas is about 85% methane and about 15% ethane. Propane and butane are used in portable heating devices such as patio grilles and camp stoves. Gasoline is largely a mixture of low-boiling liquid hydrocarbons in the C_7–C_9 range. Though some microorganisms use methane as a source of energy (by metabolizing it, not by burning it) or produce methane as a product of their metabolism, alkanes are not generally found in living systems. Carbon-containing molecules that *are* in living things often have nonpolar hydrocarbon parts that affect their properties and interactions with other molecules.

Functional groups

The possibility for interesting reactions at the temperature of life largely depends on molecules that have electrons in less stable (higher energy) orbitals. These electrons can be attracted to positive centers in other molecules and, thus, set in motion the electron and atomic core redistributions that make up chemical reactions. The orbitals that fit this criterion are π bonding orbitals and σ nonbonding orbitals. There will be π bonding orbitals in any molecule with multiple bonds. Hydrocarbons never have σ nonbonding orbitals, but there will be σ nonbonding orbitals in molecules that contain other second-period elements to the right of carbon. **Multiple bonds and the presence of atoms other than carbon and hydrogen in the molecule are responsible for the chemical reactivity of most carbon-containing molecules**. Such a center of reactivity in the molecule is called a **functional group**.

Alkenes

The simplest functional group involves π bonds in hydrocarbons. The only hydrocarbons with π bonding that we will consider here are the **alkenes**, compounds that have at least one double bond between carbon atoms in the molecule. The names of alkenes end in "-ene."

Alkynes and aromatic rings also contain π bonds. Alkynes have a triple bond between two carbons. Aromatic rings are typified by benzene, C_6H_6. The six carbons are in a ring with three π-bonding orbitals delocalized around the ring. Many biomolecules contain aromatic rings, but they are often there for rigidity rather than for reactivity. See WEB Chap 5, Sect 5.7.1 for more about the bonding in benzene.

Alkenes are more reactive than the corresponding alkanes, as you observed in Investigate This 5.22, where you found that hexene reacts with potassium permanganate while hexane, an alkane, does not. Other examples of alkenes are ethene in Figures 5.13, 5.19, and 5.20 and *trans*- and *cis*-2-butene in Figures 5.33 and 5.34. Important structural characteristics of an alkene are the planar arrangement of the atoms bonded to the carbons and the strong resistance to rotation of the double bond, which results in stable *cis-trans* stereoisomers of alkenes.

Functional groups containing oxygen and/or nitrogen

The greatest richness of chemistry in carbon-containing compounds lies in the variety of chemical linkages between carbon and elements such as oxygen, nitrogen, and sulfur. We will begin to consider various classes of compounds where these elements are found. This will not be an exhaustive survey. Functional groups based on oxygen and nitrogen that you have met before and will meet again in this text are listed in Table 5.4 with an example of each. What we want you to get here is an acquaintance that will help you recognize them and focus on their polarity, which is the key to their reactivity.

Table 5.4. Oxygen– and nitrogen–containing functional groups.
In each representation, the lines from the carbons are bonds to either hydrogen or another carbon. The ball-and-stick representations of the examples show the σ bonding framework, not σ_n or π orbitals.

Functional Group	Representation	Example
alcohol	—C—O—H	ethanol
ether	—C—O—C—	dimethyl ether
aldehyde	—C(=O)H	propanal
ketone	—C—C(=O)—C—	propanone (acetone)
carboxylic acid	—C(=O)O—H	ethanoic (acetic) acid
ester	—C(=O)O—C—	methyl ethanoate (acetate)
amine (primary)	—C—NH$_2$	methyl amine
amide	—C—C(=O)NH$_2$	propanamide

Structure of Molecules Chapter 5

5.65. Investigate This *How many $C_3H_8O_2$ isomers are there?*

Work in small groups for this investigation and then share your results with the entire class. Use your molecular models to build as many different $C_3H_8O_2$ structures as you can. Record the structures you build by making drawings showing the *connections*. One of the possible structures and its representation are:

$$\text{H}-\underset{\underset{\text{H}}{|}}{\overset{\overset{\text{H}}{|}}{\text{C}}}-\text{O}-\underset{\underset{\text{H}}{|}}{\overset{\overset{\text{H}}{|}}{\text{C}}}-\underset{\underset{\text{H}}{|}}{\overset{\overset{\text{H}}{|}}{\text{C}}}-\text{O}-\text{H}$$

How many different structures can you build? What criteria do you use to tell that they are different from one another? Experiment with each isomer you make to see what sort of shapes it can twist into without breaking the connections. Record your observations about the structures of the molecules.

5.66. Consider This *What are the functional groups in $C_3H_8O_2$ isomers?*

(a) Consider the drawings of the structures you made in Investigate This 5.65. Which of them contain one or more alcohol functional groups?

(b) Which structures contain a functional group (or groups) not shown in the table? How would you represent this group in the style shown in the second column of Table 5.4? Have you seen examples of such a group before? Explain.

(c) Are there any stereoisomeric pairs among your structures? Explain why or why not.

Alcohols and ethers

Alcohols are compounds that contain an -OH group bonded to carbon with no other reactive atoms bonded to that carbon. The presence of the -OH group is designated by an "-ol" ending. For example, the butanol isomers in Sections 5.1 and 5.2 and the retinol (vitamin A) in Consider This 5.10 are alcohols. The endings of their names reveal that these compounds contain an alcohol functional group. Many biologically important molecules, such as the sugars you used in Investigate This 5.59, contain several alcohol groups.

For alcohols, chemical reactions are often associated with the nonbonding pairs of electrons on the alcohol oxygen. These electrons form a concentrated region of negative charge, which is attracted to the positive centers of other molecules and ions. Other reactions of alcohols occur because the electron density in the sigma bonds is pulled toward the electronegative oxygen, leaving the carbon and hydrogen as positive centers. The $^{\delta-}\text{O}-\text{C}^{\delta+}$ and $^{\delta-}\text{O}-\text{H}^{\delta+}$ bonds are both polar and the carbon and hydrogen positive centers are attracted to negative centers of other molecules and ions. In Chapter 1, you saw a result of these attractions, the hydrogen bonding among alcohol molecules. The molecule as a whole is polar, with its negative end toward the oxygen. The computer-generated model in Figure 5.37 allows us to visualize the electronic distribution in the molecule and shows the charge centers we have been discussing.

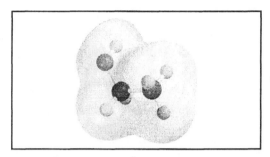

Figure 5.37. Computer-calculated electron density distribution in ethanol.

Note that localized molecular orbitals disappear in the computer-calculated electron density distributions. In Figure 5.37, there is no hint of separate regions of electron density between atomic cores. The calculations that produce these pictures are based on a model of molecular bonding that accounts for the fact that electrons cannot be distinguished from one another, so that we cannot find any particular electron. Although the localized and computer models lead to different pictures, they are complementary, not contradictory. For example, both show that there is a region of high electron density associated with oxygen. Our two localized σ_n nonbonding orbitals are represented as a single region of electron density in Figure 5.37. The localized model is very successful for understanding molecular geometry and the attractions that hold molecules together. Computer modeling is very useful for understanding molecular reactions.

Ethers are compounds that contain an oxygen atom bonded to two alkyl groups. Ethers have nonbonding electron pairs on the oxygen that can form hydrogen bonds with other molecules, but ether molecules do not hydrogen bond with one another. One consequence is that ethers boil at lower temperatures than the isomeric alcohols. The oxygen atom in ethers is a negative center, as it is for alcohols, but the partial positive charge is more spread out over the alkyl groups. As a result, ethers are not as reactive as alcohols. Sometimes it's useful to think of alcohols and ethers as derivatives of water in which alkyl groups replace one hydrogen atom (alcohols) or both hydrogen atoms (ethers).

Carbonyl compounds

Aldehydes and ketones contain a carbon–oxygen double bond called a **carbonyl group**. **Aldehydes** have two hydrogen atoms or a hydrogen atom and an alkyl group bonded to the carbonyl carbon and **ketones** have two alkyl groups bonded to the carbonyl carbon. The names of aldehydes often end in "-al" and the names of ketones often end in "-one" (long "o"). Pi-bonding electrons are not held as tightly as the σ-bonding electrons, so electrons in the carbonyl double bond are pulled even more toward the oxygen than those in the carbon–oxygen single bond of alcohols and ethers. Figure 5.37 shows the computer-calculated electron distribution in methanal (formaldehyde), the simplest carbonyl compound. The greater shift of electron density away from the carbon atom (compared to ethanol in Figure 5.37) is represented by the more extensive blue region (low electron density) at the carbon end of the molecule. As with alcohols, the positive and negative ends of the carbonyl bond, $^{\delta+}C=O^{\delta-}$, will be attracted to complementary charges on other polar molecules and could react.

Figure 5.38. Computer-calculated electron distribution in methanal (formaldehyde).

5.67. Check This *Retinol and retinal*

Explain the relationship of retinol, Consider This 5.10, to retinal, shown here and in two stereoisomeric forms in the chapter opening illustration.

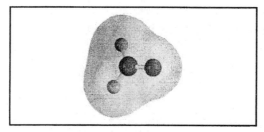

5.68. Consider This *What are the interactions between an alcohol and an aldehyde?*

(a) Can ethanol, Figure 5.37, and methanal, Figure 5.38, hydrogen bond with one another? If so, which charge center on each molecule will be interacting? Make a sketch, using 3-d structures to show how the molecules would interact.

(b) Make a sketch showing the other charge centers in each molecule interacting. If a partial bond were to form between the molecules in this interaction, which atom in each molecule would be bonded? Give your reasoning.

Chapter 5 — Structure of Molecules

Interactions between the centers of positive and negative charge in polar molecules are similar to the interactions between positive and negative ions that we discussed in Chapter 2. Since polar molecules have only partial charges, the interactions are not as strong, but, under appropriate conditions, they can lead to bond formation of the sort you imagined in Consider This 5.68(b) and thus to new molecules. We will explore more of these interactions in Chapter 6.

Carboxyl compounds

If one oxygen atom attached to a carbon atom will create a partial positive charge on carbon, two should create even more of a charge. **Carboxylic acids**, are characterized by a **carboxyl group**, $\overset{\delta+}{C}\overset{\delta-}{-O}\overset{}{\underset{O\delta-}{\|}}$, with a hydrogen bonded to the oxygen, $\overset{\delta+}{C}\overset{\delta-}{-O-H}\overset{}{\underset{O\delta-}{\|}}$ [–C(O)OH]. They have a carbon atom that is positive enough to be quite reactive. The positive center at carbon is shown in the computer-calculated electron distribution for methanoic acid (formic acid), the simplest carboxylic acid, Figure 5.39. The systematic names for carboxylic acids have an "-oic acid" ending. You have already encountered carboxylic acids on several occasions in the text, primarily in connection with acid–base chemistry. Another kind of carboxylic acid reaction forms an **ester**, another carboxyl-group-containing compound, Table 5.4, from a carboxylic acid and an alcohol. We will discuss this reaction in Chapter 6.

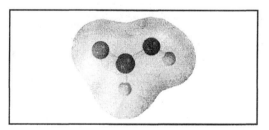

Figure 5.39. Computer-calculated electron distribution in methanoic acid (formic acid).

5.69. Check This *Identifying functional groups*

(a) What functional groups are present in retinol (Consider This 5.10)? in retinal, (Check This 5.67 and chapter opener)? How many of each group are present in each molecule?

(b) Find the functional group(s) in each of these molecules. Circle the atoms that define each functional group. Name each functional group. Give the molecular formula for each molecule.

    (i)   (ii)   (iii)   (iv)

(c) Do structures (iii) and (iv) represent the same molecule? Why or why not? How can you prove your answer? *Hint*: See Consider This 5.64(c).

5.70. Check This *Information from different representations of molecules*

WEB Chap 5, Sect 5.10.1. View and manipulate the representations of acetone and methylethanoate. Which representation is most similar for the two compounds? Explain how it is similar. Explain why the similarity in electron density representations would lead to a similarity in reactivity.

Multiple functional groups

Most biologically active compounds contain multiple functional groups. The groups are sometimes close enough together that their polarities affect one another. The amino acids found in proteins, for example, contain both a carboxylic acid group and an amine group bonded to the same carbon. Figure 5.40 shows the electron density distribution in glycine, the simplest example of these acids. Note the availability of electron pairs on the nitrogen and oxygen atoms as well as the polarization within the molecule. As we go on, you will see examples of chemical reactions of multifunctional compounds. In most cases, however, we will focus on the chemistry of an individual functional group in a compound.

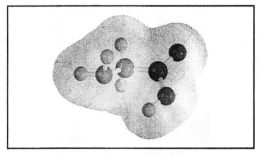

Figure 5.40. Computer-calculated electron distribution in glycine, an amino acid.

Section 5.11. Molecular Recognition

5.71. Consider This *Can you construct a "molecule" to order?*

(a) Use your molecular models to construct a pair of C_4H_8O *cis-trans* stereoisomers. Compare your solution with those of other students. How many different *structural* isomers are represented by all these solutions?

(b) Use your molecular models to construct a pair of C_4H_8O optical stereoisomers. Compare your solution with those of other students. How many different *structural* isomers are represented by all these solutions?

Chapter 5 — Structure of Molecules

Noncovalent interactions

Although we focus a great deal of attention on covalent bonds in chemistry, it is often the far weaker non-covalent interactions which control the properties of chemical and biological systems. For example, you have learned that relatively weak hydrogen bonds, typically about 5% the strength of single covalent bonds, account for the relatively high boiling points and specific heats of alcohols and water. In DNA, the accuracy and reproducibility of base pairing is enhanced by hydrogen bonding. Even weaker and more subtle interactions among mixtures of polar and nonpolar molecules are responsible for the fact that "oil and water don't mix." You will see, in Chapter 8, that these interactions help form the cell membranes that separate the inside of our cells from the outside. And, at this moment, the *cis*-to-*trans* isomerization of retinal is changing the shape of opsin molecules and causing a signal to be sent to sent to your brain (without a chemical reaction between the retinal and opsin), so that you can read this text.

Molecular recognition

You may be familiar with enzymes (many protein molecules) and their role in catalyzing cellular reactions. An enzyme is a protein molecule that is quite specific in catalyzing the reaction of only one type of molecule (out of a set of very similar molecules); the reacting molecule is called its **substrate**. What is the origin of the specificity of an enzyme for its substrate? There is a region of the enzyme called the **active site** (**receptor**), usually in a cleft in the three-dimensional enzyme structure. Functional groups from the amino acids that make up the protein are present in the active site. In order to bind in the correct orientation for reaction, functional groups on the substrate and those in the active site must be attracted to one another by noncovalent interactions. This is the way an enzyme "recognizes" its substrate.

5.72. Check This *One form of molecular recognition*

WEB Chap 5, Sect 5.9.7. Play the animation and determine which isomer fits the receptor molecular surface. Explain how this animation exemplifies the description of molecular recognition in the preceding paragraph.

Molecular recognition is not restricted to enzymes. The ways in which antibodies recognize foreign substances in our bodies, the way that one nerve cell communicates with its neighbor, and the senses of taste and smell all rely on the specific interactions of a binding site and a

> **WEB Chap 5, Sect 5.11**
> Explore these animations to see how a taste receptor and its substrates interact.

substrate. Recognition is becoming extremely important in non-biological applications as well, for example, to create ordered arrays of molecules with desirable electronic properties. The more

Structure of Molecules Chapter 5

we know about the fundamental nature of molecular recognition, the better we will be able to design new materials, molecules with specific characteristics (as in the simple examples in Consider This 5.71), and devise new technologies for analysis and sensing. For example, we may soon be able to use "artificial noses" to detect a wide range of odors with a sensitivity well beyond the human nose. In the not-to-distant future, your refrigerator may be able to tell you when it contains spoiled food and even which food it is!

Reflection and projection

The sites of reactivity in carbon-containing molecules, their functional groups, are characterized by the presence of π bonding and σ_n nonbonding electrons. In hydrocarbons, the only possible functionality is the π bond. Almost all reactions of carbon-containing molecules involve functional groups containing atoms other than carbon, such as oxygen and nitrogen. These molecules all have nonbonding electrons and many also have π bonding electrons between carbon and other second period atoms. These functional groups include alcohol, ether, carbonyl (aldehydes and ketones), carboxyl (carboxylic acids and esters), amine, and amide.

In all these functional groups, the electronegativity of oxygen or nitrogen, relative to carbon, polarizes the electron density toward the oxygen or nitrogen. Positive and negative centers are a characteristic of most functional groups and are responsible for their reactivity. Even this limited set of functional groups can undergo a great variety of reactions that depend on the reaction conditions and the other functional groups present in the system. In the next chapter, we will explore a variety of reactions, all of which are initiated by the attraction of positive to negative centers in molecules and ions. We finished here by examining how the shape of molecules and the interactions between their functional groups provide a means for molecular recognition.

Section 5.12. Outcomes Review

In this chapter, we used electron wave mechanics to further develop the bonding models introduced in Chapter 1. We found that the atomic cores in all molecules are held together by attractions between their positive charges and negatively-charged two-electron sigma bonding (σ) molecular orbitals located directly between them. Atomic cores in molecules may also have sigma nonbonding (σ_n) molecular orbitals. There are also molecules with multiple bonds between atoms, a σ bond and a pi (π) or two π bonds. In some cases, orbitals may be delocalized over three or more atomic centers in a molecule or over an entire crystal in metals.

The purposes for developing the bonding model are to understand and predict the shape and reactivity of molecules. The geometry (shape) of a molecule is determined by its framework of σ

and σ_n electrons and we discussed several ways, including using molecular models, to represent shapes. Isomers, both structural isomers and stereoisomers, are distinguished from one another by different properties and different molecular shapes. The reactions of molecules generally involve attraction of a positive site in one molecule to a negative site in another. Models that help us visualize the electron distribution in molecules show the location of these sites, which are the functional groups in the molecule.

Check your understanding of the ideas in the chapter by reviewing these expected outcomes of your study. You should be able to:

- write appropriate Lewis structures for molecules whose atomic composition you know, including multiple Lewis structures representing possible isomers [Sections 5.2, 5.5, 5.8 and 5.9].

- build molecular models based on the connectivities shown in Lewis structures of the molecules [Sections 5.2, 5.5, 5.8 and 5.9].

- for a series of compounds, correlate patterns of chemical behavior with their Lewis structures and molecular models and, based on these patterns, predict the behavior of other compounds [Sections 5.1, 5.2, 5.5, and 5.9].

- use Lewis structures to predict multiple bonding in the molecules of a compound [Section 5.5].

- draw and/or describe the σ, σ_n, and π orbitals for simple molecules containing elemental atoms from periods one, two, and three [Sections 5.3, 5.4, 5.6, and 5.7].

- draw and/or describe the sigma framework of a molecule and predict the shape of the molecule [Sections 5.4 and 5.6].

- use Lewis structures to predict whether a molecule is likely to have delocalized π orbitals [Sections 5.7 and 5.8].

- use the delocalized molecular orbital ("sea of electrons") model of metallic bonding to explain the properties of metals [Section 5.7].

- use bonding models to determine bond order in molecules [Sections 5.7 and 5.13].

- beginning with any one of the following representations of relatively simple molecules, write, draw, or build all the other representations: condensed (line) formula, Lewis structure, 3-d structure, condensed structure, skeletal structure, molecular model [Sections 5.2, 5.5, and 5.8].

- use appropriate molecular representations to show whether a pair of structures are unrelated to one another, are identical to one another, are structural isomers, or are stereoisomers (*cis/trans* and/or optical isomers) [Sections 5.1, 5.2, 5.8, and 5.9].

Structure of Molecules
Chapter 5

- use appropriate molecular representations to show whether structural isomers and/or stereoisomers (*cis/trans* and/or optical isomers) are possible for a given molecular formula [Sections 5.2, 5.8, and 5.9].
- identify the functional group(s) present in a molecular structure [Section 5.10].
- use the polarities of functional groups (from electronegativities and bond polarities and/or calculated electron distributions) to predict where molecules will be likely to interact with one another [Sections 5.10 and 5.11].

Section 5.13. EXTENSION — Antibonding Orbitals: The Oxygen Story

5.73. Investigate This   *Are O_2 and N_2 gas affected by a magnet?*

Do this as a class investigation and work in small groups to discuss and analyze the results. You need a small, strong magnet, a large Petri dish about half full of a soap bubble solution, a syringe with a short needle, and sources of nitrogen and oxygen gas. You can use an overhead projector to project your experiment and make it easier to see. Fill the syringe with nitrogen gas, insert the needle a little below the surface of the soap solution, and blow a small, hemispherical bubble of nitrogen gas about 5-6 mm in diameter on the surface of the solution. Bring the magnet close to one side of the bubble and observe how the bubble reacts to the presence of the magnetic field. Can you move the bubble with the magnetic field from the magnet? Does the other pole of the magnet give the same result? Record your results. Repeat the experiment with oxygen gas.

5.74. Consider This   *How are O_2 and N_2 gas affected by a magnet?*

Are the results for nitrogen and oxygen gas any different in Investigate This 5.73? If so, how do they differ? How might you explain any difference(s)?

This chapter has dealt mainly with many-atom molecules, as we have tried to prepare you to understand the structure and reactivity of the natural and synthetic substances you meet daily. In this EXTENSION we will examine a small molecule, the diatomic oxygen molecule. Oxygen is essential to life, as it has evolved on Earth. Reactions of oxygen and fossil fuels also provide the vast amounts of energy that drive the economy of the world. Like water, oxygen is all around us and we rarely give it a second thought. Also, like water, when examined in more depth, oxygen has surprising properties. The one that you discovered in Investigate This 5.73, attraction of oxygen to a magnet, poses problems for the Lewis bonding model.

Paramagnetism of oxygen

You found in your investigation that oxygen gas in a bubble can be pulled about by a magnet. The oxygen molecule is **paramagnetic**. Paramagnetic (*para* = near + magnet = nearly a magnet) substances have magnetic strengths between those of **ferromagnetic** substances (like refrigerator magnets) and **diamagnetic** (*dia* = through) substances (all electron spins paired) that are unaffected or slightly repelled by magnets. The magnetic properties of oxygen were discovered in the 19th century. Recall from Chapter 4, Section 4.9, that atoms with unpaired electron spins are attracted to the poles of a magnet. Paramagnetic substances have atoms or molecules with unpaired electron spins. Quantitative measurements on oxygen show that it has two electrons with the same spin, that is, unpaired spins. Oxygen molecules have an even number of electrons; other covalently bonded molecules with an even number of electrons do not have unpaired electrons and are not paramagnetic.

The Lewis structures for nitrogen, :N≡N: , and oxygen, :Ö=Ö: , both seem straightforward and predict that all electrons in both molecules will be spin-paired. Your results in Investigate This 5.73 show that nitrogen is not attracted to a magnet (it is diamagnetic), so the Lewis structure gives a correct prediction. Oxygen, however, is a strange case for which the standard Lewis rules don't fit.

Antibonding pi molecular orbitals

In Section 5.6, we discussed electron wave interference and illustrated in Figure 5.16 how reinforcing waves from adjacent sigma-bonded atoms could create a π bonding molecular orbital. When we discussed wave interference in Chapter 4, Section 4.3, we said that both constructive and destructive wave interference occur; these were illustrated in Figure 4.9. In Section 5.6 we only considered the constructive interference of electron waves. Destructive interference of electron waves from two adjacent atoms is illustrated in Figure 5.41.

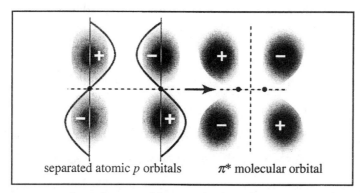

Figure 5.41. Electron wave interference (destructive) to form a π^* antibonding orbital.
Electron wave amplitude between the atomic cores is cancelled for waves with nonmatching + and − amplitudes. Nodes are shown by dashed lines. The cancellation creates a second node perpendicular to the bond axis.

The molecular orbital that is formed by destructive interference is called a **pi antibonding orbital**, symbolized as π^* ("pi star") to signal its antibonding character. You can see that the π^* orbital concentrates electron density *beyond* the two atomic cores, so its attraction to the atomic cores is in a direction that pulls them apart rather than holding them together. The π^* molecular orbitals are called antibonding because these electrons *raise* the energy of the molecule *relative to the atoms from which it is made*. Most molecules do not have any valence electrons in π^* antibonding orbitals, but two that do are essential to life. Molecular oxygen, O_2, and nitric oxide, NO, are two molecules that have electrons in π^* antibonding orbitals.

> Nitric oxide is a toxic gas. In minute amounts it plays an essential role as a nervous system signaling molecule, a function that has been recognized only since 1990.

Molecular orbital model for oxygen

Figure 5.42 is an approximate energy level diagram, similar to Figure 5.25, for the bonding and antibonding valence molecular orbitals in O_2. The two nonbonding orbitals, one on each atom, are omitted from this diagram, since their four electrons don't affect the bonding energies. The energies in this diagram are given *relative to the energies of the separated atoms*. Bonding orbitals are at lower (negative) energies and antibonding orbitals are at higher (positive) energies. There are two sets of energy-degenerate orbitals, the π and π^* orbitals.

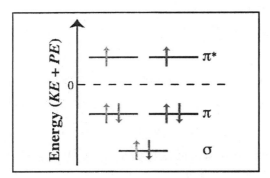

Figure 5.42. Relative energies for oxygen molecular orbitals.
Energies are relative to the separated atoms. The spins of electrons in the orbitals are shown as up and down arrows. Levels for the nonbonding orbitals, which contain 4 electrons, are at the zero of energy and are not shown.

The arrows in Figure 5.42 represent the eight valence electrons in these orbitals, with the direction of the arrow representing the spins of the electrons. In accord with our model that describes electrons by the lowest energy orbitals available to them, we can accommodate six electrons in the three lowest energy orbitals: one σ bonding and two π bonding orbitals. We need to add two more electrons to complete the molecule and the next lowest energy orbitals are two energy degenerate π^* antibonding orbitals. We place one electron in each π^* orbital, since that keeps them in different regions of space and reduces electron-electron repulsion.

Because the two antibonding electrons are in different orbitals, they do not need to be spin paired. Recall the experimental evidence from Chapter 4, Sections 4.10 and 4.11 (especially Table 4.3 and Consider This 4.62) which showed that when single electrons occupy energy-degenerate orbitals, they do so with unpaired spins. Thus, we predict that the ground state of oxygen will have two electrons of the same spin in the π^* orbitals, as shown in Figure 5.42. These unpaired electron spins explain the observed paramagnetism of oxygen.

In Figure 5.42, there are paired electrons in three bonding orbitals, a σ and two π, just like in the triple bonds you have seen previously. However, the two electrons in the π^* antibonding orbitals cancel the bonding energies of two of the electrons in the π bonding orbitals. Therefore, the *bond order*, that is, the effective number of two-electron bonds is two; molecular oxygen has a double bond. The molecular orbital model successfully explains (or rationalizes) both the paramagnetism and double-bond properties of molecular oxygen.

5.75. Check This *Molecular orbitals in nitrogen*

Assume that the energy level diagram in Figure 5.42 is also applicable to N_2. Draw a diagram like Figure 5.42 that describes N_2. Show the spins of electrons described by each orbital as up and down arrows. What is the bond order for N_2? Are the molecular orbital and Lewis descriptions compatible? Explain.

Ground-state oxygen reacts slowly

The reactivity of molecular oxygen is profoundly affected by its unpaired electrons. Ground-state oxygen, with its unpaired electron spins is quite unreactive. For ground-state oxygen, room temperature reactions go so slowly as to be almost unobservable. Combustion reactions of oxygen and fuel molecules are rapid, but they require a source of energy to get the oxygen and fuel hot, before the reaction can occur. In the laboratory, we can easily excite oxygen to a state in which all its electrons are spin-paired (the two highest energy electrons in different π^* orbitals, but with opposite spins). The room temperature reactions of spin-paired oxygen with carbon-containing compounds are rapid and readily destroy these compounds.

If oxygen had a spin-paired ground state, it is unlikely that life based on aerobic reactions would exist. The retinal molecule shown in the chapter opening illustration would probably be destroyed as fast as it could be formed. Recalling Wordsworth's thought from the beginning of the chapter, you might think that it is fortunate that the active principle assigned to the form (electronic state) of oxygen is to be *less* active. Oxygen is perhaps an extreme example of the connection between molecular bonding and reactivity, but it is a good example to keep in mind as you continue through the text and see many more examples of this fundamental connection.

Index of Terms

σ bonding framework, 5-22
σ framework, 5-22
π^*, pi antibonding orbital, 5-78
π, pi bonding orbital, 5-33
σ, sigma bonding orbital, 5-19
σ, sigma orbital, 5-18
σ_n, sigma nonbonding orbital, 5-20
3-d structure, 5-48
active site, 5-73
alcohols, 5-68
aldehydes, 5-69
alkanes, 5-65
alkenes, 5-66
atomic cores, 5-9
bent bonds, 5-30
bond angle, 5-23
bond length, 5-17
bond order, 5-42, 5-79
bonding orbital, 5-18
carbonyl group, 5-69
carboxyl group, 5-71
carboxylic acids, 5-71
cis-isomers, 5-57
condensed structure, 5-51
delocalized π molecular orbital, 5-41
double bond, 5-30, 5-42
ductile, 5-47
ester, 5-71
ethers, 5-69
functional group, 5-66
hydrocarbons, 5-65
isomers, 5-7
ketones, 5-70

localized π molecular orbital, 5-35
localized, one-electron molecular orbitals, 5-18
malleable, 5-47
metallic bonding model, 5-46
molecular orbital, 5-15
molecular shape, **5-22**
nodal plane, 5-34
nonbonding electrons, 5-20
octahedral, 5-26
optical isomers, 5-62
optical rotation, 5-61
optically active, 5-61
paramagnetic, 5-77
pi antibonding orbital, 5-78
pi bonding molecular orbital, 5-33
polarimeter, 5-59
polarized, 5-59
polarized light, 5-59
polarizer, 5-59
receptor, 5-73
sigma bonding orbital, 5-19
sigma molecular orbital, 5-18
sigma nonbonding molecular orbital, 5-20
single bond, 5-42
skeletal structure, 5-53
stereoisomers, 5-58
structural isomers, 5-56
substrate, 5-73
trans-isomers, 5-57
trigonal bipyramidal, 5-25
trigonal pyramidal, 5-22
triple bond, 5-32, 5-37

valence electrons, 5-9

| Chapter 5 | Structure of Molecules |

Chapter 5 Problems

Section 5.1. Isomers

5.1. This is the structure of 2-methyl-1-propanol, $C_4H_{10}O$.

Which of these structures is also a correct representation of 2-methyl-1-propanol? Explain your decision in each case. Use models, if necessary, for your explanation.

(i)   (ii)   (iii)

5.2. Both 1-pentanol and 2-pentanol have the same molecular formula, $C_5H_{12}O$.

(a) Draw the structure of each alcohol.

(b) Predict which of these two alcohols has the higher boiling point. Explain the reason for your prediction.

(c) Draw all alcohols with the formula $C_5H_{12}O$. *Hint:* there are eight total including 1-pentanol and 2-pentanol.

5.3. A student followed the procedure in a laboratory manual to prepare an ionic compound, a green solid with the composition, $Co(H_2NCH_2CH_2NH_2)_2Cl_3$. She dissolved the compound in water, warmed the solution, and found that the solution turned reddish purple. She evaporated the water, recrystallized the reddish-purple crystals, and analyzed them for cobalt, chloride, and 1,2-diaminoethane, $H_2NCH_2CH_2NH_2$. She found that the reddish-purple compound had the composition, $Co(H_2NCH_2CH_2NH_2)_2Cl_3$. What conclusion(s) can you draw about the relationship(s) between the green and reddish-purple compounds? Explain your reasoning.

Section 5.2. Lewis Structures and Molecular Models of Isomers

5.4. Predict the relative values for the boiling points of 1-propanol and 2-propanol.

1-propanol   2-propanol

Structure of Molecules   Chapter 5

5.5. This is the Lewis structure for propane, C_3H_8:

$$\begin{array}{c} H\ H\ H \\ | \ \ | \ \ | \\ H-C-C-C-H \\ | \ \ | \ \ | \\ H\ H\ H \end{array}$$

   (a) Use your model kit to make this model, and then draw a representation of this structure in three dimensions.

   (b) What is the shape (geometry) around the central carbon atom?

   (c) Are there any other isomers for propane, C_3H_8?

5.6. This is the Lewis structure of n-pentane, C_5H_{12}.

$$\begin{array}{c} H\ H\ H\ H\ H \\ | \ \ | \ \ | \ \ | \ \ | \\ H-C-C-C-C-C-H \\ | \ \ | \ \ | \ \ | \ \ | \\ H\ H\ H\ H\ H \end{array}$$

   (a) How many valence electrons are represented in this structure?

   (b) Including electrons in atomic cores, what is the total number of electrons implicitly represented?

5.7. The boiling points of the three pentane isomers are given in Table 5.2. Make molecular models of these compounds and cover them with plastic wrap or aluminum foil to represent the surface of each molecule. Explain why n-pentane has the highest boiling point and neopentane has the lowest boiling point of this set of isomers. *Hint:* Consider the amount of surface each molecule has to interact with its neighbors in the liquids.

5.8. In Check This 5.11, you were asked to explain the boiling point data (Table 5.1) for the four $C_4H_{10}O$ alcohols. Draw the three ethers that also have the formula $C_4H_{10}O$ and predict which one will have the lowest boiling point. Explain the reasoning for your selection. *Hint:* See Problem 5.7.

5.9. In Check This 5.7 you wrote Lewis structures and made models of the four isomeric amines having the formula C_3H_9N. Predict which one will have the lowest boiling point and explain the reasoning for your selection. *Hint:* Consider all the possible intermolecular interactions in the liquids.

Section 5.3. Sigma Molecular Orbitals

5.10. Which electrons, core or valence, are generally responsible for chemical bonding? Explain why.

5.11. Explain the difference(s) between an atomic orbital and a molecular orbital.

5.12. Explain the difference(s) between a bonding and a nonbonding molecular orbital.

5.13. Why do two hydrogen atoms combine to form a hydrogen molecule (H_2)? Explain using

(a) a Lewis structure for bonding.

(b) the MO model for bonding.

5.14. Why don't two helium atoms combine to form a helium molecule (He_2)? Explain using

(a) a Lewis structure for bonding.

(b) the MO model for bonding.

5.15. Explain why it is possible for molecules to exist, if there is always a repulsive force between positively charged nuclei?

5.16. Identify the number of valence electrons in each of the following:

(a) Na (b) CO_2 (c) NH_3 (d) Se

5.17. Consider Figure 5.4, Relative energies for a two-nuclei-one-electron molecular orbital system. It has been redrawn here with curves 1-4 labeled.

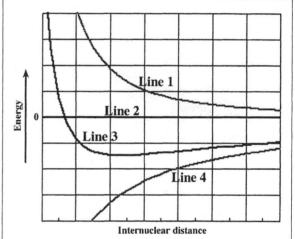

(a) What information does line **1** give you as two nuclei move towards each other?

(b) What information does line **2** give you as two nuclei move towards each other?

(c) What information does line **3** give you as two nuclei move towards each other?

(d) What information does line **4** give you as two nuclei move towards each other?

Section 5.4. Sigma Molecular Orbitals and Molecular Geometry

5.18. In certain gas-phase reactions, methane can lose a hydrogen cation, H^+, to form the methide anion, CH_3^-.

(a) Write the Lewis structure for the methide anion.

(b) How would you describe the geometry of the sigma orbitals in the methide anion?

(c) How would you describe the geometry of the methide anion?

(d) Are your answers to parts (b) and (c) the same? Explain why or why not.

Structure of Molecules Chapter 5

5.19. Sodium amide, $NaNH_2$, is a white crystalline ionic compound.

(a) Write the Lewis structure for the amide anion, NH_2^-.

(b) How would you describe the geometry of the sigma orbitals in the amide anion?

(c) How would you describe the geometry of the amide anion?

(d) Are your answers to parts (b) and (c) the same? Explain why or why not.

5.20. (a) How many valence electrons are there in the Lewis structure for IF_5?

(b) How many total electrons, including all of the core electrons, are there in a molecule of IF_5?

(c) Write a Lewis structure for IF_5.

(d) The geometry of the fluorine atoms in the IF_5 molecule is square pyramidal (a pyramid with a square base). Why is this different from the molecular shape of PF_5, in which the geometry of the fluorine atoms is trigonal bipyramidal, as shown in Figure 5.11(b)?

5.21. For each of the following molecules or ions, write a Lewis structure, including all non bonding pairs of electrons.

(a) I_3^-

(b) BF_4^-

(c) SF_4

(d) XeF_4

(e) PF_6^-

5.22. For each molecule or ion in Problem 5.21, explain why these are the observed geometries (shapes).

(a) I_3^- is linear.

(b) BF_4^- is tetrahedral.

(c) SF_4 is shaped like a "see-saw", /\ , with S at the fulcrum.

(d) XeF_4 is square planar, with the Xe at the center of the square.

(e) PF_6^- is octahedral with the P in the center of the octahedron.

Section 5.5. Multiple Bonds

5.23. Which second-period elements are capable of making double bonds under normal conditions? Why are the others not included? Explain the reasoning for your choices.

5.24. Which elements never make double bonds? Explain the reasoning for your choices.

5.25. For each molecule, write a Lewis structure and determine the molecular geometry for each carbon atom. *Hint:* each molecule has a multiple bond. Molecular models may be helpful.

(a) C_3H_6 (propene has one carbon-carbon double bond)

(b) C_3H_4 (propyne has one carbon-carbon triple bond)

(c) C_3H_4 (allene has two carbon-carbon double bonds)

(d) C_2H_4O (ethanal has one carbon-oxygen double bond)

(e) C_2H_3N (acetonitrile has one carbon-nitrogen triple bond)

Section 5.6. Pi Molecular Orbitals

5.26. What is(are) the difference(s) between a sigma (σ) and a pi (π) orbital?

5.27. Why do all molecules have sigma bonding orbitals, but not necessarily pi-bonding orbitals?

5.28. If two carbon atoms are bonded by a sigma bonding orbital, why is a second bond between those two atoms a pi orbital rather than another sigma bonding orbital?

5.29. Why are molecules with multiple bonds generally more reactive than similar compounds without multiple bonds?

Section 5.7. Delocalized Orbitals

5.30. Consider the following molecules and ions. Circle each one that contains a π bond. For those molecules with a π bond, place a D next to ones where the π bond is delocalized and an L next to ones where the π bond is localized. Building models may be helpful.

(a) H_2CCH_2 (d) NH_3 (g) CH_4

(b) H_2CO (e) CO_2 (h) SO_2

(c) NO_3^- (f) $CH_3C(O)OH$ (i) H_2O

Structure of Molecules   Chapter 5

5.31. Consider the hydrogen carbonate anion, $(HO)CO_2^-$, and the ozone molecule, O_3.

(a) What is the total number of valence electrons in each species?

(b) Draw the Lewis structure for each species.

(c) How many σ bonding orbitals are there around the central atom in each species?

(d) What geometry is predicted for each species?

(e) How many π bonding orbitals are there around the central atom in each species?

(f) How many localized and how many delocalized π bonding orbitals are there in each species? Explain how you arrive at your answer.

(g) What are the bond orders for carbon-to-oxygen in $(HO)CO_2^-$ and oxygen-to-oxygen in O_3? Use sketches of the delocalized orbitals, if any, to illustrate your answer. What, if any, similarities are there between the delocalized orbitals and the bond orders in these two species? Explain.

5.32. Consider the nitrate ion, NO_3^-, the nitrite anion, NO_2^-, and the nitronium cation, NO_2^+.

(a) What is the total number of valence electrons in each ion?

(b) Draw the Lewis structure for each ion.

(c) How many σ bonding orbitals are there around the central atom in each ion?

(d) What geometry is predicted for each ion?

(e) How many π bonding orbitals are there around the central atom in each ion?

(f) How many localized and how many delocalized π bonding orbitals are there in each ion? Explain how you arrive at your answer.

(g) What is the bond order for nitrogen-to-oxygen in each ion? Use sketches of the delocalized orbitals, if any, to illustrate your answer.

5.33. A single Lewis structure for the nitrate ion doesn't represent the distribution of the π bonding orbital over all four atom cores, an alternative that has been extensively used is to draw more than one Lewis (or 3-d) structure:

$$\left[\begin{array}{c} \ddot{O} \\ \ddot{O} - N = \ddot{O} \\ | \\ \ddot{O} \end{array} \longleftrightarrow \begin{array}{c} \ddot{O} = N - \ddot{O} \\ | \\ \ddot{O} \end{array} \longleftrightarrow \begin{array}{c} \ddot{O} - N - \ddot{O} \\ \| \\ \ddot{O} \end{array} \right]^-$$

The double-headed arrows are used to indicate that the best representation is some intermediate structure (that is impossible to write).

(a) If all three of these structures existed separately, they would have identical energies. Explain why.

(b) If the structures have identical energies, then we would expect an intermediate structure based on them to have equal contributions from each structure. What fraction contribution would each make to the intermediate structure? What is your reasoning?

(c) The delocalized π bonding structure, Figure 5.23(a), led us to the conclusion that all three of the nitrogen-oxygen bonds in nitrate can be thought of as $1\ ^1/_3$ bonds. Show how the same conclusion can be reached here, based on equal contributions from these three structures to the proposed intermediate structure.

5.34. **(a)** We have said that molecules are likely to have delocalized pi orbitals, if there is more than one energy-equivalent way to write localized pi orbitals. Write the Lewis structure for carbon dioxide, as you did in Consider This 5.38. Would you expect the molecule to have delocalized pi orbitals? Explain why or why not.

(b) The molecular orbital picture for carbon dioxide, Figure 5.24, shows two pi orbitals delocalized over all three atomic cores in the molecule. Is this consistent with your answer in part (a)? Explain why or why not.

(c) Another way to represent the localized pi orbitals in carbon dioxide is shown in this molecular model. (The sigma nonbonding electrons are not represented in the model.) Is there more than one equivalent way to model these pi orbitals? Explain why or why not and construct a model (or models) to demonstrate your explanation.

(d) Is your answer in part (c) consistent with the delocalized pi orbital picture for carbon dioxide? Explain why or why not.

5.35. When some material is connected between the terminals of a battery, there is an electrical potential difference between the two connections to the battery. If electrons can enter the material at the negative terminal and leave at the positive terminal to complete the electric circuit, we call the material an electrical conductor. Metals are good electrical conductors. How does the delocalized molecular orbital ("sea of electrons") model of metallic bonding help you understand the high conductivity of metals? Present your reasoning clearly.

5.36. You can often break a piece of wire or thin strip of metal by bending it back and forth repeatedly. This action causes dislocations of atoms in the crystal like those you see represented in Figure 5.26. Explain how this effect can eventually lead to breaking the piece of metal.

Structure of Molecules

Section 5.8. Representations of Molecular Geometry

5.37. The number of isomers possible for any particular molecular formula depends on the geometry about the atoms that make up the molecule. For each of the following compounds, use your models to figure out how many isomers are possible if the four bonds about the central carbon atom (shown in **bold**) have a square planar molecular shape? How many isomers are possible if the four bonds about the central carbon atom have a tetrahedral molecular shape?

(a) **CH$_3$**Br, (methyl bromide, a fumigant)

(b) **CH$_2$**Cl$_2$, (dichloromethane, a solvent used in semiconductor processing)

(c) CH$_3$**CH**(OH)CH$_2$OH, (propylene glycol, a shampoo additive)

5.38. Skeletal structures for several molecules are given. Draw the condensed representations for each molecule. What is the molecular formula in each case.

(a) aspirin

(b) nicotine

(c) vitamin C

(d) cholesterol

5.39. **Conformations** are forms of a molecule that differ only in that there have been rotations about single bonds. Identify which of the following pairs of structures are **conformers** (*i.e.*, the same molecule) and which are actually different molecules that are not interconvertible by any combination of rotations about single bonds or motions of the molecule as a whole. Building molecular models might be helpful in some cases.

(a)

(b)

(c)

Chapter 5

Structure of Molecules

(d), (e), (f) [stereochemistry structures]

5.40. Rotations about single bonds are rapid. This allows molecules easily to assume different conformations. (See problem 5.39.) For each of the following molecules, can you find a conformation that allows all the carbons to lie in a single plane? Draw 3-d structures to illustrate your answers. *Hint*: molecular models are extremely valuable in answering this question.

   (a) butane, $CH_3CH_2CH_2CH_3$

   (b) 2-methylpropane, $CH(CH_3)_3$ (common name: isobutane)

   (c) 1,3-butadiene, $CH_2CHCHCH_2$

   (d) 1,2-propanediene, CH_2CCH_2 (common name: allene)

5.41. For each of the following molecules, can you find a conformation (See problem 5.39.) that allows all the carbons AND all the hydrogens to lie in a single plane? Draw 3-d structures to illustrate your answers.

   (a) butane, $CH_3CH_2CH_2CH_3$

   (b) 2-methylpropane, $CH(CH_3)_3$

   (c) 1,3-butadiene, $CH_2CHCHCH_2$

   (d) 1,2-propanediene, CH_2CCH_2

5.42. For each of the following molecules, can you find a conformation (See problem 5.39.) that allows all the atomic centers to lie in a single plane? Draw the Lewis structures first to help answer this question. Draw 3-d structures to illustrate your answers.

   (a) methanol CH_3OH (wood alcohol)

   (b) hydrogen peroxide H_2O_2 (used as a disinfectant and to bleach hair)

   (c) hydrogen cyanide   HCN   (poisonous gas)

   (d) nitric acid $(HO)NO_2$ (useful industrial acid)

   (e) nitrous acid   (HO)NO   (nitrite salts are suspected carcinogens)

Structure of Molecules Chapter 5

Section 5.9. Stereoisomerism

5.43. Stereoisomers (isomers that share the same connectivity of their bonds, but differ in their 3-d shapes) can be subcategorized as optical isomers (those stereoisomers that bear a mirror image relationship to one another) and those that are not optical isomers. Identify which of the following pairs of structures are **identical**, which are **optical isomers**, and which are **other types of stereoisomers**. Building models could be very helpful.

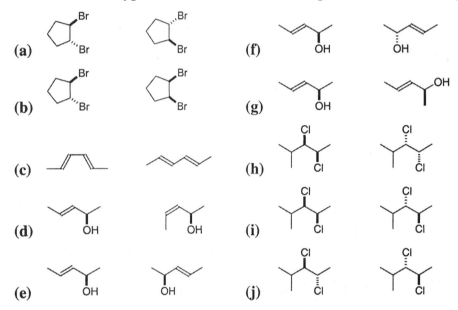

5.44. (a) Complete this table by filling in the blanks with either "the same" or "different."

Property	Structural Isomers	Stereoisomers	
		Optical Isomers	Other
Molar mass			
Boiling point			
Density			
Bond connectivity			
Chemical reactivity			

(b) For one of these data columns, all the entries are "the same." By definition, two compounds cannot be isomers if <u>all</u> of their properties are the same. What is the physical property that allows us to distinguish one optical isomer from its mirror image optical isomer?

5.45. Use your molecular models to make as many models of isomers of C_5H_{10} as you can. How many are there? (If it is easy for you to visualize molecular structures, like Lewis or condensed structures, written on a piece of paper, you can do this problem with a pencil and paper.) How does your result compare with the number of C_4H_8 isomers you obtained in Investigate This 5.55? What conclusion(s) can you draw about the relationship between number of carbons and number of isomers? Does this conclusion make sense? Explain why or why not.

Section 5.10. Functional Groups -- Making Life Interesting

5.46. **(a)** Use your molecular model kit to construct a model of cyclopentane. (See Investigate This 5.45.). How flexible is the ring?

(b) Remove one of the CH_2 groups and reconnect the ring to make a four-carbon ring, cyclobutane. How easy is it to do this? How flexible is the ring?

(c) Remove another CH_2 group and reconnect the ring to make cyclopropane. How easy is it to do this? How flexible is the ring? Would you expect cyclopropane to be more reactive, less reactive, or about the same as cyclopentane? Give the reasoning for your answer.

5.47. Use your model kit to make a cyclopentane model.

(a) Replace one of the hydrogens with an –OH group; the model now represents cyclopentanol. Is there more than one structure for cyclopentanol? Why or why not? If so, make separate models of each one. Are any of them optical isomers? How can you tell?

(b) Add a second –OH group on the carbon adjacent to the first –OH group. The new model is 1,2-cyclopentanediol ("di" = two). Is there more than one structure for the diol? Why or why not? If so, make separate models of each one. Are any of them optical isomers? How can you tell?

5.48. Table 5.4 catalogs a number of important functional groups. Use the table to answer the following questions.

(a) Which functional groups contain oxygen?

(b) Which functional groups contain nitrogen?

(c) Which functional groups in the table are most likely to ionize? (You may have to refer back to Chapter 2.) Explain your reasoning.

Structure of Molecules   Chapter 5

5.49. Can you discover a mathematical relationship between the number of C's and the number of H's in an alkane? [Because this is a rather abstract question, we will give you an example to illustrate what we mean. If you wished to know the number of toes in a crowded room, you could simply count the number of feet and multiply by 5. Mathematically, the number of toes is a function of the number of feet. Note that this relationship is independent of the number of feet in the room (assuming no three-toed aliens or other exceptions)!]

(a) Can the number of H's in an alkane be described as a function of the number of C's? You will need to examine a number of alkanes to be certain that you have discovered a general relationship. Your relationship should fit branched as well as straight-chain alkanes.

(b) Can the number of H's in an alk**ene** be described as a function of the number of C's?

(c) Can the number of H's in an alk**yne** be described as a function of the number of C's?

5.50. Certain terms used in the nomenclature of carbon-containing compounds sound very much alike but do not mean the same thing at all. Clearly explain the differences between these pairs of words and draw a specific compound illustrating each.

(a) alkane   alkene
(b) alcohol   aldehyde
(c) ether   ester
(d) amine   amide
(e) carboxyl   carbonyl

5.51. Draw the Lewis structures and structural formulas for each of the following compounds. Draw bond dipoles on the structural formulas. Identify the shape of each compound. Identify if the compound is polar or nonpolar.

(a) H_2CO (formaldehyde)

(b) $HO(O)CCH(NH_2)CH_2C(O)OH$ (aspartic acid)

(c) $H_2NCH_2CH_2NH_2$ (1,2-diaminoethane)

5.52. (a) In Problem 5.2, you drew structures for isomeric alcohols with formula $C_5H_{12}O$. There are six other structures that have the formula $C_5H_{12}O$. Draw their structures and identify the functional group(s) in these molecules.

(b) Predict which one of the six molecules in part (a) will have the lowest boiling point. Explain the reasoning for your choice.

Section 5.11. Molecular Recognition

5.53. In a double helix of DNA, the nitrogenous base cytosine is held to the nitrogenous base guanine by hydrogen bonds. (See Investigate This 1.33.) Dashed lines in the diagram represent hydrogen bonds.

(a) How many hydrogen bonds can form between, thymine and adenine? Use these two structures for thymine and adenine to draw a diagram showing the hydrogen bonding that can take place between thymine and adenine.

(b) Thymine does not pair with guanine. Offer a possible explanation for this observation.

(c) DNA has the ability to replicate (make copies of itself) accurately and reproducibly. How can this be related to base pairing? Explain your reasoning.

5.54. **WEB** Chap 5, Sect 5.11.1-3. Work through these pages before answering the following questions.

(a) Explain the relationship of the activity in Sect 5.11.3 to the one you did in Check This 5.72.

(b) Make a model of aspartame. A ball-and-stick model is shown rotating in Sect 5.11.1 and here is a 3-d representation of the molecule:

$$\begin{array}{c} \text{HC–CH} \\ \text{HC} \quad \text{CH} \\ \text{HC=C} \\ \text{O} \quad \text{CH}_2 \\ ^+\text{H}_3\text{N} \quad \text{C} \quad \text{C}\cdots\text{H} \\ \text{H} \triangleright \text{C} \quad \text{N} \quad \text{C=O} \\ \text{CH}_2 \quad \text{H} \quad \text{OCH}_3 \\ ^-\text{O(O)C} \end{array}$$

An important point to note about the structure of this dipeptide (aspartyl phenylalanine methyl ester) is that the four atoms in the center of the structure, O=C–N–H (shown in red), lie in a plane and rotation about the C–N bond is very restricted. This bond acts much like a double bond because interactions between the pi bond in the carbonyl and the nonbonding electrons on the nitrogen atom delocalize the electrons over the three second period atoms and make their structure rigid. Two possible ways to construct the six-membered ring are shown in these photographs. Use whichever one you wish.

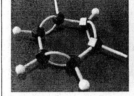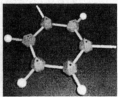

(c) As sweet receptor, cut an appropriate size hole in a piece of paper and color its edge red, blue, and green as in sweet receptor shown in Sect 5.11.3 (or simply label the edge with charges). Show that your aspartame model complements the receptor, as does the correct phenylalanine isomer in the web activity.

(d) Exchange the –H and –NH$_3^+$, on your model (left-hand side of the structure above) and try part (c) with the new isomer. Will this isomer still taste sweet? Explain why or why not.

(e) Undo the exchange in part (d) and then exchange the –H and one of the other groups attached to the central carbon at the right-hand side of the structure above. Try part (c) with the new isomer. Will this isomer still taste sweet? Explain why or why not. Discuss the similarities and differences between your observations in parts (d) and (e).

Chapter 5 — Structure of Molecules

Section 5.13. EXTENSION -- Antibonding Orbitals: The Oxygen Story

5.55. The double bond in O_2 and a triple bond between two carbons have almost the same bond length, 121 and 120 pm, respectively. If bond lengths get shorter as bond order increases, why is the double bond between oxygen atoms the same length as the triple bond between carbon atoms? *Hint*: Recall the trend in atomic sizes and the reason(s) for it as you go across a period of the periodic table.

5.56. Fluorine (F_2) is the most reactive of the halogens. According to the Lewis model, each fluorine atom of this covalent compound has three nonbonding pairs of electrons.

(a) How many valence electrons are there in diatomic fluorine?

(b) Assume that the molecular orbital diagram shown in Figure 5.42 is also applicable to molecular fluorine. Draw a diagram like Figure 5.42 that shows how the valence electrons are assigned to the various molecular orbitals. (Remember that two nonbonding sigma orbitals are not shown in the diagram.)

(c) What is the bond order of the F-to-F covalent bond that you derive from your results in part (b)? Explain clearly.

(d) How does the molecular orbital bonding model correlate with the Lewis model for fluorine?

General Problems

5.57. (a) Write a Lewis structure for the cyanate anion, OCN^-. Can you write more than one satisfactory structure?

(b) What geometry does(do) your Lewis structure(s) predict for the cyanate anion? How does the structure and geometry of this ion compare to the structure and geometry of the carbon dioxide molecule?

(c) Would you expect the cyanate anion to have delocalized π electrons? Explain the reasoning for your response.

(d) Answer these same questions for the isocyanate anion, ONC^-.

(e) Would you expect the cyanate or isocyanate anion to be the more stable isomer? Give the reasoning for your choice.

5.58. (a) At room temperature, beryllium fluoride, BeF_2, is a white crystalline solid and boron trifluoride, BF_3, is a gas. What conclusions can you draw about the bonding in these compounds? Explain.

(b) Write the Lewis structure for BF_3. What geometry do you predict for BF_3? What figure in the chapter helps you explain your choice of geometry? Explain. *Hint:* This compound does not obey the octet rule.

(c) Under certain conditions BeF_2 molecules can be observed in the gas phase. Write the Lewis structure for BeF_2. What geometry do you predict for BeF_2? What figure in the chapter helps you explain your choice of geometry? Explain. *Hint:* This compound does not obey the octet rule.

Index of Terms

δ (a small amount)	1-28	amount	1-58
ψ (electron wave function)	4-65	amplitude	4-14
π (pi bonding orbital)	5-33	anion	2-18
σ (sigma bonding orbital)	5-19	aqueous	1-5
σ (sigma orbital)	5-18	atom	1-13
ν (wave frequency)	4-15	atomic core	1-19
λ (wavelength)	4-14	atomic cores	5-9
Δ [change (final − initial)]	1-53	atomic mass unit	3-15
σ bonding framework	5-22	atomic number	1-15
σ framework	5-22	atomic number	3-15
π* (pi antibonding orbital)	5-78	Avogadro's number	1-62
ΔE (energy change)	1-53	balanced chemical equation	2-23
α–helix	1-48	ball-and-stick model	1-18
σ_n (sigma nonbonding orbital)	5-20	basic	2-61
3-d structure	5-48	basic	2-63
A (mass number)	3-16	bent bonds	5-30
absorption spectrum	4-12	beta particle	3-25
acidic	2-61	Big Bang Theory	3-20
acidic	2-63	binding energy per nucleon	3-42
active site	5-73	biomolecules	1-22
alcohol	1-64	Bohr quantum model	4-34
alcohols	5-68	boiling point	1-36
aldehydes	5-69	boiling point	1-9
alkali metal	4-6	bond angle	1-19
alkali metals	2-21	bond angle	5-23
alkaline earth metal	4-6	bond dipoles	1-31
alkaline earths	2-21	bond length	1-19
alkanes	5-65	bond length	5-17
alkenes	5-66	bond order	5-42
alkyl	1-64	bond order	5-79
alpha particle	3-25	bonding orbital	5-18
amino acid	1-47	Brønsted–Lowry acid	2-63

Brønsted–Lowry base	2-63	denatured	1-49
bulk properties	1-17	density	1-7
C (molar heat capacity)	1-68	deoxyribonucleic acid	1-50
calorie	1-56	diffraction pattern	4-16
cancellation	4-16	dipole moment	1-31
carbon cycle	2-77	dipole-dipole attractions	1-38
carbonyl group	5-69	dispersion forces	1-37
carboxyl group	2-67	dissolved	2-6
carboxyl group	5-71	DNA	1-50
carboxylate ion	2-67	double bond	5-30
carboxylic acids	2-67	double bond	5-42
carboxylic acids	5-71	ductile	5-47
cation	2-17	electric dipoles	1-28
chain reaction	3-45	electromagnetic radiation	4-20
chemical equation	2-22	electromagnetic spectrum	4-21
cis-isomers	5-57	electromagnetic wave theory	4-19
collision-induced fission	3-45	electromagnetic waves	4-20
compound	1-18	electron	1-13
concentration	2-42	electron	3-15
condensation	1-9	electron configuration	4-67
condensation	3-20	electron spin	4-53
condensed phase	1-9	electron wave function	4-65
condensed structure	5-51	electron-dot model	1-18
conjugate acids and bases	2-71	electronegativity	1-32
continuous spectrum	3-8	electronegativity	4-58
core electron	1-14	electron-shell model	1-14
Coulomb's law	2-25	electrostatic attraction	1-12
coulombic repulsion	3-21	element	1-13
covalent bond	1-19	elementary charge	3-15
critical mass	3-46	endothermic	2-8
daughter nucleus	3-32	energy	1-53
de Broglie wavelength	4-38	energy change	1-53
degenerate energy levels	4-66	entropy	2-12
delocalized π molecular orbital	5-41	entropy	2-8

equilibrium	2-35	hydrocarbons	5-65
equivalence point	2-74	hydrogen bond	1-41
ester	5-71	hydronium ion	2-60
ether	1-64	hydroxide ion	2-62
ethers	5-69	hydroxides	2-66
evaporation	1-54	induced-dipole attractions	1-37
excited state	4-35	intensive property	1-70
exclusion principle	4-53	ion	2-17
exothermic	2-8	ion	3-19
extensive property	1-71	ion–dipole attractions	2-19
favorable factor	2-39	ionic crystal	2-18
favorable factor	2-8	ionic solution	2-18
formula unit	1-60	ionization energy	4-9
formula unit	2-43	isomers	5-7
freezing	1-9	isotopes	3-17
frequency	4-15	isotopic signature	3-62
functional group	5-66	joule	1-56
gamma radiation	3-25	ketones	5-70
gas	1-8	kinetic energy	4-45
ground state	4-35	kinetic energy	4-74
groups	1-13	lattice	2-24
half life	3-32	lattice energy	2-25
halogens	2-21	Le Chatelier's principle	2-79
heat	1-71	Lewis structure	1-21
Heisenberg uncertainty principle	4-43	limiting reactant	2-53
		line formula	1-18
hertz	4-23	line spectra	3-8
heterogeneous	2-6	liquid	1-7
homogeneous mixture	2-6	localized π molecular orbital	5-35
homonuclear bond	4-57	localized one-electron molecular orbitals	5-18
hydrated ions	2-19		
hydration energy	2-28	London forces	1-37
hydration layer	2-19	macroscopic	1-17
hydrocarbons	1-38	malleable	5-47

mass	1-58	nonbonding electrons	5-20
mass number	3-16	nonpolar molecule	1-28
mass-energy equivalence	3-37	nuclear binding energy	3-41
melting point	1-9	nuclear fission	3-39
metallic bonding model	5-46	nuclear fusion	3-21
miscible	2-11	nuclei	3-15
mol	1-59	nucleic acids	2-66
molar heat capacity	1-68	nucleon	3-16
molar mass	1-60	nucleon	3-42
molar mass	2-43	nucleus	1-13
molarity	2-43	nucleus	3-15
mole	1-59	octahedral	5-26
mole	1-62	octet rule	1-21
mole	2-43	optical isomers	5-62
molecular formula	1-18	optical rotation	5-61
molecular model	1-17	optically active	5-61
molecular orbital	5-15	orbital	4-44
molecular shape	1-19	oxyacids	2-64
molecular shape	5-22	oxyanion	2-64
molecule	1-17	paramagnetic	5-77
momentum	4-38	parent nucleus	3-31
monatomic	2-20	Pauli exclusion principle	4-53
N (number of neutrons in a nucleus)	3-16	periodic table	1-13
		periodic table	4-6
n (principal quantum number)	4-66	periodicity	4-6
negative ion	2-18	periods	1-13
net ionic equation	2-34	PET scan	3-30
neutrino	3-25	pH	2-61
neutron	1-13	phases of matter	1-6
neutron	3-15	photoelectric effect	4-28
noble gases	4-8	photon	4-30
nodal plane	5-34	pi antibonding orbital	5-78
node	4-14	pi bonding molecular orbital	5-33
nonbonding electrons	1-21	Planck's constant	4-26

polar molecules	1-28	sigma bonding orbital	5-19
polarimeter	5-59	sigma molecular orbital	5-18
polarized	5-59	sigma nonbonding molecular orbital	5-20
polarized light	5-59		
polarizer	5-59	single bond	5-42
polyatomic	2-20	skeletal structure	5-53
positive ion	2-17	solid	1-7
positron	3-25	solute	2-6
positron emission tomography	3-30	solution	2-6
potential energy	4-46	solvation	2-8
potential energy	4-76	solvent	2-6
precipitate	2-32	space-filling model	1-30
principal quantum number	4-66	specific heat	1-56
products	2-22	specific heat	1-69
protein	1-47	spectator ions.	2-34
proton	1-13	spectrograph	3-8
proton	3-15	spectroscope	3-8
protonation	2-60	spectroscopy	3-6
quantized	4-26	spectrum	3-6
quantum	4-26	speed of light	3-37
quantum model	4-34	speed of light	4-22
radioactive decay	3-31	spin	4-53
radioisotope	3-31	spin-paired electrons	4-54
reactants	2-22	spontaneous nuclear fission	3-44
receptor	5-73	stable isotopes	3-61
reinforcement	4-16	standard temperature and pressure	1-8
reorganization	2-12		
reorganization	2-39	standing waves	4-41
reversible	2-36	stereoisomers	5-58
s (specific heat)	1-69	stoichiometry	2-51
saturated solution	2-36	STP (standard temperature and pressure)	1-8
Schrödinger wave equation	4-65		
shell model	4-56	structural isomers	5-56
side group	1-47	substrate	5-73

superimposition	4-16	u (wave velocity)	4-15
supernova	3-23	uncertainty principle	4-43
temperature	1-70	unfavorable factor	2-39
tetrahedral	1-26	unfavorable factor	2-8
tetrahedral angle	1-26	unit charge	1-13
tetrahedron	1-26	valence electron	1-14
thermal energy	1-71	valence electrons	5-9
total energy	4-47	valence shell	1-16
total energy	4-76	vaporization	1-9
trans-isomers	5-57	viscosity	1-74
traveling waves	4-41	wave equation	4-65
trigonal bipyramidal	5-25	wave mechanics	4-42
trigonal pyramidal	5-22	wave velocity	4-15
triple bond	5-32	wavelength	4-14
triple bond	5-37	Z (atomic number)	3-15